Modern Manufacturing:
Information Control and Technology

Advanced Manufacturing Series
Series editor: D.T. Pham

MODERN MANUFACTURING

INFORMATION CONTROL AND TECHNOLOGY

Edited by

Marek B. Zaremba and Biren Prasad

With 176 Figures

Springer-Verlag

London Berlin Heidelberg New York
Paris Tokyo Hong Kong
Barcelona Budapest

Marek B. Zaremba, PhD, MSc
Département d'Informatique
Université du Québec à Hull
101 St-Jean-Bosco, Hull, Québec, Canada J8Y 3G5

Biren Prasad, PhD
CERA Institute
5421 South Piccadilly Circle, West Bloomfield, MI 48322, USA

Series Editor

D.T. Pham, PhD
University of Wales, College of Cardiff
School of Electrical, Electronic and Systems Engineering
P.O. Box 917, Cardiff CF2 1XH, UK

ISBN 3-540-19890-3 Springer-Verlag Berlin Heidelberg New York
ISBN 0-387-19890-3 Springer-Verlag New York Berlin Heidelberg

British Library Cataloguing in Publication Data
A catalogue record for this book is available from the British Library

© Springer-Verlag London Limited 1994
Printed in Great Britain

The publisher makes no representation, express or implied, with regard to the accuracy of the information contained in this book and cannot accept any legal responsibility or liability for any errors or omissions that may be made.

Typesetting: Camera ready by contributors
Printed and bound at the Athenæum Press Ltd., Gateshead
69/3830-543210 Printed on acid-free paper

Preface

Manufacturers worldwide are faced today with important and in many respects unprecedented challenges resulting mainly from international competition, changing production processes and technologies, shorter product life-cycles, market globalization, and environmental requirements. Fundamental to meeting these challenges is the understanding and control of information across all stages of the Computer Integrated Manufacturing (CIM) process. Manufacturing competitiveness is a balancing act. There is no single technical solution that can be copied easily or bought from another successful company. A well-orchestrated process, using the capabilities of information technology to their full extent, is required to achieve corporate goals and objectives.

This book offers a solid tutorial and reviews the state of art in the information oriented aspects of Computer Integrated Manufacturing and Intelligent Manufacturing Systems. The book deals with modern manufacturing knowledge that is sufficiently general to provide at least partial intellectual support for a number of specific CIM cases, applications, and the introduction of new products and systems. Particular emphasis is placed on the impact of new software engineering technologies, object-oriented approach, database design, hierarchical control and intelligent systems. A number of fundamental techniques and methods for achieving manufacturing realization are extracted from modern manufacturing theory and the currently successful practices.

The structure of the presentation reflects four fundamental aspects of manufacturing: planning, modeling, design and control. Thirteen chapters of the book are grouped together in six parts. The first part addresses general issues of modern manufacturing paradigm and enterprise modeling. Chapter 1 begins with an introduction of fundamental notions of modern manufacturing. It lists major elements of Intelligent Information Systems and discusses main barriers hindering the development of a product. Eight areas that determine the performance of enterprise competitiveness are examined. Chapter 2 deals with enterprise modeling. Enterprise modeling system presented in the chapter provides a comprehensive approach to model the business and engineering processes as well as to simulate dynamically the operations within a manufacturing enterprise.

Three next three chapters are focused on object-oriented approach. Chapter 3 discusses object-oriented information modeling paradigms - a description of a distributed information system model that includes both information (data, knowledge, and processes) and functional specifications. Chapter 4 provides a more detailed presentation of database technology - an essential ingredient of a CIM system. Particular emphasis is placed on relational and object-oriented databases, which provide technology responding to the needs of the next generation distributed database systems. Object-oriented approach to design of process parameters is explored in Chapter 5. A framework of a star topology system is described, where the central node is a blackboard and all the surrounding nodes are objects representing different design perspectives.

Hierarchical structures in the context of production management are the topic of the next two chapters. Chapter 6 describes a framework for building a hierarchical system to model large scale production systems and for execution of such a system to effectuate optimal planning, decision and control. Hierarchical control approach to solve complex managerial problems is discussed in Chapter 7. First, the problem of production planning for multi-product CIM systems is dealt with for both linear and non-linear cases. Then, the problem of optimizing the operation of an autonomous manufacturing/supply chain and a composite inventory-marketing problem are considered. Finally, an integrated multi-product dynamic model for long-term strategic planning is given.

The next part of the book examines topics relating to modeling and control of Flexible Manufacturing Systems. Chapter 8 addresses the implications of information control and technology in integrated control and management of the shop-floor operations. Heuristic and knowledge based approaches are emphasized. Modeling of manufacturing systems using Petri nets theory is discussed in Chapter 9. The modeling techniques are analyzed from the standpoint of system performance evaluation and optimization, including deadlock handling and efficiency prediction. Automated synthesis of performance-oriented Petri net models is examined. Chapter 10 deals with the problem of fault tolerance in computerized manufacturing systems. Explicit introduction of fault tolerance at the specification and the design stage of modern manufacturing systems is proposed. A case study of the Volvo Uddevalla computerized assembly plant is presented.

Chapters 11 and 12 deal with the issues of Computer-Aided Design in its broad aspects. Chapter 11 presents computer-aided process planning tasks as the link between CAD and CAM (Computer-Aided Manufacturing) systems. Both fully automatic planning and group technology approaches are looked upon based on the results of two BRITE/EURAM projects. In Chapter 12, computer-aided design process based on features is discussed. Features are used as vehicles to introduce functional aspects into CAD techniques.

Finally, Chapter 13 describes advantages and difficulties of implementing Just-In-Time technology in manufacturing systems. Upon a discussion of the JIT environment, three case studies are presented.

The editors wish to thank all the chapter authors for their contributions to the book. Many of the authors are well-known scholars in modern manufacturing and information technology areas. Many of them are respected instructors and long-time colleagues of the editors. Their names are listed in the list of contributors. Our thanks also go to Miss Imke Mowbray and Mr. Nicholas Pinfield, editor at Springer Verlag for their help with the book production. We are thankful to Suchitra Mathur and Dayana Stetco, who helped edit the manuscript, and our family members: Mrs. Pushpa Prasad, Miss Rosalie Prasad, Miss Gunjan Prasad, Miss Palak Prasad, and Janina, Renata and Chris Zaremba for their love and support.

Marek B. Zaremba
Biren Prasad

The Contributors

G.M. Acaccia

Industrial Robot Design Research Group
University of Genova
Via all'Opera Pia 15/A
I - 16145 Genova, Italy

A. Adlemo

Department of Computer Engineering
Chalmers University of Technology
S - 412 96 Göteborg, Sweden

L.A.R. Al-Hakim

School of Management
David Syme Faculty of Business
Monash University
P.O. Box 197
Caulfield East, Victoria 3145
Australia

S.-A. Andréasson

Department of Computing Science
Chalmers University of Technology
S - 412 96 Göteborg, Sweden

Z. Banaszak

Department of Robotics and Software Engineering
Technical University of Zielona Gora
65-246 Zielona Gora, Poland
This work was completed when the author was on a leave at:
Department of Mathematics
Kuwait University, Kuwait

M. Callegari

Industrial Robot Design Research Group
University of Genova
Via all'Opera Pia 15/A
I - 16145 Genova, Italy

A.W. Chan

Institute for Advanced Manufacturing Technology
National Research Council
Montreal Road, Bldg. M-50
Ottawa, Ontario, Canada K1A 0R6

Y.-T. Chen

Department of Industrial Engineering
The University of Iowa
Iowa City, IA 52242-1527, U.S.A.

K.A. Jørgensen

Department of Production
University of Aalborg
Fibigerstraede 13
DK - 9220 Aalborg 0, Denmark

G. Kapsiotis

Intelligent Robotics and Control Unit
Department of Electrical and Computer Engineering
National Technical University of Athens
Zographou 15773
Athens, Greece

A. Kusiak

Department of Industrial Engineering
The University of Iowa
Iowa City, IA 52242-1527, U.S.A.

R.C. Michelini

Industrial Robot Design Research Group
University of Genova
Via all'Opera Pia 15/A
I - 16145 Genova, Italy

R.M. Molfino

Industrial Robot Design Research Group
University of Genova
Via all'Opera Pia 15/A
I - 16145 Genova, Italy

R. Nagi

Department of Industrial Engineering
State University of New York at Buffalo
342 Bell Hall
Buffalo, NY 14260, U.S.A.

H. Nordloh

Bremer Institut für Betriebstechnik und angewandte
Arbeitswissenschaft an der Universität Bremen,
Hochschulring 20
D-28359 Bremen, Germany

B. Prasad

CERA Institute
5421 South Piccadilly Circle
West Bloomfield, MI 48322, U.S.A.

J.-M. Proth

INRIA-LORRAINE
Technopôle de Metz 2000
4 rue Marconi
57070 Metz, France

M. Schulte

Universität des Saarlandes
Lehrstuhl für Konstruktionstechnik/CAD,
Postfach 151150
66041 Saarbrücken, Germany

A.S. Sohal

School of Management
David Syme Faculty of Business
Monash University
P.O. Box 197
Caulfield East, Victoria 3145
Australia

R. Stark

Universität des Saarlandes
Lehrstuhl für Konstruktionstechnik/CAD,
Postfach 151150
66041 Saarbrücken, Germany

S. Tzafestas

Intelligent Robotics and Control Unit
Department of Electrical and Computer Engineering
National Technical University of Athens
Zographou 15773
Athens, Greece

F.B. Vernadat

INRIA-LORRAINE
Technopôle de Metz 2000
4 rue Marconi
57070 Metz, France

Ch. Weber

Universität des Saarlandes
Lehrstuhl für Konstruktionstechnik/CAD,
Postfach 151150
66041 Saarbrücken, Germany

Contents

1 Modern Manufacturing

Biren Prasad

1. INTRODUCTION

Modern Manufacturing and "world-class manufacturing" are often used interchangeably. Establishing Manufacturing Competitiveness is considered equivalent to being a world-leader in "world-class manufacturing." In other words, manufacturing competitiveness means sustained growth and earnings through building customer loyalty -- by creating high value products -- in very dynamic global markets. During the last decade, a number of modernization programs have been launched by many industries, such as in Newport News Shipbuilding, NEC, IBM, GM, FORD, dealing with how a company does its business, focussing on business processes and information usage in the manufacturing environment. Programs, such as enterprise re-engineering, multidimensional integration of engineering, manufacturing and logistics practices have resulted in far reaching changes and profit. These changes could not have been possible without process tools like Concurrent Engineering (CE) and Continuous Acquisition and Life-cycle Support (CALS -- earlier known as "Computer-aided Acquisition and Logistics Support) in action [Ross, 1992]. Attributes such as empowerment, flexibility, total quality management, agility, fast-to-market, accountability, teamwork, and integration are inherent in the notion of modern manufacturing.

2. Computer Integrated Manufacturing:

Most research & development efforts toward automation for modern manufacturing have been independently developed. The result is that the areas of automated application are self-assertive. At the present time, highly automated areas in manufacturing include Computer Aided Design (CAD), Computer-aided Process Planning (CAPP), Computer-aided Manufacturing (CAM), Manufacturing Resource Planning (MRP), and Computer-integrated Inspection techniques (CII). With the dependency on computational and logical techniques, the emphasis has been on integrating the existing CAD, CAM, CAPP, and CII systems to provide a computer-integrated manufacturing (CIM) environment (see Figure 1).

Today, CIM systems are merely being applied to integration and processing (storage and automation) of data and processes (hard uncertainty and hard information). The communication part of CIM design is information. This information includes many different categories of product images, CAD data, CAM data, CAPP, CII, design specifications, the history of production, and interface information in various forms, including electronic, text, raster images, videos and their mixture, as well as many different types of paper formats. They are seldom used for processing knowledge (in addition to data and processes), and for solving problems related to decision and control, even though there has been an increasing interest in subjects such as knowledge based systems, expert systems, etc. The latter slew of tools is more of a cognitive nature and is the basis of decision making used in CIM and CE.

3. Concurrent Engineering

Concurrent Engineering is considered one of the key concepts which enables companies to attain world-class stature. Today, it is an organizational keyword and promises to reach the level of fame once held by Management-by-objectives. The term refers to the same organizational process described as "simultaneous engineering" and sometimes as the "multi-disciplinary approach" or, "concurrent design." Here, everyone contributing to the final product from conceptual to support levels is required to participate in the project from its very inception. CE and Computer Integrated Manufacturing go hand in hand. Together, they are called the Intelligent Information System. CIM plus CE equals IIS (See Figure 2).

Information System. CIM plus CE equals IIS (See Figure 2).

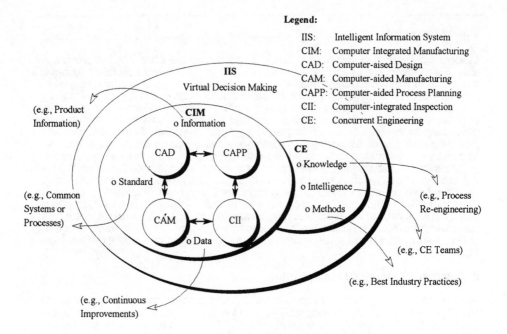

Figure 1: Current Level of Automation

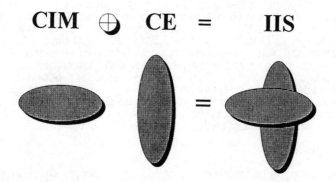

Figure 2: Intelligent Information System: What is it?

4. Intelligent Information System

The need for a CIM system has long been recognized. Most people consider manufacturing information-intensive. Many more believe that intelligent handling of information transfer through computer technology can produce a better Computer Integrated Manufacturing system since it reduces the need for manual intervention. CE brings forth the necessary value system such as the culture for embedding a procedural discipline in CIM operations and in enterprise-level communications. The key to IIS success is understanding the obstacles to CE in existing CIM processes and identifying opportunities for improvement. The identification of improvement opportunities and the implementation of effective product development process control strategies can be facilitated by the systematic collection and monitoring of relevant in-process metrics.

4.1 Major Elements

Business methodology and the role of the engineer are changing today. The people, the environments, and the cultures have all changed. The dominant material of society is information, and the dominant process is mechanization. Almost everything is mass produced or mechanized. Today a watch would cost thousands of dollars if it were built using the traditional "crafts manufacturing" technique. Now, with "mass production," we can manufacture accurate digital watches for a price of $2 to $5. The technological needs are also demanding a more specialized group of people in the usual Product Development Team (PDT). It is beyond the realm of a single person or a group to optimize a new design with all its multidisciplinary requirements. There are too many factors or disciplinary skills involved for any one person to handle. Besides the knowledge requirements in each disciplinary field, the depth and levels of skills required from these individuals are also very high. As a result, companies employ a large highly specialized work force, and use various tools to produce new system or component designs. The major elements of Intelligent Information System for product development are:

o   *Exchange of Data:* This is accomplished by networked computer systems to encourage the exchange of data among CE team members. It can quickly relay design changes, iterations, reviews, and approvals to CE participants.

o   *Requirements Management:* It deals with attaining a balance between product and process requirement management. Management process starts with understanding the requirements, interfaces, plan of manufacture for an existing design to extending support for a new design, if envisioned, while meeting the artifact's functional goals.

o   *Team Make-up:* This entails including in the Product Development Team (PDT) manufacturing support personnel with "x-abilities" expertise or the use of substitute "x-abilities" tools during the early stages of design rather than being called upon when the design is set in stone.

o   *Empowerment:* This entails empowering multidisciplinary teams to make critical decisions early on, instead of later, at major design reviews, when the cost of change is much more dramatic.

o   *Modeling:* It entails the use of various models which electronically represent, in convenient forms, information about the product, process, and the environment in which it is expected to perform.

o   *Tools & Technology:* It uses tools (such as rapid prototyping, Design for X-ability, simulation, etc.) to visualize product and process concepts quickly and in a format that is understandable to all participating IPD teams. It applies these tools to experiment a number of product and process options that are available or feasible.

o   *Information Sharing Architecture:* This fosters more effective information shared among the many different personnel and CE teams involved in a consistent manner (such as recording of design history, division of responsibility, and maintenance of agendas). Mozaic -- the Auto-trol Technology's object-oriented CE architecture -- is an example of the modularity and architecture that comes from object-oriented technologies. Objects represent the real-world product & process decompositions, tree structures, and their ease of communication across platforms that make them ideal for collaborative environments.

o   *Cooperative Problem Solving:* It means sharing problem solving information, so that instead of a single individual, the whole team can make joint decisions. Functions that focus and facilitate collaborative discussions involve the use of a sound analytical basis such as recording of design rationale and electronic critiquing of designs, and planning and execution of design changes.

o   *Intelligent Decision Making:* Design decision making can be viewed as a process of creating an artifact that performs what is expected (specified as requirements) in the presence of all sorts of constraints and operating environments that govern its behavior. Depending upon the cognitive knowledge about a product available to a decision maker, design decision may range from cognitive to progressive.

Although, IIS has a major impact on productivity, there are some barriers that inhibit the progress.

4.2 Major Barriers

Despite the fact that many of the aspects in design, engineering, prototyping, production, and manufacturing can be assisted and automated by computer aided tools, each activity is mostly dealt with in its own domain. There are several barriers that hinder the development of a product. In a joint study with ARPA's (formerly DARPA's) Initiatives in Concurrent Engineering (DICE) program, many large companies, including General Motors, General Electric, Lockheed Missiles and Space Company and Boeing Aerospace, have experienced similar barriers while implementing modern manufacturing concepts. Some major classes of barriers are (see Figure 3):

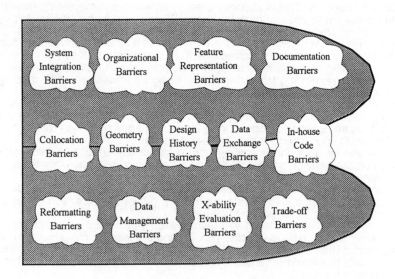

Figure 3: Major Barriers to Manufacturing Competitiveness

o   *Organizational Barriers*: These types of barriers are man-made and generally appear due to cultural and social differences of the people involved. There are three sub-classes of organizational barriers as shown in Figure 4. They are caused by (a) organization structure, (b) conflicting agendas, and (c) tradition. Some examples of organization structures are division of labor, culture bureaucracy, hierarchy, etc. Organizational charts divide work into segments and thus it tends to create boundary lines. Invisible walls develop since managers are evaluated on the performance of their individual segments. Examples of conflicting agenda include results versus performance metrics, capital dollars versus expenditures, delivery dates versus project schedules, etc. Most employees do not like changes which will modify the status-quo that a manger has so comfortably developed over the years. They tend to be uncooperative, not because they love the invisible walls, but because they are innocent victims of their past customary practices. Examples of tradition include standard operating procedures, culture, Not Invented Here (NIH) syndrome, "why fix something if it isn't broken," etc. Further, since it does not pay-off personally, they have very little interest in wanting to change such traditions.

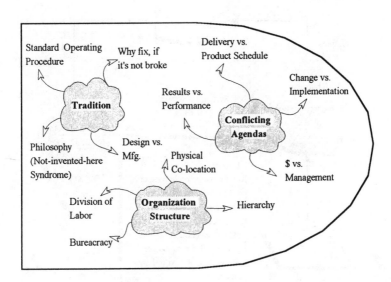

Figure 4: Organizational Barriers

o   *Geometry Barriers:* This results from the differences between underlying mathematical representations of various CAD systems. For example, solid modeling techniques are widely used for defining mechanical components and products. They are represented in various geometrical forms, and often serve as a potential tool to integrate design with manufacturing. Contemporary solid modeling systems provide good support for analytical procedures that can be used for interference checking and mass properties computations, such as volumes, etc. However, geometrical forms do not explicitly specify the functional requirements and allowable variations of a product that are important to manufacturing, assembly, and quality evaluations.

o   *Data Exchange Barriers*: Most CAD systems have their own proprietary formats for representing CAD entities. To communicate with another CAD system, design information must be converted to a neutral file or use a system to system translator. Currently there are several neutral data formats for transferring CAD information. These include IGES, EDIF, and DXF formats. In neutral file format, there are at least three levels of interpretation. First, data is written using the parent CAD system to a neutral file. Second, data is transferred and read-in by the receiving CAD system. Third, data is then reformatted from its neutral format to a new proprietary format. Many new data translation tools are now in use to aid this process such as PDES or STEP. Unfortunately this process is not robust. Numerous categories of errors, such as inaccuracies, incomplete or extraneous information, missing data, ripples, and instabilities, occur during the translation process (see Figure 5).

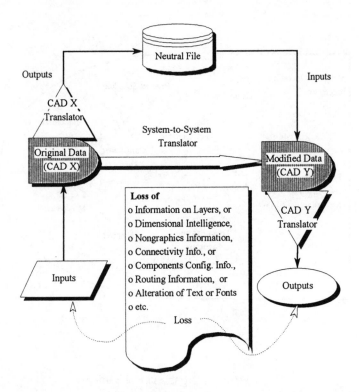

Figure 5: Types of Output Errors after a Typical Data Translation Process

o   *Features Representation Barriers:* Many CAD models include no description of parts functional features and of their interactions with other parts. The lack of modeling facilities for functional requirements creates gaps in integrated product development (IPD) for concurrent engineering. Similarly, the lack of tolerance specification facilities in current CAD environment represents a serious deficiency for the implementation of a process-rich CIM system.

o   *Reformatting Barriers:* Most of the programs that are used to support product development require the creation of models of the products such as CAD solid models, FEM/FEA models, NC models, etc. (See Figure 6). Information that is required at various stages of product development needs to be reformatted to suit the needs of the application, even though the type of data remains the same. This is because each model describes the same entity and the same object independently. This makes the control of data tedious and error prone. Time spent translating, correcting, and rebuilding the data model is costly and adds little value to the design-for-manufacture process.

o   *Trade-off Barriers:* One of the major shortcomings of present practices is the lack of adequate tools for early product trade-off. Due to the crude nature of early product data and stringent time allocations, most trade-off studies at the beginning of the design cycle are either done quickly or are ignored. Enough design checks are not performed and it is most likely that design, when passed on for downstream operations, remains unchecked or incomplete. There are minimal or no tools to compare design alternatives during the early product development. It is cheaper to solve any problem early on during the design stage rather than in later phases. This is because cost of changes in later phases can be many times more. Furthermore, the early handling of problems leads to fewer problems at later stages. It is therefore advantageous to carry on more iterations and tradeoffs during the early part of the design cycle than later. This is essential to contain any possible change requests during later phases.

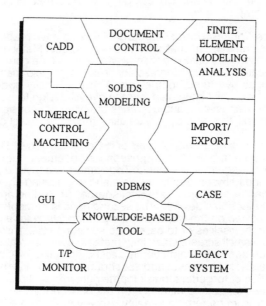

Figure 6: Reformatting Barriers

o   *Data Management Barriers:* In most companies, even team members keep their day's work files separate from each other and store them in their own private directories. There is a lack of group filing or network sharing. Even if a product data management (PDM) system is used, there are problems with maintaining current product versions and revisions. If someone needs to share data, he needs to ask for a copy or know the security codes or passwords of the owner to retrieve it. Advances in electronic communication and standards are now being readily used to access information on a network. But incompatibility of hardware and operators' errors are most common. Other problems that arise frequently are the difficulty in properly describing entities in the data base, the speed of transmission and translation of data among different functions, activities, and subsystems.

o   *Design History Barriers:* Tracking the history of design is normally not an integral part of a product development process. It is typically done after the facts, when work is over. Thus, there is no way to find out when the person returns after a break, where he left off, and what parameters and values he was using at that time. He puts down as notes what he remembers at that time. There is no log for tracing out the design history. Notes are subject to errors, which makes such records unreliable and less useful. The problem is somewhat contained when the group is small. In a small group, one person may be responsible for all his work and little or no interface is required. When the group is large and multi-disciplinary, or when the product is complex, a single member cannot possibly complete all the assignments. Co-workers often come in as resources to provide supplementary knowledge, or the expertise required for the job. In such a scenario, it is difficult to work without a common history, and to track progress as a natural part of the teamwork environment.

o   *In-house Code Barriers:* Most organizations have made significant investments in commercial analysis and design codes. Many have also developed their own niche of customization over and above the commercial codes (such as application-specific integration or linkages) to help them gain performance advantages over their competitors. These specialty codes typically remain isolated or loosely integrated into the overall design-to-manufacture process. Frequent maintenance and upgrades are often not considered the engineers' job functions, though they may be using them for product optimization.

o   *Documentation Barriers:* A user of a Computer-Aided Design and Drafting (CADD) system has often two needs for document outputs. The first is the need to manage multiple media. Today there are "hybrid printers" that serve as a bridge between graphic-plotters and engineering-copiers. These printers can accept input from a variety of sources, including aperture cards, hard copy, or electronic devices. The second need is perception. Even if a design can be viewed on the workstation, designers still want a hard copy in hand, either for themselves or to pass on to someone else. They cannot see everything on a small screen. Often they prefer to consult someone or annotate the design on paper. If information is passed on in the form of paper, it is error-prone, subject to delays, and susceptible to traps like not having the right information or forgetting to ask for the right information. Collaborative visual techniques for setting meetings on the network can solve this problem but this is far from becoming a reality. However, there are some performance problems in real time setting to show all visual effects quickly and inexpensively.

o   *Collocation Barriers:* In a multi-disciplinary organization, it is not possible to collocate all co-workers. People are often geographically distributed by virtue of being responsible for multiple tasks, or being part of different organizational or reporting structures. In a matrix organization, they usually do not report to the same manager. The team thus experiences problems in exchanging and sharing of data and information. As electronic communication based networking becomes more advanced, geographic separation will cease to be a concern when forming workgroups.

o   *X-ability Evaluation Barriers:* This comes from the inability to use CADD for Design for Assembly (DFA) or Manufacturability (DFM) type of x-ability evaluations. Most work relates to qualitative evaluations relative to Design for X-abilities. The current CAD system does not evaluate the degree of difficulty associated with a given mechanical system and assembly (MSA). There are some quantitative methods based on empirical data or index of difficulty, but none are normally tied to CADD.

o   *System Integration Barriers:* In designing an information system for an enterprise, one is immediately confronted with inconsistencies between data that describe (a) product design and parts family, (b) process engineering, (c) the know-how involved in the manufacturing processes, (d) plant floor (programmable logic controller; PLC) level and data that describes, and (e) management and business operation. These leads to at least five types of data series that are present: (i) one used for CAD environment (fortunate if there is only one CAD system), (ii) another for process planning, (iii) a third for manufacturing processes, MRP, etc., (iv) fourth for PLC operations and, (v) a fifth series for business and management operations.

All these forms are supposed to be homogeneous, i.e., the pertinent data forms must (a) describe the common entities the same way, (b) use the information which is required without translation, and (c) share a common data representation schema. In actuality, these data sets are often heterogeneous (describe the same object independently), produce duplicate information, or are incomplete. For large organizations, it is not easy to allow a complete level of integration that is based on a single data structure. Most industrial data that exists today is partially integrated. Without integration, it is difficult to improve operational efficiency or be globally competitive. Without integration and automation, it would be impossible to produce complex products such as automobiles or aircraft in a competitive fashion.

Competition has driven engineers to include terms such as time compression, total quality management, teamwork, quality function deployment, and Taguchi into their vocabularies [Hauser and Clausing, 1988]. Intelligent Information management (CE + CIM) coupled with automation efforts is vital in maintaining a competitive position in today's marketplace. Despite recent advances in computational and communication technology, it is still not an easy task to win competitiveness effectively.

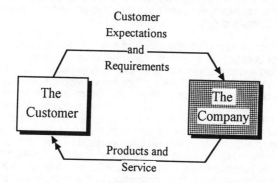

Figure 7: Basic Premises of Manufacturing

5. Areas of Manufacturing Competitiveness

A basic premise of manufacturing refers to the best transformation of customer expectations and requirements into useful products and services (Figure 7). A large number of companies across Europe, America, and Japan was recently studied [Ferdows and De Meyer, 1986], with the focus on manufacturing strategies and competitive priorities. Manufacturers who used to be able to differentiate themselves because of a lock on raw materials, technical knowledge, capital, or innovation, have found that manufacturing is a vulnerable market. When products come to market anybody can copy their salient features. The competitive edge, if any, is usually short lived. Technology by itself cannot create long-lasting competitive advantage. What is difficult to duplicate is how technology is deployed into one's processes. The improvements made through deployment of technology can provide a real competitive advantage. Engineering schools and researchers tend to ignore the process management factors and look exclusively at the technological solutions (e.g., CAD/CAM, CAE, CIM, etc.), while the industrial researchers think of computers and all off-the-shelf tools as commodities that anyone can buy and use. Most truly successful companies (both in US and Japan) believe that process management techniques are the product of decades of "corporate learning" that others cannot buy or copy. The Japanese seem to be far ahead in mastering the technology and structuring it to fit their únique environments [Whitney, 1992]. Coming to terms with the Japanese market was one of the challenges Americans and Europeans' had to meet to narrow the competitive gap. The only thing competitors cannot buy is their own unique process or their own unique organizational culture. This can be a blessing in disguise or a curse depending upon how one looks at it. For most American automotive industries, the production process is rooted deeply in the way teams design and manufacture the products -- thus inflexible, while the Japanese seem to have a better handle on it. With regards to culture, Americans seem to be more open than the Japanese whose strong cultural ties facilitate joint collaboration and teamwork. Thus, Americans seem to fall short on both ends. The studies have also shown some subtle differences in the way an enterprise looks at its processes. While the Europeans still focus on quality improvements and operational efficiency during the process of manufacturing, the Japanese focus on flexibility while continuing improvements in quality, dependability, cost, and productivity [Wilson and Greaves, 1990]. To the Japanese, flexibility in manufacturing means a rapid and efficient process of introducing changes in production volumes and product mix [Wheelright and Hayes, 1985]. In the field of manufacturing, the Japanese focus on a process of rapid development of new products and on becoming innovators of new process technologies. Achieving perfection in process flexibility is not simple. Locals of the Toyota factory in Japan can receive their car built to their specification within two days of placing the order. Such devastating process flexibility cannot be attributed to merely an edge in technologies. It has been observed [Putnum, 1985] that the success of the Japanese is largely due to progress in process management and continuous refinements, examples being the well known Kanban system of production control, Taguchi method of quality control, etc.

The recognized decline in the productivity of many U.S. companies has been a strong stimulant for companies to search for ways to improve their operational efficiency and to become more competitive in the marketplace. Many have changed their attitudes towards customers, their production processes, and their internal management approaches, whereas others continue to search for the reasons of their dwindling profits. Successful companies have been the ones who have gained a better focus on eliminating the waste, normally designed into their products, by understanding what drives product and process costs and how can value be added. They have chosen to emphasize high-quality production rather than high-volume production. With increasingly pervasive global competition, manufacturing excellence is becoming as fundamental a competitive weapon as engineering excellence. Significantly, what we are seeing is the completion of the definition of what it takes to be a World Class Manufacturing Company.

Manufacturing competitiveness is a balancing act. There is no single technical solution that can easily be copied or bought from other successful companies. A well orchestrated process, not just a program, is required to achieve corporate goals and objectives. The shorter lives of products today simply do not leave time to fix problems, correct design errors, or iterate many times to redesign products for lowering costs or improving quality. A company is considered to have reached a world class manufacturing status if the worth of products and services far outweighs the process and methodologies expended to produce it. Such a company gets the product right the first time.

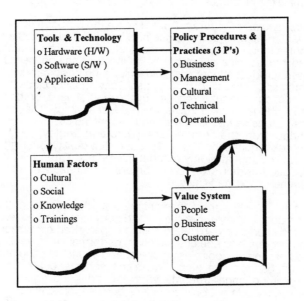

Figure 8: Nature of An Enterprise Business Operations

5.1 Products & Services

The nature of an enterprise business operation that distinguishes it from others is governed by four major controlling factors: tools and technology, value system, human factors, and policy practices/procedures (see figure 8). Today, the nature of technology, people, and market is changing rapidly, and the methods and practices of the past, and in particular of the Iron or Craft age of mass production, are no longer valid. A recent trend is towards an "invisible enterprise" where key functions and "virtual enterprise" are all subcontracted to suppliers. The time is not far when such enterprises will serve as a glue tying all functions or processes back to the value system. There are still many enterprises in which at least one of the following life-cycle function is performed outside (a) design (b) manufacturing (c) marketing or (c) customer support. For example, an enterprise may choose to design its product in-house, manufacture it somewhere else, and rely upon a third party (say value added reseller, VAR's) for marketing and servicing their products. The single rifle-shot approach of trying to get everything right through marketing has not worked and will not work. People gain a collective perception of a product by its actual field use, its value to their lives, and by relative comparison to other products. It is not just a matter of increasing quality in the sense of statistical conformance, but of creating products that are superior in both functional and cultural performance and technological contents. Cultural performance denotes the customers' human aspects that make a product effective in the context of easing their lives, profession, and work values.

Products and services are one side of the balance shown in figure 9. The attributes that determine the outcomes of this side of the balance include:

o   *Functions and Features*: Does it serve the purpose? Does it have features which are simple, easy to use and handle, and versatile? The product attributes that are important to the customer must be enumerated and translated into technical counterparts so that they can be measured by engineers. These product attributes should include the basic product functions that the customer assumes will be provided. They should include the essential level of product performance required by the marketplace, or the level necessary to lead technologically if that is the product strategy. They should also include those attributes that will attract and delight the customer, and differentiate this product from the competitors [Dika and Begley, 1991]. Manufacturing success requires the ability to produce top quality, often individually customized, products at highly competitive prices. Often they have to be mass-produced or delivered to customers in batches as small as one unit with extremely short lead-times.

o   *Quality*: Quality is more than shiny paint. It ranges from the visual (such as good door fits, style, color, etc.) and nonvisual (such as reliability, producibility, and scores of x-ability), to customer perceived excellence of functions and features (such as ride and handling, performance, drivability, noise-vibration and harshness, door closing efforts and sound, heating and air conditioning system performance, etc.). The myth that higher quality translates into higher costs is no longer true. Many have found that overall production of higher quality products not only costs less but also reduces cost-overruns. In many cases it has been shown that higher quality products yield as much as 40% higher return on investment than lower quality products do [Whitney 1988]. Quality is not an option; it is a given fact, which is to say it is the way of doing business. It is implicit that each of the design alternatives in consideration are able to meet the customer's quality requirements. Quality should not be involved when making tradeoffs. If an alternative can't meet the quality target, it should not be taken into consideration. On the other hand, if the product functions far exceed its expectations at a cost penalty, the customer will not consider it to be a value. The product should not be over-designed in areas where the customer is insensitive [Dika and Begley, 1991].

o   *Cost*: As quality is the price of admission in the competitive marketplace, cost is viewed as the "ticket of survival." Cost is influenced by many factors, and the relationship is more complex. For example, an increase in quality, and for that matter any x-ability, has an adverse effect on local cost. Cost reduction cannot be achieved by keeping status-quo. Cost reduction means looking at alternate concepts, materials, and process driven designs. Fixed costs (direct labor and materials) as well as variable costs (indirect labor and materials), and those due to waste in all forms, complicated manufacturing system, inventory costs, equipments (e.g. single-purpose machines), facilities and layout are all ingredients of "process-driven costs". Reducing waste and process driven costs helps achieve higher profit margins and offers the opportunity for the manufactures to price their products more competitively.

13

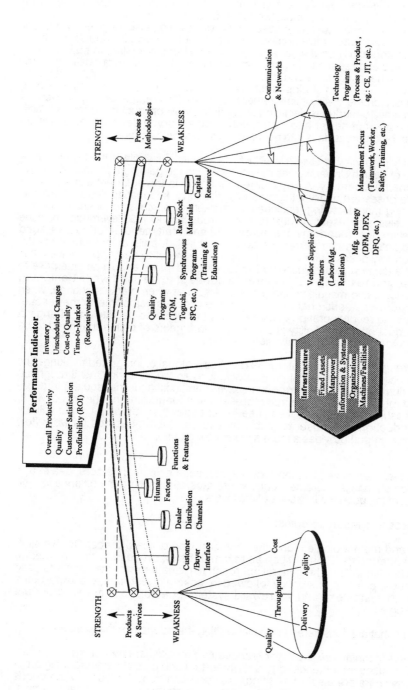

Fig 9: The Balancing Elements for Manufacturing Competitiveness

o   *Delivery (fast-to-market, responsiveness)*: In today's global marketplace, time is becoming a major competitive force. Quality used to take this place but not anymore. Today customer sees quality or anything else as given or guaranteed. Every automotive company wants to get its new car models on showroom floors in record time. Airlines spend billions of dollars in maintenance facilities and repairs to cut the non-flying downtime. Overnight delivery carriers are handling more and more packages with services faster than anytime before. Even food chains (such as pizza parlors) are competing for home delivery on the basis of time rather than taste. The range for responsiveness and time-to-market is declining every time a new product is introduced. With the advent of newer and faster technologies, some day companies will be able to compress days into hours and minutes into seconds.

o   *Agility (economy of ease or flexibility)*: A prestigious study at the Iacocca Institute of Lehigh University has defined agility as the paradigm for manufacturing in the next century. A key element of agility is the agile enterprise. Agile enterprises are totally integrated organizations. Information flows seamlessly among manufacturing, engineering, marketing, purchasing, finance, inventory, sales, and research. Just as mass production in 1950 leveraged "machine intelligence" (economy of automation), and craft manufacturing leveraged people's skills and dexterity (economy of skills), agile manufacturing is the economy of ease with which a company can react to new opportunities. It is not a technological solution that, once plugged in, will enable an enterprise to respond to change. It is merely a concept to be used while designing products and manufacturing processes so that when markets change, when customers needs change, agility in the process can come to the rescue, allowing reconfiguration of the set-up with calculated penalty and change in product offerings. Agility is part of measurements of merits that need to be considered in product realization.

o   *Throughput*: A global enterprise requires a global or integrated thinking. The former CAD measures of productivity, such as "engineering CAD throughput" or number of drawings creation in a week are local (thus inadequate) throughput measures. It is not a true representation for maximizing the flow of work. Measures on "number of engineering changes," or "ramp up time to volume production" are closer to global measures. Such measures on work-flow (through-put) provides a more accurate big picture.

Besides, there are several secondary weights, such as human factors, dealer and distribution channels, customer & buyer interface, that can tip the balance in their favor depending upon the locations of the levers.

5.2 Process and Methodologies:

Process and methodologies form the "content" side of the balance. On this side, value is added to the inputs (such as raw materials), process or the knowledge (such as human skills). The general approach to management , the concepts used to market the product, the policy followed in making investment, the importance assigned to customers, and the reward system for employees are all important methodologies.

The factors that are considered influential in tilting the balance on this side are:

o   *Lean Manufacturing or Synchronous Organizations*: Lean or synchronous, often used interchangeably, is an important and crucial manufacturing process. There are a set of 17 things that one can do to be lean. Synchronous organization improves efficiency through such things as systematic elimination of waste, error-proofing, just-in-time inventory, work place organization, and 13 others. Lean manufacturing means developing an environment which is conducive to synchronous principles, elimination of wastes such as bottle-necks of information or material movements, smooth flow of work, rationalize committee structure, etc.

o   *Manufacturing Strategy*: This includes strategies such as DFM, DFA, DFQ, DFX, etc. (see figure 10)

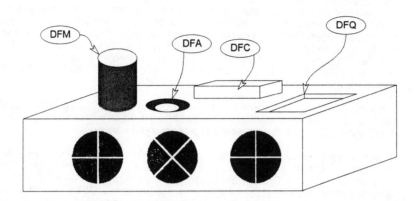

Figure 10: Examples of Competing Manufacturing Strategies

o   *Manufacturing Automation Protocol (MAP)*: The proliferation of stand-alone manufacturing tools made it difficult, or impossible in some instances, for one machine to talk to another on the factory floor. MAP was subsequently developed by General Motors and its industry partners to facilitate and simpli-fy communications between the growing number of machines and computers in today's automotive factories. MAP allows equipment to share data or "talk" to each other for improved efficiency in Computer Integrated Manufacturing (CIM) factories.

o   *Management Focus*: Companies will adopt a management style that allows more leeway to workers (empowerment), pushing decision making to the lowest appropriate levels, and greater flexibility to managers. It will reduce the reliance on corporate headquarters and begin to establish intra- and inter-corporation projects, with multidisciplinary focus teams with a great range of expertise. There will be emphasis on "the process" and continuous improve-ment, long range thinking, and total customer-orientation in the projects that are undertaken. There will be more and more decentralization of activities and better coordination between the groups involved.

o   *Use Common System*: Organizations must develop technologies that use common systems to produce a variety of products, each with its own unique features and distinctions. This means that processes and technologies are not developed with a single solution in mind; they are developed to be used in a variety of applications. Technology in itself should not be the goal of an organization. Rather, an organizational objective should be to develop tech-nologies that have broader applications throughout the SBU for solving exist-ing and anticipated problems.

o   *Deploy Common Systems*: Organizations must deploy common systems -- standardized tools, methods, and practices to design and develop products, tool-rooms, die design, fabrication and construction, and business support infrastructure.

o   *Vendor/Supplier partners (Labor Management Relations)*: Manufacturing enterprises can, and most likely will, have borders going well beyond the confines of an individual company subsidiaries. This postulate is supported by the success of Japanese Keiretsu and is the key to the agile manufacturing concept [Goldman and Priess, 1991]. The latter is manifested in many recent international developments resulting in cooperative agreements such as U.S. congressional and presidential directives in USA and Japan to mandate transfer of technology from federal research facilities to private sectors such as CRADA (Cooperative Research & Development Agreement) and U.S. National Science Foundation and Department of Commerce's strategic partnership initiatives.

o   *Strategic Sourcing*: Most companies are adopting policies to optimize outside supplier base, develop partnership with strategic suppliers to develop mutual trust, and enter into various discounts arrangement concepts.

o   *Data Communication and Networks programs*: The sleuth of Electronic Mail (E-mail) and Ethernet network (such as Local Area Networks (LAN), Wide Area Network (WAN), Broad band) systems and a new generation of workstations are providing direct links between work-groups such as design, analysis, manufacturing, testing, and the CE team users. Improved communication features allow data to be transferred rapidly at a higher rate in a one-to-two order of magnitude faster than what was feasible few years ago. Networks like LAN can quickly transmit huge data files almost instantaneously to a large number of interconnected users. Programs on these networks thus enable the rapid transmission of design information between work-groups and teams facilitating the CE approach.

o   *Electronic Data Interchange (EDI)*: one of the biggest barriers to effective CE programs has been the lack of a common graphics exchange standard. Much of the design generated by a CAD/CAM supplied by one vendor cannot be recognized directly by another CAD/CAM system. This has been alleviated at least partly by the introduction of exchange standards. Initial Graphics Exchange Specifications (IGES) is the most common approach that is used currently to allow dissimilar systems to talk to each other. In this approach, the transmitting system translates data into a second language -- a so called IGES neutral file -- that can be sent to different systems. Translators at the receiving end reformat the data into appropriate native forms. Today many standards are being explored such as PDES and ISO STEP standards. With such standards the initial product definition data-transfer capability of IGES has been extended to include feature-based and object-oriented representations. They are currently being expanded to many application areas, such as sheet metal forming, finite element, numerical control, process planning, etc. As this range of functionality and application interface expands further, it will be easier for engineers, designers, and manufacturing personnel to transmit data regardless of individual brands of software and hardware systems that are used.

o   *Technology Programs*: This includes programs such as just-in-time (JIT) manufacturing, Pull system, Concurrent Engineering, Relational data base systems. There is no universal form that can fit all. These and other automation programs must be tailored to the needs of the individual company.

o   *Raw Stock or Materials*: Raw materials are defined as any truly variable cost of producing a unit of a product.

o   *Synchronous Programs*: The objectives of synchronous programs are quite focussed -- they aim at productivity improvement, performance improvement (such as quality and other x-abilities), and elimination of popular waste. These programs are often directed towards materialist cost improvements, reduced manpower through process-driven design concepts (such as simplified manufacturing), inventory cost reduction through the combination of techniques such as in-line sequencing, just-in-time delivery, defect free supplier, and improved material handling. Programs include training and education.

o   *Quality Programs* (such as Quality Circles, TQM, Taguchi, SPC, Six Sigma Program, etc.): One of the principles of Statistical Process Control (SPC) is that inspection, once a part is produced, even if it is done on a statistical basis, is wasteful. By the time a part comes off the end of a production line and is inspected, many bad parts will have been produced. It is likely that the special cause that produced the bad part will not be easy to identify. What needs to be done is the measurement of the parameters within the process on a real-time basis in order to be sensitive to the cause of the variation and address it. Six Sigma program is named after the statistical figure of six times the standard deviation measurement. Like many earlier ones, these programs require re-engineering of the organization and how it operates. Quality programs should not only be directed towards minimizing defects in production, but also augmenting the capability of the products to monitor their own operations (SPC to determine the level of variability in the process, on-line factory information and performance feedback system to monitor the flow of the product through manufacturing processes, and corrective actions to eliminate those variations). It requires bench-mark schemes, rationalized schemata for products and processes, and a company wide strategy for implementing team communication networks. The success will depend upon the check and balances of the operations, for example, potential failures detected at any level need to be relayed back to team leaders. Through the chain of commands it ought to be conveyed to the design team for error-proofing the process so that it does not occur again. Such quality programs will not only create good products but will also ensure continuous improvement guaranteed over the life of the product.

Figure 11 shows three examples of how such a process and capability work in sync to meet both the company and the customer interests. The processes are "manufacturing strategies," "Agile/Modular/Flexible Production system," and the "quality programs." As shown in Figure 11, the company interests are represented by one or more of the teams (multidisciplinary, multi-functional or core competency). In the case of "manufacturing strategies" process example, teams are shown to take the initial step in determining the appropriate strategies, such as JIT, CE, Poka-yoke, etc, which are suitable for the problem at hand. This means that they would be helpful in either enhancing the product values to the customer, or meeting the company interests (such as profitability). In the case of the "quality programs" example, the process is reverse. The customer (in terms of VOC) provides the basic needs and wants, which in turn are translated into appropriate quality programs to be used in product design and development. Many of the processes and capabilities described in this section will have similar scenarios. Like the examples in Figure 11, they will be leveraging the strengths of each other in protecting both the company and customer interests.

6. Performance Indicators:

Figure 12 shows a list of eight areas that determine the performance of an enterprise competitiveness. Each indicator provides a measure of a company's efficiency in the world marketplace. The points on the full circle show their current state of performance. Depending upon whether the performance is to be minimized or maximized the corresponding point is shown inside or outside the circle. The performance indicators must show profit at the current conditions as well as point the CE teams toward the new product development or technology insertion. New accounting measures (such as, Activity-based-costing (ABC) and Goldbratt's theory) are often used.

COMPANY

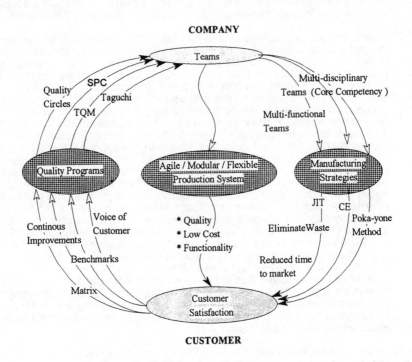

Figure 11: Examples of Process and Methodologies for Meeting the Company and Customer Interests

o *Productivity (gain or loss):* Productivity means overall gain or loss. A higher level of productivity in one specific department or discipline is not a good measure. Productivity means creating concepts that positively impact on both the upstream and the downstream operations.

Productivity (P) = T / OE;

where T is a throughput and OE is operating expenses.

o *Customer Satisfaction:* Customer satisfaction means meeting the customers' needs, at the right time, and in the quantity, price, and performance they want. The cornerstone of these performance measures is the customer. It goes without saying that if the customer does not want to buy the product, improvements in cost, weight and investment don't really matter. At the same time, if the customer becomes disappointed with the workmanship of the product or encounters problems over its life, he or she will not buy it again.

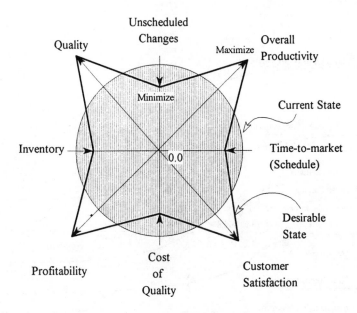

Figure 12: Performance Indicators of an Enterprise Competitiveness

o   *Unscheduled Changes:* The success of rapid product realization depends
    upon the team's ability to handle unscheduled changes. Unscheduled
    changes occur in many ways: some are avoidable some are not. Avoidable
    changes are typical of products thrown over the wall before they were ready
    for manufacturing. Once the parts are sent back to originating team, un-
    scheduled changes have to be squeezed in between work. Unavoidable
    changes occur when circumstances change, people move, and the steps are
    no longer valid. Unwanted changes are caused by changes in product lines,
    product functionality, technology, etc. Though number is an important
    measure, unscheduled changes can be very serious. For example, if errors
    are detected late (say during a downstream operation), it might be very costly
    to implement them.

o   *Inventory (I):* Inventory includes all assets including property, plant, and
    equipment, but excluding value-added parts. The new definition broadly
    stated includes any item which the company could sell, not just the finished
    products. By including capital assets in the inventory category, teams are
    forced to focus on the way they are utilizing all of the investments under their
    control.

o   *Cost of Quality:* There are two contributory elements which effect the cost of
    quality: (a) cost to ensure quality (c-t-e-q) and (b) cost to correct quality
    (c-t-c-q). They are shown in Figure 13.

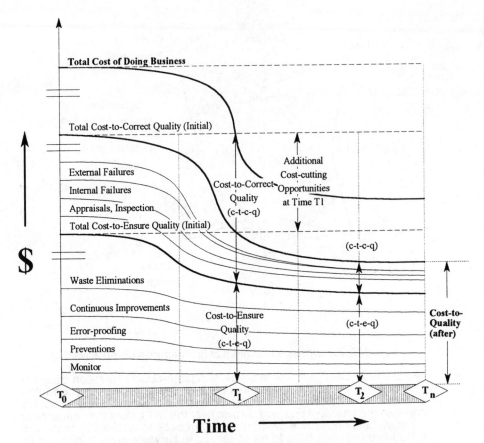

Figure 13: Cost-cutting Opportunities through Preventions

Cost to ensure quality is the cost of doing things right (e.g. choosing the right process), the cost of doing right things (e,g. choosing right actions), and the cost of preventing mistakes (such as anticipating problems). Prevention costs are the costs associated with those activities that ensure or build quality during designing, implementing, and manufacturing products and services. The cost to correct quality is the cost incurred because of doing things wrong (e.g., choosing wrong process), the cost of doing wrong things (e.g. choosing wrong actions), and the cost of inspections to discover mistakes committed earlier. Appraisals are the costs associated with evaluation, measurement, and audit to assure "conformance to quality requirements." Other types of cost-to-correct-quality are internal and external failures. Internal (or external) failures are the costs associated with a product or service that does not meet the quality requirements (such as building codes) prior to transfer (or after transfer) to the customer. Most cited product quality indicators attempt to measure the parts per million (PPM) level of conformance. This does not, however, account for criticality -- for example a 1 dollar part failure may result in a thousand dollar component failure if one part is encapsulated into another. Another measure of overall effectiveness is to track cost of quality, both "cost-to-correct-quality" and "cost-to-ensure-quality."

C-t-q Effectiveness = = [{cost-to-ensure-quality} / {cost-to-quality}] *100

where, cost-to-quality is the sum of two parts.

If the C-t-q effectiveness number is close to 100, the company is doing things more right than wrong. The effectiveness number thus provides an analytical aid to decision making or to track quality improvement opportunities.

o   *Profitability (ROI):* The return on investment (ROI) is defined as the ratio of throughput (T) minus the operating expenses (OE) to inventory costs, i.e.

$$ROI = = [\{T\text{-}OE\} / \{I\}]$$

where, throughput is defined as
Throughput (T) = Net Sales - Raw Materials

where, Net sales (or volumes) are defined as the irreversible transfer of product to the consumer. Such a definition of sales does not allow the transfer of goods in a consignment from a manufacturer to a dealer to be counted as a sale. OE is computed using all normal operating expenses plus direct labor and factory overhead. By grouping direct labor and factory overhead in an OE category, there is little reason for teams to overbuild their inventory. Direct labor is recognized as a fixed cost.

o   *Time-to-market:* This is a measure of the time duration required to design and develop a marketable product (from concept through rate production).

Additional performance indicators that are being used are in the areas of delivery, risk management, and teamwork communication. The key to understanding customer satisfaction is the recognition that there are two basic type of activities: support and value-added. While support tasks are necessary for internal planning and control, they consume the teams' effort and time while providing no direct benefits to the customer.

6.1 Cost-drivers:

There are three main cost drivers during the entire product life from conception to grave: company costs, user costs, and society costs (see Figure 14). Company costs are the costs of activities required in planning, design development, assembly, production, distribution, and servicing a quality product. It includes all costs that are incurred from needs to delivery up to the point when the product is shipped to the customer. User costs are the costs to the users of those activities that are performed by the user from the time when the product is delivered or shipped, to its disposal when the ownership ends. The society costs are the costs that are inflicted on the society from the time the product is in user custody until it is disposed off or recycled safely. The ability to recycle the material, or its impact on the environment, is the major contributing cost to society. Many of these costs are intangibles and cannot be measured or quantified accurately.

7. Agile Virtual Company

The term agile emphasizes "economy of flexibility or ease" and virtual applies to "economy of teamwork or cooperation" and "economy of seamless flow of information." The trio contributes to the "economy of change." Together they translate the intellectual power of the company into new and constantly evolving products, processes, and services. An agile virtual company is thus built on the intellectual strength and cooperation of the concurrent teams to constantly produce new products quickly to meet evolving demands. Its products will be customized to the needs of its customers. When the "agile virtual company" is in place, authority will be diffused (and concurrent) throughout the enterprise, and decisions will benefit from

22

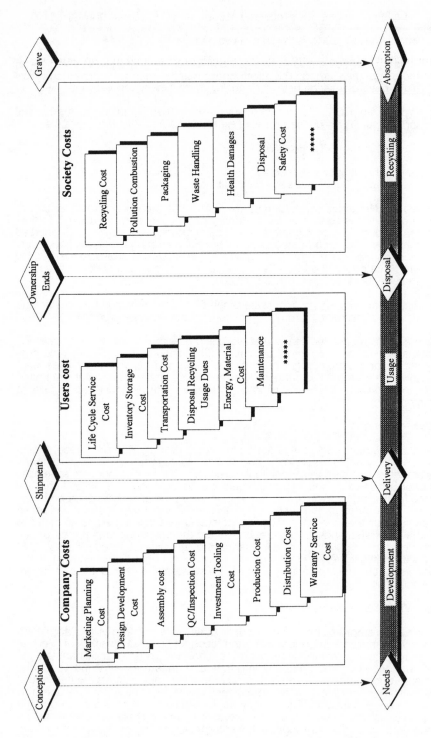

Figure 14: Life Cycle Cost Drivers

seamless flow of information among diverse life-cycle work-groups, e.g. engineering, manufacturing, marketing, etc.

Major technical breakthroughs may not have to be shifted into an "agile virtual" paradigm, just like none had to be shifted for lean or synchronous manufacturing trails. A common understanding of the critical factors that influence the quality, time to market, customer satisfaction, speed, efficiency, and productivity is essential. A technological system push can help the concurrent teams within the "virtual enterprise" to make the best decisions on how to respond quickly to change. In the "agile virtual" world, partnership might dominate the marketplace because no single company is likely to have all the intellectual strengths to be able to compete on its own.

On the manufacturing side, the definition of "flexible" resides in the material handling and production equipment, systems, and the workforce. Figure 15 compares five manufacturing traits -- craft, mass, lean, agile, and virtual. They are compared against a number of organizational terms that are essential to manage change. These terms are:

7.1 Organizational Terms:

o   *Reconfigurability:* the ability to easily upgrade a product or process to quickly react to changes in the market conditions.
o   *Scope and Applications:* the extent to which this change can be applied. It defines the aspects of the product life cycle or production chain over which the change can be managed.
o   *Management participations and Commitments:* A steady level of participation and commitment from the beginning to the end of the project activities.
o   *Partnering:* A strategic alliance approach where closer relationships are established with suppliers and customers.
o   *Interface-compatibility:* ability to interface with a work-group (say a partner) through standardized policy, procedures, and manufacturing practices.
o   *Plug-in-compatibility:* ability to plug into any hardware and software environment, irrespective of the brand name and applications.
o   *Communication-networks:* ability to provide inter-operability (robust communication network) capabilities up, down, within, and across organizational lines in a multi-vendor environment.
o   *Empowerment:* the freedom of decision making as low as possible in employee ranks, at the point of maximum information.
o   *Policy, Procedures, and Practices (3Ps):* ability to electronically use 3Ps to prevent the possibility of undesirable conditions that could possibly contribute to a problem.
o   *Motivation and Rewards*: A system of rewards that are based on both teams' success and individual creativity.
o   *Responsiveness:* ability to reduce both the time it takes in each life cycle operation and its cumulative total.
o   *Team-cooperation:* from everyone including supplier-base, so that simultaneous development of the product, process and business operations can take place.

This book deals with modern manufacturing knowledge that is sufficiently general to provide at least partial intellectual support for a number of specific CIM and CE case histories, applications, and introduction of new products and systems. A number of fundamental techniques and methods for achieving manufacturing realization are extracted from the currently successful modern manufacturing theory and practices. These are presented in the fifteen chapters of this book.

Manufacturing Traits	Craft Manufacturing	Mass Manufacturing	Lean Manufacturing	Agile Manufacturing	Agile-Virtual Manufacturing
Economy-basis	Economy of skill	Economy of scale	Economy of Kaizen	Economy of Flexibility	Economy of Cooperation
Organiz.-Focus	Short-term	Short-term	Medium-term	Long-term	Long-term
Reconfigurability	Performs within a fixed production cycle	Performs within a fixed production cycle	Performs both within & across production cycle that exists.	Reconfigures a factory for unplanned production reqts.	Partners with V/C to exploit an unexpected market opportunity.
Scope & Applications	Takes account of specific aspects of components manufacturing.	Automates the specific aspects of components manufacturing.	Takes account of entire production chain from customer to disposal.	In addition to all elements of "lean" scope -- includes process re-engineering & corporate strategies.	In addition to all of "agile" scope, includes partnership, teamwork,cooperation, collaboration and information highway links
Decision-making Style	Top-down (1-d) Directive Management Style	Top-down (1-d) Directive Management Style	Mostly Top-down (1-d) Directive Management Style	Top-down, Bottom-up (2-d) Supportive Managment Style.	Bottom-up, Top-down (2-d) Goal-oriented Decision Making
Mgt. Participation & Commitment	Bottom-up driven, no definite actions.	Little management commitments & actions	Bottom-up driven, some mgts. commitments & actions	Top-down driven, medium level mgt. commitment & actions.	Both bottom-up & top-down driven a large mgt. commitment & actions
Partnering: strategic Approach to developing Products	Personal strength & skill	Machine's capabilities, technology utilization & insertions	Teamwork & collaboration within groups	Temporal partnerships of workteams, cooperation across groups, suppliers & customers	Teamwork & cooperation across groups, intra-organization, business relationships & teamwork
Interface-compatibil	None	None	Small	Medium	Large
Plug-in Compatibility	None	Little	Small	Distributed Computing	Cooperative Computing
Communication Network	Through voice & paper	Through machine language and instructions	Through information network or pipeline	Through seamless network across & within organization	Meeting-on-the network: Free echange of ideas up, down, & across
Empowerment	No empowerment	No empowerment	Transfer of decision making to top level managers	Transfer of decision-making to mid-level managers	Transfer of decision-making as low as possible in employee ranks (such as
Policy, Procedures, and Practices	Extremely rigid, complex procedures, organization dictated most of the steps	Less rigid and more flexible than "craft manufacturing"	Less rigid and more flexible than "mass manufacturing"	Less rigid and flexible than "lean manufacturing"	Extremely flexible, simple procedures, teammates dictate most of the steps
Team Cooperations	No team cooperation, high personal contrib.	Low team cooperation, medium individual contri.	High team cooperation, low individual contributions	High team cooperation & individual contributions	Very high team cooperation and less of "hidden" goals.
Motivation & Reward System	Is based on individual creativity & contributions	May be little better than "craft manufacturing"	May be little better than "mass manufacturing"	More of team-based and less of individual-based	A reward system that weighs both team and individual contributions

Figure 15: Differences of Manufacturing Traits

8. REFERENCES

Dika, R.J., and Begley, R.L., " Concept Development Through Teamwork - Working for Quality, Cost, Weight and Investment", SAE Paper # 910212, *Int'l Congress and Exposition, SAE*, Feb. 25-March 1, 1991, pp. 1-12., 1991, Detroit, MI.

Ferdows, K., and De Meyer, A., 1986, "Towards an Understanding of Manufacturing Strategies in Europe - A Comparative Survey 1985-1986", *Insead Research Paper*, 1986.

Goldman, S. and Priess, K., Editors, *21st Century Manufacturing Enterprise Strategy: An Industry-Led View*, Volume I, Iacocca Institute, Lehigh University, Nov., 1991.

Hauser, R., and Clausing D., 1988, "The House of Quality", *Harvard Business Review*, May-June 1988, pp. 63-73.

Putnum, A.O., 1985, "A Redesign for Engineering", *Harvard Business Review*, May-June 1985, pp. 139-144.

Ross, E.M., 1992 " CALS: Enabling the New Manufacturing Paradigm", *CALS Journal*, Winter 1992, pp. 29-37., 1992.

Wheelright, S.C., and Hayes, H., 1985, "Competing Through Manufacturing", *Harvard Business Review*, Jan-Feb. 1985, pp. 99-108.

Whitney, D.E., 1988, "Manufacturing by Design", *Harvard Business Review*, July-August, 1988, pp. 83-91.

Whitney, D.E., 1992, "State of the Art in Japanese Computer-Aided Design Methodologies for Mechanical Products -- Industrial Practice and University Research", *Scientific Information Bulletin*, Office of Naval Research Asian Office, NAVSO P-3580, Vol. 17, No.1, Jan.-March 1992.

Wilson, P.M., and Greaves, J.G., 1990, "Forward Engineering - A Strategic Link between Design and Profit, *Mechatronic Systems Engineering*, Volume I, pp. 53-64, 1990.

2   Enterprise Modelling*

Albert W. Chan

1 Introduction

In response to the development of new technologies and the competitive pressures of a global economy, the nature of manufacturing is changing. Companies once organized along hierarchical lines, using traditional equipment to produce standardized products for local markets, are being replaced by flexible organizations and company networks employing advanced, reconfigurable technologies to produce quality products in order to meet specific customer requirements in international markets. Technology is advancing so rapidly that many manufacturers are already operating in an environment of continuous, and often dramatic, change.

While most manufacturers in virtually all industries are experiencing severe market pressures, the companies that survive and excel will be those that are able to quickly bring new products to customers who demand ever-increasing levels of performance, value and quality. And they will do so in a global environment. As business opportunities will not last long, competitiveness will have to depend, above all, on timing. Companies must be able to rapidly identify new market opportunities, establish supply and customer networks, master an appropriate production technology, and have the right production facilities available to launch into production and capitalize on the window of opportunity. This may lead to the formation of virtual companies around the world based on advanced manufacturing technologies, modular production systems and organization structures, networks of alliances among suppliers, ·producers and customers.

World-class manufacturers recognized this challenge, and have begun grooming the agile, efficient and effective new-product development techniques necessary for the accelerating pace of innovation. Concurrent engineering concepts are being exploited which may also lead to virtual reality systems, allowing manufacturers to design, develop and test products and processes jointly with their internationally-based suppliers and customers. One class of techniques which is gaining prominence for

* NRC No. 37110

meeting these challenges is to utilize fully integrated computer-based tools to model, analyze and re-engineer the organization and processes in a manufacturing enterprise.

This chapter presents enterprise modelling as an important candidate of such a technique. A particular Enterprise Modelling System developed at the National Research Council of Canada is described. The relevance and potential applications of such a system to modern manufacturing are discussed.

2 Enterprise Modelling and Integration

Manufacturers are continually being reminded that they must change in order to survive in the global competitive economy of the 90's. This message of change is usually accompanied by a bewildering array of buzzwords and acronyms, such as total quality management (TQM), just in time (JIT), continuous improvement, concurrent engineering (CE), agile manufacturing, business process re-engineering (BPR), etc. Yet when faced with a barrage of ideas for modern manufacturing, how does a manufacturing enterprise decide what needs to be done? Rarely is the cry for change accompanied by a set of concrete steps for identifying opportunities for improvement, or a plan for implementing good solutions. Decision-makers need good tools to guide them to the right answers, ones that will incorporate much of the knowledge about how the business currently works, and are able to examine what the impacts of any changes might be. One important tool to help a manufacturer to understand the implications of the different approaches, strategies and options is simulation modelling.

Computer modelling and simulation is a powerful system analysis technique for evaluating and fine-tuning new and existing processes in a manufacturing system, and for providing a basis for tools to analyze organizational structures and the flow of information and control within an enterprise. It is equally useful for analyzing the operating characteristics of existing decision-making rules such as for inventory and scheduling systems, and for verifying control strategies for automated systems. In contrast to optimization techniques, simulation provides a descriptive model predicting how a manufacturing system will behave under given conditions. It is useful for comparing alternative process structures, facility layouts and cell configurations, and for studying the logistics and operation of material handling systems such as automated guided vehicles, conveyors and robots, as well as storage components such as automated storage and retrieval systems. Via statistical analyses, mean values and statistical distributions for activities can be utilized to simulate the dynamic interactions of processes. Subsystems can be modelled in detail, providing statistics for an aggregate model addressing the performance of the overall manufacturing system.

Enterprise modelling provides a holistic approach to model the business and engineering processes as well as to simulate dynamically the operations within a manufacturing enterprise. A proliferation of "islands" of information due to the piecemeal growth of organizations, lack of in-house standards for information handling and inflexible data processing technology has impeded improvements in efficiencies and productivity. While there are commercial MRP and simulation packages for planning material requirements and modelling of the operations on a production shop floor, an enterprise-wide model is more suitable for process engineering and concurrent engineering studies. By capturing the information exchanges and interactions between business processes such as marketing and purchasing, and engineering processes such as product design, production and maintenance, the impact of changes in one process on various performance measures of an enterprise can be studied. As well, the effect of changes in external variables, such as market share fluctuations, consumer preferences and demands, and material costs, and the relationship with vendors, suppliers and customers can be modelled. For example, the competitive advantage derived from the *extended factory* approach via establishing a family type relationship with a small selected group of suppliers/vendors can be ascertained.

Enterprise integration is concerned with improving and managing the performance of complex cross-functional processes among the functional activities in a given enterprise. It encourages continuous process improvements and the application of computer technology and information exchange standards to achieve maximum efficiency. The level and sophistication of the enterprise integration depends on the degree of efficiency and effectiveness with which the integration team collectively operates toward its goals. An integrated enterprise is characterized by its goals, teaming, flexibility, and people. A modelling tool to facilitate the design and re-engineering of business and manufacturing processes, and to study the impact of changes in a design process throughout the life cycle of a product before the design process is finalized, will therefore be an important asset to an integrated manufacturing enterprise. The expense of using the factory floor as a test bed for new manufacturing approaches and process changes can be obviated.

AMICE, a project undertaken by a European consortium under the ESPRIT (European Strategic Program for Research and Development in Information Technologies) program, has done some significant conceptual work in defining an open systems architecture for computer integrated manufacturing (CIM-OSA) [1,2]. A framework has been set up for the four components (Function, Information, Resource and Organization) of an enterprise. An enterprise model can be considered the basis of an integrated information system at three levels (conceptual, technical and implementation) and from three different views (data, function and organization) [3]. Data modelling emphasizes on the conceptual description of objects and their relationships and serves to identify the most important objects within the enterprise, but it does not directly result in an electronic data processing implementation model. There are two basic approaches to modelling: data-oriented approaches focus more on the modelling of static objects and their relationships, whereas function-oriented approaches focus more on modelling the dynamics. They can both facilitate the creation of simulation models.

Nelson [4] uses an Information System Framework (ISF) to support multi-state representation of entities, processes and events that characterize the behavior of an enterprise. Mertins, *et al* [5] follow an object-oriented approach by defining three main object classes to represent the different aspects of a manufacturing enterprise in Integrated Enterprise Modelling (IEM). While there has been much conceptual and theoretical work done to advance the methodology for enterprise modelling, only a small portion of the work has begun to be implemented. The enterprise modelling work discussed in this chapter follows a more pragmatic approach in that the models developed for an enterprise will be created with specific objectives in supporting decision making for the user. Rather than developing a model for deeper understanding and for analyzing the enterprise in general, specific processes, resources or operating practices will be modelled for process re-engineering, for example. Through the use of mostly drag-

and-drop operations and a friendly user interface, a user can experiment with alternatives by easily making changes to process definitions, resource allocation, decision rules, etc. Impact of these changes throughout the entire enterprise can be evaluated. The goal is a flexible, user-oriented decision support tool.

3 The Enterprise Modelling System (EMS)

3.1 Objectives

A prototype Enterprise Modelling System (EMS) has been jointly developed by the National Research Council of Canada and SIMCON. SIMCON is a consortium of companies collaborating with the Institute for Advanced Manufacturing Technology on Computer Integrated Manufacturing and related research. The main objective of the EMS is to provide a comprehensive set of tools for the creation of structural and process models of manufacturing enterprises. The scope of these models encompasses the business as well as production operations within an enterprise, with capabilities specifically aimed at continuous process improvement and evaluation of decision-making alternatives. These models of manufacturing enterprises can be applied to help address issues such as process re-engineering, concurrent engineering, information infrastructure design, and system integration.

There are significant advantages for an organization to clearly define and maintain up-to-date models of all processes, together with the flow of information and material between them. First, through modelling of the processes, people within the organization can gain valuable insight into the process details. Second, process models lead to opportunities for continuous process improvement by providing the ability to predict how changes in the process structure or changes within the steps of a process will affect the process yield. They help to position an organization to quickly respond to changes in the global environment, thus maintaining or actually gaining competitive advantage.

3.2 Potential Users

The Enterprise Modelling System can be used by application developers to create generalized sector-specific models for, say, the electronics, aerospace and automotive industries, or specific models for a particular organization. The application developer is expected to have a good understanding of both the functionalities of the modelling tool and the behavior of the system being modelled. Users in this group include industrial engineers, systems analysts, and process re-engineering specialists who have a good understanding of enterprise-wide processes.

Each enterprise model created by an application developer will, in turn, be operated by an end-user, such as a process specialist, for the purpose of improving the performance of the individual processes in the enterprise. The model will be run in a systematic manner to compare alternatives and to produce the information on which management and operational decisions can be based. Users in this group include decision-makers and designers such as corporate planners, product planners and designers, process engineers, manufacturing engineers, plant managers, production schedulers, maintenance engineers and sales staff, many of whom could be on a multidisciplinary team with enterprise-wide missions and concerns.

3.3 System Overview

The prototype Enterprise Modelling System has been built to provide a general framework, together with templates for modelling a range of activities found in an enterprise by including a suite of modelling functionalities and decision support capabilities. Selected modules have been developed in detail to illustrate how the tool can be used to set up a model of a manufacturing system. Business processes such as hiring and purchasing have been modelled, providing components for applications in business process re-engineering.

In addition to static process analysis tools, which include the cost and time charting for a process, the EMS provides capabilities for the user to analyze processes dynamically by gathering the statistics of interest while a simulation is running. Where multiple processes in an enterprise interact either in precedence relationships or sharing of resources, simulation can provide an added dimension to system analysis. The build-up of bottlenecks, the cumulative effects of delays, and the dynamic impact of various corrective measures can be monitored and examined using the EMS. The integration of process re-design features and simulation capabilities provides a powerful tool for analyzing complex manufacturing systems.

3.4 Object Oriented Approach and Expert System

The Enterprise Modelling System was developed following the object oriented approach and implemented in Smalltalk-80 using a discrete event simulation methodology. Whereas programs written in traditional programming environments using procedural languages are difficult to modify and extend, object oriented technology provides advantages in flexibility, portability and reuseability. Since the Enterprise Modelling System is aimed at providing a framework where only some of the modules will be fully developed to illustrate how the framework can be used to create a customized model, there is a likely need for the user to develop additional modules using the templates provided. Accordingly, it is crucial that the software developed be modular, reuseable and easily maintained, facilitating the modelling of large and complex systems.

In order to provide the flexibility for an Enterprise Model to support a wide range of decision-making issues, artificial intelligence concepts such as production rules and forward chaining inferencing have been incorporated in an object oriented environment. This facilitates the implementation of decision rules that can be easily modified, added or deleted. By including expert system capabilities, such as for rule-based inferencing, which are aimed at supporting the evaluation of decision-making alternatives, the Enterprise Modelling System becomes well-suited for examining the design and internal makeup of many decision-making processes found in an enterprise. How human creativity can be most efficiently incorporated into the decision-making process, thereby multiplying the benefits of re-engineering business processes and automating production processes, can be assessed. The EMS can also facilitate the establishment of a manufacturing strategy for an enterprise. This should be a dynamic process which allows periodic updates to the strategy by taking into account the changing conditions of external variables, such as competition, consumer taste, economic environment, and unpredictable occurrences in the marketplace. This unique combination of object orientation and expert system technology found in the EMS offers powerful potential to those organizations which are striving to be responsive to changing customer demands, and which are determined to improve their competitive edge in a global market.

4 EMS Features and Capabilities

4.1 Multiple Views of the Enterprise

The Enterprise Modelling System provides three different views of an enterprise to the user -- the organizational view, the resource view and the process view. Each of these views has a static as well as a dynamic aspect. The organizational view displays a hierarchy of business units which comprise the enterprise. The resource view facilitates creating new resources, copying the attributes of an existing resource to a new resource, and assigning responsibilities to a resource by allocating a set of activity template models. The process view provides an editable representation of activities and processes in the form of a network. The activities in a process are performed by the resources, which provide the link between the organization view and the process view.

The user builds an enterprise model by first creating an organizational structure consisting of business units or resource groups. This is followed by the allocation of the appropriate resources to the business units. The next step involves defining processes as hierarchical networks of sub-processes and activities linked together by data paths. The user can also create various process segments and aggregate them to form higher level processes. The user interface allows the user to create several versions of a process and then compare them using scenario analysis.

4.1.1 Organization View

In the Enterprise Modelling System, the organizational structure of an enterprise is represented as a hierarchy of generic *business units*. A *business unit* is any type of organizational or functional unit consisting of one or more resources (both human resources and equipment) for performing either manufacturing or non-manufacturing operations described through a set of processes. The behavior of each *business unit* is determined by a series of activities carried out by its resources in order to achieve the goals of its various processes. In a manufacturing context, a generic *business unit* can be a production department, a work cell or a workcenter. The user enters the hierarchical relationships of the *business units* in the form of a tree structure, and specifies the resources, processes and activities for each *business unit* via browsers. Figure 1 gives an example of the organization view.

The efficiency and effectiveness of an organization is dependent upon the way its resources are organized to handle the activities carried out in the enterprise. Specifying the hierarchical relationships between resources and groups of resources defines the infrastructure of the enterprise. Technological changes tend to repeatedly challenge the organizational structure, leading to restructuring within an enterprise. The Enterprise Modelling System provides capabilities for the modelling of the enterprise structure as a suite of business units and the relationships between them.

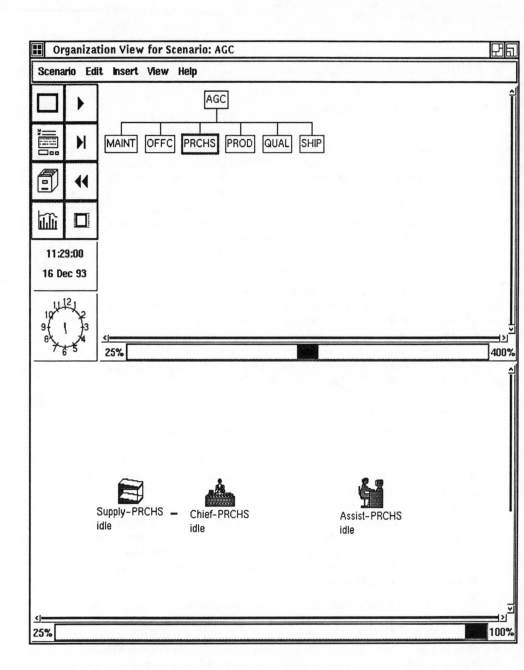

Figure 1. Example of Organization View and Resource View

4.1.2 Resource View

The Enterprise Modelling System categorizes resources in an enterprise into resource types for human resources (such as managers, computer personnel, and clerical personnel) and for equipment (such as material handlers, manufacturing workcenters, and storage buffers). Each resource type is represented by a user-defined icon and can have associated with it a unit cost and a list of user-specified capabilities, which are stated in terms of generic activities that can be performed by that resource type. The user can add to the library of resource types by creating new types or modifying existing types. As well, the day-by-day, hour-by-hour availability of a type of resource can be specified via a resource calendar.

In defining the configuration of a manufacturing enterprise, the user will specify the number of instances of each resource type inside each business unit. Each instance will inherit the capabilities and availability of the resource type by default, but the user can customize each instance, for example, by removing inherited capabilities or augmenting the list of capabilities. Associated with each individual resource are a resource calendar (which determines the availability of the resource), fixed and variable costs of utilizing the resource, statistical probabilities for breakdown and repair (for equipment), and other descriptive information for resource identification.

An important feature provided by the EMS is the association of a user-specified rule base with any resource. Where decision making is needed in the Enterprise Model, such as scheduling production, selecting the next task, or dispatching a material handler, either mathematical procedures, optimization algorithms, or rule-based inferencing can be deployed. This capability offers a host of opportunities for the investigation and comparison of decision support concepts.

During simulation, status information of the resources in any given business unit can be monitored by the user, while cost information and utilization statistics are automatically tracked by the Enterprise Model. Simulation results can also be charted for detecting trends. Figure 1 includes an example of the resource view.

4.1.3 Process View

The Enterprise Modelling System keeps a library of the processes found in an enterprise. Each process has a goal and is defined in terms of subprocesses and activities. A business process can have one or more standard operating procedures for achieving its goal. Similarly, a manufacturing process can have alternate process plans. The process view (Figure 2) provides a graphical representation of a process with specific features to facilitate process redesign, such as changing the process flow, and adding or deleting a process step.

An activity is a low level task that can be performed by one or more resources. Associated with each activity is a priority, time for performing the activity, and a list of resources needed to perform the task. Following the ISO reference model for shop floor production [6], activity templates have been defined for generic activities such as transform, transport, store, retrieve and verify. The activities and subprocesses in a process have precedence relationships, which can be described via a network representation of the process. Inputs and outputs of each activity are specified in terms of physical or informational objects, such as fixtures, parts, forms, etc. The average and

34

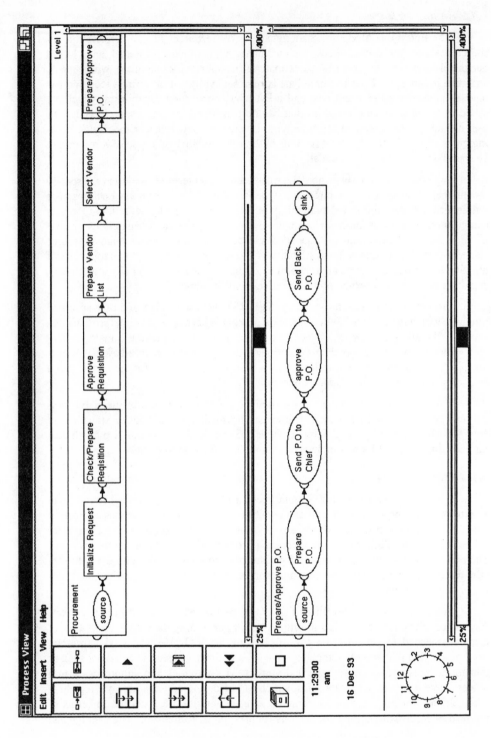

Figure 2. Example of Process View

total time and cost of the component steps of a static process can be charted for easy evaluation.

A process browser (Figure 3) is provided by the EMS for the user to organize the processes defined for the enterprise. Processes are categorized, and a cross-reference table displays the processes against the resources and business units in the enterprise. Through a filtering mechanism, the user can easily view the unit or resources responsible and contributing to a given process. Conversely, all those processes in which a given unit or resource is involved can also be displayed.

During simulation, service requests will trigger the execution of subprocesses and tasks of those processes being simulated following an event-driven approach. When there is more than one task competing for the same resource, decision-making rules based on various selection criteria, such as priority or task time, are used to determine the coordination of tasks. Random variations of the time taken for each task, and the random occurrences of breakdowns, delays and rejects can all be accommodated using statistical distributions. As well, process performance statistics, such as costs, throughput time, production, etc. are tracked, and can be presented as required.

4.2 Scenario Analysis

Another functionality of the Enterprise Modelling System facilitates the managing of simulation scenarios and process scenarios, both real and hypothetical, by allowing the user to record, name and track the genealogy of scenarios for the purpose of evaluating alternatives. By designing processes in the early stages of product development, the user can compare various scenarios by decomposing them into component processes and studying the concurrency of the associated activities. Since the concurrency of processes is increased by reducing the dependency among their activities, the degree to which the concurrency of activities can be effected will depend on the way high level processes are designed and aggregated into activity groups. The Enterprise Modelling System provides the tools for experimenting with various activity groupings with the intention of reducing dependencies among them, thus facilitating process re-engineering.

As a decision support tool, the EMS can be used to help set up various decision-making procedures to align with certain strategies or missions. Alternatives can be compared and decision rules fine-tuned by utilizing the scenario manager to perform scenario analyses. A friendly user interface invites the experimentation of innovative ideas, and a well-structured scenario management system will permit the user to freely exercise his ingenuity rather than being inhibited by the onus of keeping track of a large number of scenarios and their interrelationships.

4.3 Performance Measures, Statistics and Costing

The EMS provides capabilities for the user to gather the statistics of interest while the simulation is running. Performance measures relevant to specific processes, resources and activities can be defined, monitored and plotted graphically. The basis of a cost tracking system, enabling the value-added costs of parts to be tracked throughout the production process, has been set up. In addition to cost factors, the consideration of other performance measures such as process yield, cycle time, and resource utilization can also be monitored. Since an important objective of modern manufacturing is to re-

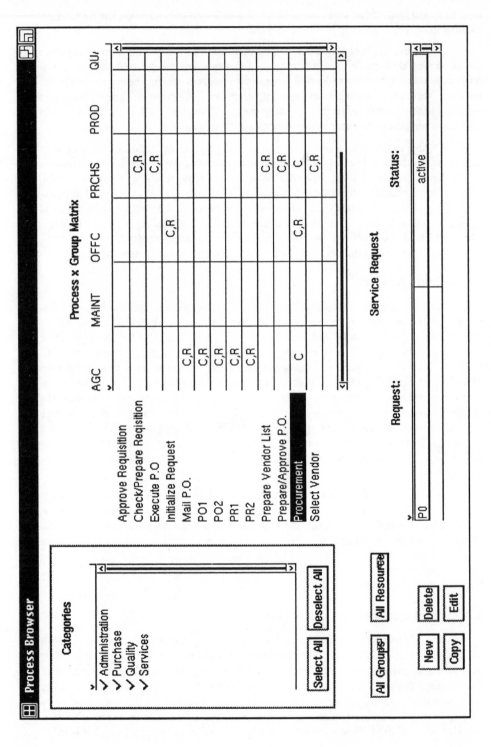

Figure 3. Example of Process Browser

duce the total product development time, these performance indicators form the bases for comparing two or more scenarios for a process.

In a global market environment, it is often necessary to establish company and product performance goals using competitive benchmarking. While attempting to re-engineer its manufacturing and business processes, an enterprise may need to compare the performance of its processes or products with comparable activities of its competitors, and more importantly, of the best world-class companies, and set performance targets accordingly. By measuring the effectiveness of the various aspects related to manufacturing, such as the product design process, the communication with suppliers, the distribution and servicing systems, and the methods for selecting equipment and manufacturing technologies, a company may decide to revise its manufacturing strategies. The resulting changes in market shares and profits can also be predicted. The EMS supports these important considerations.

4.4 Support for Decision Making

With information related to its organization structure, resources and processes captured in an Enterprise Model, an enterprise is well positioned to study its decision-making processes and to improve on its decision-making strategies. In facilitating the experimentation with different decision-making schemes, such as in scheduling manufacturing operations, allocating financial and human resources, plannning maintenance strategies, etc., the EMS can accommodate mathematical procedures, optimization algorithms, as well as rule-based inferencing. As users of the Enterprise Modelling System are not required to be familiar with Smalltalk, the decision-making sub-system [7] has the ability to provide explanations of each production rule in plain English, rather than in Smalltalk syntax. The system can also explain its reasoning by giving a list of those production rules that have been fired, in the order they were fired, to arrive at a conclusion. When the system requires certain data for a hypothesis, it will prompt the user for the data, and the user can, if desired, query why the information is needed.

Decision rules can be specified for activities or resources, permitting rule-based expert system capabilities to be incorporated into any number of decision-making objects and invoked where it is required. It is expected that the decisions for next-task selection for a resource, for selection of a particular resource amongst a group of resources to carry out an activity, for activity branching in process definitions, and for controlling starting and stopping conditions for activities are all candidates for utilizing the capabilities of the rule-based decision-making subsystem. As rules are kept separate from the underlying model for the manufacturing system itself, decision making can be distributed among the entities of the system, rather than being centralized, so that the decision-maker can capitalize on opportunistic reasoning. Cooperation between decision-makers at various levels can also be investigated. In order to allow for the easy modification of rules to reflect changes in the system or to test various scenarios, the methods of specifying the decision-rules used in enterprise modelling should be highly flexible. A rule editor has been designed for rule creation, modification and deletion, the use of which requires no familiarity with programming.

The object oriented paradigm adopted in the EMS not only allows artificial intelligence to be made available to any object requiring decision-making capabilities in the system, but it also permits intelligence and capabilities to be inherited through a class hierarchy. In modelling a real-life enterprise, this provides the flexibility for special knowledge, decision-making rules of thumb, and management authority to be propagated through selected parts of the organization structure if the modeller so desires. For any specific instance in this hierarchy, the user can override the inherited behavior in order to customize an individual or group at any level. Similar applications can be found in the modelling of equipment and machine controllers, where rule-based control logic, such as the routing of automated guided vehicles or robotic manipulation procedures, can be effected through an intelligent control hierarchy for shop floor equipment.

Via the decision support capabilities provided by the EMS, an Enterprise Model can be used as a test-bed for comparing decision-making strategies and fine-tuning those decision-rules related to the selected strategy. The interactivity provided by the rule editor and the explanation facilities enable the decision-maker to test out any innovative ideas he may have before actual implementation, while at the same time allowing him to iteratively formulate his decision rule set and to get immediate feedback from the model regarding its implications.

5 Application to Modern Manufacturing

This section explains why the Enterprise Modelling System is well-suited to be applied to a range of concepts relevant to modern manufacturing, and describes how the techniques to implement some of these key concepts can be formulated, assessed, and refined using a simulation modelling tool for the entire enterprise.

5.1 Information Infrastructure Design

In the light of pressures to globalize operations and of new competitive requirements such as increasing product quality and decreasing time to market, manufacturing companies must try to better manage the interdependencies both within their own departments and with external agencies. Information technology is being applied to solve business and strategic challenges associated with cross-functional integration, coordination and control of mutually dependent value chain activities, and team development across organizational and geographical boundaries. Factors that need to be taken into consideration include the extent of the organization's need for networking resources to exchange information among multiple business units, and the requirements to share data elements among business units or with external firms.

An awareness of the information technology capabilities which can be offered by an appropriately designed information infrastructure can positively influence process design. For example, information technology has made it possible for employees scattered around the world to work as a team. Knowing that standardized computer-aided design systems with common data structures for the design process will enable engineers for the same manufacturer to share and exchange complex three-dimensional designs across continents, might affect the structure of a product development process. The role of information technology and the accompanying information infrastrucure in a process should be considered in the early stages of its design or re-engineering. In the broadest sense, all of information technology's capabilities involve improving coordination and information access across organizational units, thereby allowing for more effective management of task interdependence.

The Enterprise Modelling System can be used to evaluate alternative designs of the information infrastructure using various criteria, such as the likelihood that a design will satisfy the chosen design objectives, the simplicity of the design, the role of buffers or intermediaries, the degree of control by a single individual department, the balance of process resources, and the generalization of process tasks so that they can be performed by more than one resource. Modelling the information infrastructure in a manufacturing enterprise will also allow the benefits and impact of information technology capabilities to be assessed. For example, unstructured processes can be transformed into routinized transactions. The capture and dissemination of knowledge and expertise in order to improve a process can be facilitated by the decision-making subsystem of the EMS. The detailed tracking of task status, inputs and outputs is also enabled. Different functional units within a process can be connected, thus eliminating the need to communicate through an internal or external intermediary. Increasingly, manufacturing companies are concerned with coordinating activities that extend into the previous or next company along the value-added chain. An information infrastructure allows the implementation of inter-organizational linkages such as via the use of electronic data interchange (EDI).

In short, the Enterprise Modelling System not only facilitates predicting the effectiveness of an information infrastructure design, but it also provides an environment to compare different designs and to fine-tune a selected design of an information infrastructure before final implementation.

5.2 Process Modelling

Process modelling is the structured and formal description of processes through a defined representation framework or model. Whereas a model aims at reducing the complexity of understanding or capturing the interaction with a phenomenon by limiting the detail that does not influence its relevant behavior, a process model is an abstract description of an actual or proposed process, involving selected elements of the process that are considered important to the purpose of the model. Process models can be used to facilitate human understanding and communication about processes as well as to support process improvement and management through providing a basis for designing and analyzing processes.

In high level process design, activities needed for products and services are defined and arranged so that desired output may be obtained at a predetermined rate and quality level. In manufacturing, these activities range from purchasing and processing of raw materials to assembly, packaging and shipping of products. A process design can be evaluated in terms of its effectiveness, efficiency and adaptablility. A good design is one that will yield quality products on a consistent basis, is economical, and can accommodate changes in operating conditions. Process flow modelling is a tool for achieving such design objectives.

To do a detailed process analysis and design, one can start with the familiar process flowchart, which consists of blocks representing activities connected by links representing flow of entities through the blocks. The process view of the Enterprise Modelling System provides helpful features for the formulation of process flowcharts, and provides a means to document and maintain an up-to-date description of the processes for an enterprise. Once entered, the processes can be analyzed and evaluated according to various performance measures. Examples of commonly used performance measures accommodated by the EMS are: lead time, queue time, cycle time, yield load, backlog, utilization factor and cost variance.

Process flow modelling is useful in predicting the effect of productivity improvement measures such as combining similar operations, minimizing or eliminating nonvalue-added activites (e.g., material storage, transportation, inspection and rework) and implementing a new method. The models can also be used as a training tool for supervisors and new employees.

5.3 Business Process Re-engineering

Traditionally, organizations viewed improvements in terms of automating tasks or building upon existing business practices. Individual tasks or functions within a department were optimized independent of the overall business objectives. These companies implement "changes" to improve productivity by tinkering, streamlining and automating. They talk quality and change initiatives, but keep doing business in basically the same old ways, only making incremental improvements in productivity. In contrast, re-engineering is a mindset rather than a step-by-step process, requiring cross-functional teamwork by aligning a company's resources with its mission. Rather than thinking of solely providing a product, this mindset requires companies to strive to provide a total solution to the customer. Companies that were once strictly manufacturers must now consider providing a total solution package for the customer, including such elements as diagnostics, after sale service monitoring and maintenance. They must nimbly deal with markets that are moving targets, allowing the customer to define quality, breaking with tradition and questioning the value of every current activity. They continue to re-engineer themselves, recognizing that flexibility is more important than size, and delivery time and quality are replacing low price as a customer incentive.

The re-engineering of core business processes views the enterprise not as a sequence of functional activities, but as a set of core processes, each of which consists of interrelated activities, decision, information and material flows, which together determine the competitive success of a company. These core processes cut across functional, geographic, business unit and even company boundaries. The organization is reoriented from performing as a cluster of suborganizations -- each pursuing its own, often conflicting objectives -- to integrating activities within a number of core processes, where each core process focuses on achieving one or two overall objectives of competitive success. Redesigning core processes can simultaneously improve performance along the multiple dimensions of time, quality and cost.

Whereas traditional quality improvement techniques seek to continually fine-tune existing methods, business process re-engineering is concerned with radically altering these methods. Organizations need both re-engineering and quality improvements to stay competitive. Quality and re-engineering are distinct in motivation, objective, technique and result, and the business circumstances in which they are applied are also different. Quality improvement is necessary, but ineffective at a business-wide level due to its functional activity viewpoint. Re-engineering alters the process, eliminates waste, and redefines the steps and the jobs, by focusing on the outcome rather than the task. By taking a high level, holistic view, it also exposes process shortcomings and the inhibiting "rules" of the organization -- the often unarticulated beliefs about how a process should work. Whereas information technology is incidental to the success of most quality improvement programs, it is essential and fundamental to re-engineering programs.

The multiple views of the enterprise provided by the Enterprise Modelling System constitute a powerful toolset for business process re-engineering. Manufacturing and business processes can be modelled, documented and re-engineered. The related needs for restructuring the organization, the implications of resource re-allocation, and the corresponding impact on performance measures can all be predicted and assessed using the EMS. Static process analyses, dynamic simulation modelling, as well as "what-if" scenario experimentation, can be accommodated. The graphical representation of processes and the friendly user interface allow the re-engineering specialist to visualize and evaluate various redesign alternatives without being encumbered to manage an excessive amount of process details.

As an example of re-engineering, a manufacturer may redesign its engineering parts-purchasing process as follows. While previously the responsibility of purchasing specialists, part-purchasing is now carried out by engineers who will identify potential sources of supply for the components they need, and then select suppliers on the basis of price and delivery date. The engineers are supported by an information system that they are encouraged to update. The redesign will likely reduce the effort as well as the elapsed time required to place a component order.

Reduced cost and improved responsiveness follow from the elimination of multiple steps in the work flow in functional organizations. Instead of the work passing up and down hierarchies and across organizational boundaries, it stays with an individual or a team throughout the process. The members of the team are empowered to make decisions, actively becoming involved in all aspects of a given process or solution, leading to better motivation and a sense of satisfaction than people performing simple, repetitive tasks. When a single team holds end-to-end responsibility to a product or document, accountability and quality increase. Each team must operate with a common set of performance measures, focusing on throughput instead of individual tasks. This approach makes use of employees from all relevant departments and functions, providing benefits in reduced product development time, getting higher quality merchandise to stores and customers faster, motivated employees, greater accountability and lower costs. These benefits all add up to a considerable competitive advantage for an enterprise.

5.4 Concurrent Engineering

Concurrent engineering is a systematic approach to the integration of the design of products and the processes for design, manufacturing, delivery and support. It is a methodology that encourages continued process improvements. Its goals are to improve quality, reduce time-to-market and cost. From the time the need for a product is recognized, the variables that will influence customer satisfaction throughout the entire product life cycle will be explicitly considered. The idea of a value chain helps to focus on where value is added, how unnecessary steps might be eliminated, and how complicated processes might be simplified. During the conceptual design phase, consideration will be given to the product's manufacture. Thus product design will take into account the design of the manufacturing process, involving activities such as production plannning, equipment selection, facility design and process planning. In contrast to the traditional sequential, iterative and distributed design practices, the concurrent engineering approach requires a parallel, interactive and cooperative team approach to product and process design. In large-scale manufacturing systems involving a large number of resources and design activities, the clustering of those activities in the design process which might be scheduled simultaneously is an important step toward achieving concurrency [8], and can lead to savings and increased quality through simplification of the entire manufacturing process.

Computer tools are becoming increasingly helpful, and are often required, to manage the complex relationships between design, manufacturing and business processes. The Enterprise Modelling System provides support in organizing the enterprise structure, modelling alternatives, communicating design intent, simulating performance, creating virtual teams, and storing product definitions and best practices for future retrieval. In order to achieve significant reductions in product realization times, it is desirable to perform a considerable amount of process modelling and simulation to ensure that the right information, skills and resources are available at the right place at the right time. The concurrent engineering approach to process development involves identifying the communication requirements and segmenting the processes into clusters having mostly concurrent and independent activities. The process browser and process view of the EMS provide specific features to support this approach. While the initial ideas leading to cycle-time compression may come from the minds of experts, value will be added when the EMS is used to work out the details and elaborate on preliminary ideas, in order to identify any unanticipated shortcomings through static and dynamic process analyses in an iterative manner.

Concurrent engineering begins and ends with the customer. In using the EMS, all the key value-added functions from the initial concept through customer use can be included. Processes should be examined as a value chain by identifying who the suppliers, customers and value-adding steps are. Communications between all the key processes should be reviewed to see if the handoffs from one to another can be improved. The process framework can be used as a way of letting individual technical specialists understand what they need to know about other functions. Best practices can be benchmarked and supported, allowing an individual process to better tune its results for higher throughput, quality, customer satisfaction and profits. The EMS will allow the concurrent engineering team to set up appropriate metrics to track the impact of systems and processes on customer-oriented goals and higher market shares, rather than on internal measures. The work of individual teams and process owners should be synchronized so that an integrated system including the main areas of product/process design, manufacturing and customer support are planned to work together.

5.5 Fast Innovation

Under the time pressures in a global competitive market, a manufacturer can achieve time advantages by using organizational techniques similar to those employed by flexible manufacturers. The effects are the same -- an increased ability to accommodate variety, improved productivity and shortened response times. The structure of an organization should facilitate rapid new product design and introduction. For example, rather than aiming for significant product improvements for each introduction cycle, a fast innovator can plan for comparatively less improvements with each new product introduction, but introduce new products more frequently. While the traditional approach is to organize new product development and introduction programs by functional centers, it may be inadequate to meet the time demand. An integrated system structured to introduce new products rapidly requires that all development resources for one product be well-coordinated in their objectives, and can be structured by including marketing, design, manufacturing and sales in a modelling study using tools such as the Enterprise Modelling System.

Significant advantages can be gained for being a fast innovator. These include faster realization of cost reductions, dramatically improved quality, lower development costs and the fact that the latest technology can be used closer to the time of introduction. Externally, the organization can assume a position as technological or idea leader, and higher prices can be realized from having a fresher product offering to satisfy the customer. For example, a tableware manufacturer is reported to have turned its business into a design and fashion-intensive business by changing design every 12 weeks, rather than every 12 months as is done by most of its competitors. It was working with major retailers to take advantage of consumer desires for fashion-oriented table and kitchenware. The features provided by the EMS to model information and material flows can be very helpful for studying and fine-tuning the stringent requirement for tremendous flexibility to achieve short design cycles.

Fast innovators also have the advantage of new approaches for marketing new products. For example, rather than depend on extensive market research and testing to define the features, performance and cost specifications, companies with rapid development and introduction cycles can experimentally market new products. Due to the short lead times, the new product can be introduced and then modified quickly as the realities of the market become clearer. By employing an enterprise-wide modelling tool such as the EMS to re-engineer the new product development and introduction process, time-based companies can obtain considerable results in flexibility and responsiveness.

5.6 Pull System of Production

In today's highly competitive global marketplace, corporations are becoming leaner which requires increased effectiveness from a limited resource pool. Not only is product cycle time a focus of attention, but the inventory requirements of a manufacturing system must also be scrutinized for possible reduction. Agile manufacturing is the ability to quickly respond to unanticipated market demands. It requires the ability to manufacture highly customized products in batch sizes as small as one. This is a shift from the push-to-market model to a dynamic, highly customized quick-to-respond, pull-by-market model.

A "pull" production philosophy bases production on the tug of customer demand -- no more, no less. This philosophy leads to continuous flow manufacturing which is to build to demand. To meet such an objective effectively, a manufacturing facility must be flexible, efficient, and provide an environment of constant improvement. An effective pull system will have benefits such as reduced manufacturing cycle times, inventories, rework and scrap.

Simulating a manufacturing system using the Enterprise Modelling System provides a powerful means to analyze bottlenecks. In addition to impacting manufacturing cycle times, a bottleneck drives an enterprise's inventory needs as well. If a process or bottleneck is not integral to the requirements of the customer (i.e., value-add), the process is waste and should be eliminated. It is only after determining that it is value-add that it should be considered for improvement. If a facility cannot meet current customer demands, then manufacturing improvements should be directed toward increasing capacity. If, however, a facility can meet the rate of customer demand, improvements should be directed toward increasing flexibility. A bottleneck can be improved utilizing methods such as defect prevention, setup and operation time reduction, preventive maintenance, and process flow improvement, and the EMS provides capabilities to facilitate the experimentation and implementation of these methods.

The Enterprise Modelling System can be used to support the deployment of a new manufacturing philosophy. In the initial phase of pull system deployment, work-in-process is controlled to a level that will meet customer demand and compensate for current process deficiencies such as scrap, rework, long changeover/setup time, down time, etc. The second phase is then directed toward attacking these process deficiencies. As these processing problems are eliminated, work-in-process can be reduced, resulting in decreased manufacturing cycle times and increased flexibility.

There are important implications to the pull system of production. A process-oriented management philosophy implies that processes must be improved before results can be achieved, requiring closer linking of the organizational hierarchy. The visibility of the pull system will allow management to become aware of any obstacles to improvement and offer an atmosphere of support and open communication. As well, the focus will be on identification and elimination of problems at the source. Because of the nature of a pull production system, a bottleneck workstation has the potential to shut down the entire work cell. This leads to an increased sense of urgency on the factory floor and can be the catalyst for increased cross-training, standardization, documentation and preventive maintenance at the constraint operations. The EMS provides a test bed for an enterprise to explore new production approaches and to assess the potential benefits before implementation.

6 Concluding Remarks

This chapter has included an overview of enterprise modelling and has presented its relevance to modern manufacturing. A particular modelling tool called the Enterprise Modelling System was described. How such a tool can be applied to gain valuable insight into important manufacturing system concepts such as information infrastructure, process modelling and re-engineering, concurrent engineering, system integration, etc. were discussed. A prototype of the EMS has been developed and is being field-tested in various industrial sectors.

As quality, innovation and service are assuming more and more importance, the traditional functional management is no longer adequate. Instead, manufacturing enterprises should perceive their activities as a set of processes that cut across the conventional, departmental demarcations. Process management gives business the ability to dramatically improve response times, service and quality. The combination of advanced technology and increased global competition is driving a need for more frequent, even continuous, improvements. In contrast to traditional productivity improvement approaches, re-engineering the factory requires fundamentally rethinking the entire manufacturing process, from suppliers through to customers. Re-engineering's purpose and goal is to achieve order-of-magnitude improvements over traditional methods. The Enterprise Modelling System provides a framework for taking an integrated look at the organizational structure and the business and engineering processes in the entire enterprise, thereby allowing many process redesign and concurrent engineering ideas to be tested and analyzed before implementation. As well, once a manufacturing approach or strategy has been established, implementation details can be fine-tuned and operational performances assessed.

In today's global markets, a timely response means competitive advantage. Time is a vital element in successful innovation, which, in turn, is key to the long-term vitality of all enterprises. Companies must substantially reduce the time required to conceive, develop, and introduce new products and services. Fast innovators can experiment with their customers as they fine-tune their innovations. They can introduce a version that is their best guess and quickly adjust it to reflect consumers' reactions. The cost of a delay at any stage of the innovation cycle is actually the cost of lost opportunity in the sales cycle. It has been estimated that in a sales window of five years, for example, a six-month delay in the introduction of a new product has a bigger impact on profit then even a 50 per cent increase in development cost. Demand for a given product depends on how well that product meets customer requirements, which is reflected in the quality of the product. Speed and quality of innovation are the key factors in the potential return from a product. Having up-to-date models of an organization's structure and processes is the key to fast response and continuous improvement. With the help of tools such as the Enterprise Modelling System, manufacturing enterprises are in a strong position to accelerate the deployment of modern, world-class manufacturing techniques and to improve their competitive edge in the world market.

Acknowledgement

The author would like to acknowledge the contributions by H. Atabakhsh, U. Graefe, and A. Pardasani of the National Research Council of Canada in the Enterprise Modelling project.

References

[1] Jorysz, H.R. and Vernadat, F.B., "CIM-OSA Part I: Total Enterprise Modelling and Function View," *Int. J. CIM*, Vol. 3, No. 3, 1990, pp. 144-156.

[2] Jorysz, H.R. and Vernadat, F.B., "CIM-OSA Part II: Information View," *Int. J. CIM*, Vol. 3, No. 3, 1990, pp. 157-167.

[3] Hars, A. and Scheer, A.W., "Enterprise-Wide Modelling - the Basis for Integration," *Engineering Systems With Intelligence: Concepts, Tools and Applications*, Tzafestas (ed.), Kluwer Academic Publishers, 1991, pp. 541-548.

[4] Nelson, D.A., "Modelling Enterprise Dynamics," *Dynamic Modelling of Information Systems II*, Sol. H.G. and Crosslin, R.L. (editors), Elsevier Science Publishers, B.V., 1992, pp. 309-327.

[5] Mertins, K., Sussenguth, W. and Jochem, R., "Planning of Enterprise-Related CIM Structures," *Advances in Factories of the Future, CIM and Robotics*, Cotsaftis, M. and Vernadat, F. (editors), Elsevier Science Publishers, B.V., 1993, pp. 67-76.

[6] "Reference Model for Shop Floor Production Standards - Part I: Reference Model for Standardization and Methodology for Identification of Standards Requirements," ISO Technical Report TR-10314-1, 1990.

[7] Atabakhsh, H. and Chan, A.W., "ExTool: An Interactive Object-Oriented Expert System Toolkit for Distributed Decision Making," Proceedings of the 1993 Canadian Conference on Electrical and Computer Engineering, Vancouver, Canada, September 14-17, 1993, pp. 233-236.

[8] A. Kusiak and J. Wang, "Management of Design Projects in a Concurrent Engineering Environment", Proceedings of the International Conference on Object-Oriented Manufacturing Systems, Calgary, Canada, May 3-6, 1992, pp. 66-71.

3   Object-Oriented Information Modelling

Kaj A. Jørgensen

1 Introduction

Industrial manufacturing companies are significant consumers of information technology for industrial applications. The competitiveness of industrial companies compels them to utilise new research results as quickly as possible and to combine these with their own experiences. Information technology is available in many different forms, and it is essential to make the right choices when trying to take advantage of the new technology.

It is essential, for well functioning companies, to manage their information resources as well as possible. Therefore, computer based information systems are important in order to achieve satisfactory results.

We have seen various efforts to produce more or less sophisticated information systems for industrial companies, but, even though a lot of resources have been spent on development, implementation and utilisation of existing standard software packages for production management, the results have not been satisfactory for the companies. The users are often disappointed. They have not been able to obtain the support they need to carry out their task in an efficient way.

The information systems are often *unreliable,* because it is not safe to *reuse* existing systems or parts of systems. To apply a piece of an existing system in a new environment is often not possible, or it is very difficult to predict the result of doing it, to foresee all the consequences.

The productivity and the extendability of systems development is low. Often, it is necessary to redevelop significant parts of existing systems when they have to fulfil new purposes.

Furthermore, *the efficiency of producing new systems is inferior.* A high efficiency is only possible to obtain if extensive investments in expensive development tools are offered. It is not yet possible to build any kind of professional system by combining sophisticated, inexpensive library components on a high level.

Design of information systems is often a very abstract discipline which is rather difficult for non professionals to participate in, and, during a design process, decisions are made for conditions that do not exist, yet. The consequences of these decisions are sometimes very unpredictable, and wrong decisions could be very expensive.

Often, it is not realised how important it is to make careful planning before the actual programming is initiated. *There are many examples of information systems which function badly because preliminary analysis and design considerations have been omitted.*

Many existing methods for information modelling are not sufficiently detailed with regard to advice about how to build the basic information structures and they are not focusing very much on how entity types should be defined. Also, many methods do not clearly focus on how to deal with aspects of the process where loose ideas, inexact specifications, participant's opinions, etc. are incorporated.

The existing methods for information modelling are all based on rather old and simple ways of describing entity relationships. Only a few suggestions have been made for enriching this type of information, and to extend the entity relationship models with other sorts of semantic data.

One solution to these problems is that the companies must develop systems which are more individually accommodated to their own needs. The managers should be able to obtain exactly the information which is important for their tasks, and the companies' own employees usually have the best knowledge of the company routines, the basic structures of the company, the products, the production processes, etc. They are not always aware of their company's problems and how they can be solved, but, if the company's own employees are able to participate more actively in developing the decision system and the underlying information system, it will, undoubtedly, be making the company stronger and more competitive.

An important advantage of a project, where the company develops its own information system, is that also the information analysis will be performed by the company employees, and, therefore, the knowledge and the results will be retained within the company.

2 Information Systems Development

If a company decides to develop its own information system, it is important to utilise the development resources as efficiently as possible. The established software engineering techniques are rather expensive because the development process takes a long time and many development tools are inefficient. However, the new technologies for developing management information systems, appear to be very promising, and many companies see the new opportunities as solutions to their problems.

Unfortunately, these efforts have often lead to poor results. Certainly, it is possible, with a suitable development tool, to develop smaller information systems with rather simple information structures, but, developing larger systems, with highly integrated and complex information structures, requires a detailed methodology and demands some education and practice.

It is important that the companies perform their development processes continually. When such a process is initiated, it should *not* be necessary to start from scratch, every time. It should be possible to reuse most of the existing systems, in which the company dependant system structures are available as a platform for further development.

It is of great importance to get a good start and a lot of resources can be wasted in the early stages of a software development project. It is always more expensive to perform fundamental redesigning in a late phase of a project.

If a company can reuse and extend its previously developed system, in a reliable way, it will be easier to obtain a satisfactury start of the development project. Most of the fundamental company related structures and algorithms of the system are already developed, and program segments for manipulating the information are ready for use.

2.1 Information Modelling

An information system, designed to avoid all, or most, of the previously mentioned problems must be based on a new methodology for systems development. It is necessary to create a flexible model of the system structure on a higher abstraction level, and, from this, be able to extract the components of the software system. A system model, describing all the company's information structures and program modules, is termed a *repository*.

Because management information systems usually contain a substantial amount of persistent information (a database) supporting the functions in an organisation (a company), *knowledge engineering* and *information modelling* are the most central activities in producing the repository. It has played an increasingly important role in the development of information systems for industrial applications.

Descriptions of information system models will often include specification of both function and information, but many methodologies focus mainly on function modelling methods. Function modelling is a process driven approach while information modelling is data driven. (Ref. [13], [14], [21], [23], [24] and [33]) Recent research results appear to indicate that what is really needed is a well balanced mixture of both approaches.

A *model* is an intentionally simplified description of something. It has a specific purpose, e.g. communication and manipulation. *Modelling* is a design activity which will result in a system model. By means of modelling, it is possible to introduce the necessary *abstraction levels* in the process.

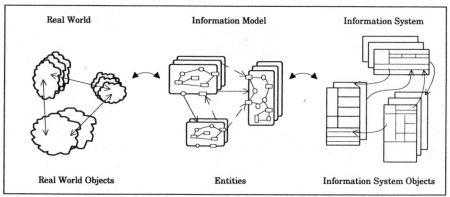

Figure 1   Information Modelling and Information Systems Development

An *information model* is a kind of link, between the real world and the information system (figure 1). It is a description of the real world and, at any time, it will define the designers' view of the real world. This view is established by means of a set of decisions which are based on the designers' communication with other participants in the project, contractors, supervisors, end-users, etc.

Furthermore, the model serves as a platform for generation and implementation of the information system. Therefore, the model will also contain detailed descriptions which are directed towards the construction of the information system.

Development projects are often performed by *prototyping,* by which a sequence of different versions of the system are designed. Prototyping is especially suitable when an efficient development tool is available. Furthermore, if such a development tool is used to perform modelling, prototyping and modelling can be combined to provide an ideal framework for system development.

Modelling and prototyping can be combined in *prototyping by modelling.* In such a project, the modelling activity will be the main activity, compared to the design of the target system. It is actually the model which is prototyped, and the target system will be produced more or less automatically.

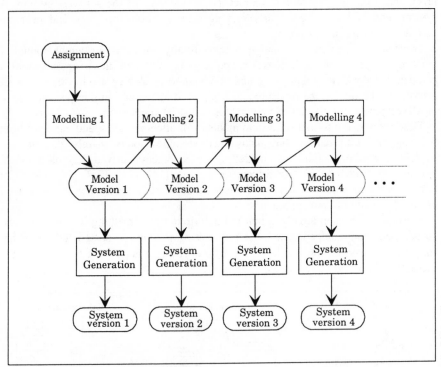

Figure 2   Prototyping by Modelling

New methodologies must, obviously, be prepared to support by means of development tools and, thereby, be encouraging prototyping by modelling. The methodology which is presented here is aimed at fulfilling these criteria.

2.2 Entities and Entity Types

An information system contains information about *organisms, things, activities, associations,* etc. (in the following, termed *real world objects),* of *physical* and *logical* systems. The information components will be termed *information system objects,* e.g. products, customers, orders and employees. Thus, *Real World Objects* are represented by *Information System Objects.*

In order to underline that a model is created, the descriptions of the real world objects are termed *entities.* Thus, *Information Model Entities* correspond to *Real World Objects* and *Information System Objects.*

Real world objects are defined by a set of *properties,* which makes it possible to identify them (see, feel, hear, etc.) and to distinguish individual objects from each other in the environment where they occur. Representations of properties in the information model are termed *attributes.* Thus, *Attributes of Entities* correspond to *Properties of Objects*

Usually, objects are not completely different. They have certain resemblance, to each other, and could, therefore, be grouped according to these. A definition of such a group is termed an *object type.*

One of the most important design tasks in the process is to consider how to define *types of entities.* Usually, this can be achieved in several ways, because every real world object can be regarded as belonging to several object types, e.g. a particular person, should he be described as an employee, an engineer, a database designer, a man or just a person. Or, should a piece of iron be described as a product, a (spare) part, a component or a screw. As indicated, it is always a matter of definition as to which type to apply.

In order to assure that the information system is constructed on a solid foundation it is important to find the *invariant characteristics.* An object type is a definition of all the invariance of a group of real world objects. In the information model, an entity type can be regarded as a *template* from which all entities of this type can be created.

3 Information Modelling Methodology

In this section a methodology for information modelling is presented. The modelling methodology is created in order that designers, with only a basic experience of software development, are able to carry out the design activities. However, it is recommended, that the information modelling is guided by an experienced *information engineer.* The present approach mainly applies to the situation in which the design work is being performed by a *project group,* of users with different backgrounds participating in the development of the information model.

It is assumed that the basic decisions have been taken about the purpose of the information system and what the primary functionality of the system should be. In the case of management information systems, such decisions are usually based on a clear impression of the management system of the company, preferably a detailed description of the functions, plan periods, tasks, information, organisational responsibility, etc. If this is only available to a minor degree, the obligations of the information modelling activity are increased, and it is expected that development of the information model, to some extent, will resolve matters in the higher order systems.

It must be remembered that it is always a matter of definition, by the project group, as to which types should be selected. It is important to take decisions which are verified in the project group, and, to some extent, among the users. Also here, they can contribute, significantly, with their knowledge of functions within the organisation.

In discussions about entity types, it is often experienced that members of the organisation have different opinions about commonly used concepts in the organisation. Information modelling motivates the participants to choose precise definitions.

For this methodology, it is assumed that a development tool is available in order to support the model designer and that this tool is well suited for prototyping.

3.1 The Object-Oriented Approach to Information Modelling

The methodology is partly based on the object-oriented paradigm, which has shown an increasing importance in recent years. This paradigm is developed in order to provide software systems with a more natural and widely accepted appearance (Ref. [5], [6], [7] and [11]). It is characterised as a special system view where the system components (sub-systems) are *active, living objects*, containing information, performing operations and communicating with each other. The objects may work in parallel and will communicate, on a *client - server* basis, by passing messages to each other.

The object-oriented approach to the development of information systems will automatically prepare it for implementation as a *distributed system* where the load can be balanced between the available computer resources. Furthermore, it is possible to allocate dedicated facilities for special purposes i.e. *graphical* or *mathematical manipulation* at one end and high performance *database operations* at the other.

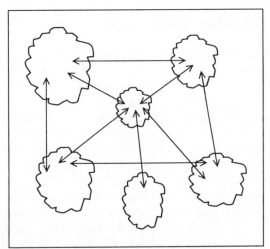

Figure 3   Communicating living objects.

For each object, special interest is paid to the *interface* which specifies the functionality of the object, a set of *public services* which can be activated from other objects. The interface determines the shape of the object to a sufficient degree for the clients which will make use of it.

Encapsulated in the objects are *private internals* which, among other things, describe the characteristics of the object and define its current state. These internals are *hidden details* which are not necessary for the clients to know. At the same time, it is *protected from outside interference*. Furthermore, it gives the designer an important flexibility because it will be possible to choose different implementation solutions without effecting the defined functionality of the object.

The object-oriented approach provides a strong contribution to the initial stages of the information system design process by providing the classification technique for the definition of entity types.

The key components of an information model is a set of entity types. Therefore, in the first phase of the development of the model, the efforts are concentrated on identifying the entity types of the model. Afterwards, the entity types are developed in greater detail and the information structures and logical relationships are developed (Ref. [20]).

3.2 Initial search for entities and entity types

When trying to define the entity types of an information model, one must keep in mind that the model, to a great extent, is a mapping of the real world. Therefore, it is natural to start looking for the basic entity types among the observed physical and logical objects of the environment in which the information system will function.

This environment will be, e.g. a company, an organisation, a management system, a production system. Thus, typically, the purpose of the information system is to store, maintain and share vital company information for members of the organisation or to support management functions in the company.

In prototyping, the information analysis is integrated in the design activities. Even if the development is started from scratch, the initial analysis need only be performed very roughly. The concerned entities, and possible entity types, will often be called by names and, therefore, the first step should be to prepare a list of names of what the project group identifies as the physical and logical real world objects to be modelled.

An information model developed in relation to a synthetic CIM Factory will be used as an example in the following. The model will be presented step by step and it will illustrate some main phases in the modelling process.

In such a company, a list would most certain include such names as: PRODUCTS, EMPLOYEES, MANAGERS, WORKERS, MACHINES, EQUIPMENT, TOOLS, BUILDINGS, VEHICLES, INVENTORIES, SUPPLIERS, CUSTOMERS, etc. Both the names of individual physical objects and names of groups of objects would be listed.

Some data files in the company could be mentioned: CUSTOMER ORDERS, PROJECTS, SOFTWARE SYSTEMS, INVOICES, CNC PROGRAMS, etc. Some of these objects could be regarded as logical objects. Other logical objects could be: CONTRACTS, ACCOUNTS, POTENTIAL CUSTOMERS, DELIVERY DATES, PRODUCT IDEAS, MARKET IMAGE, ECONOMIC RESULTS, etc.

During the information model design, all these names, and names of many other objects, would be evaluated and, if not excluded, they would possibly be entered in the model as entity types, entities or attributes of entities. At this point it would not be clear what these names represent.

3.3 The foundation for organising entity types

The fundamental relationships between entity types are determined by the two basic abstraction mechanisms, generalisation and aggregation (Ref. [28] and [29]). More precisely, they are defined by two orthogonal evaluations: 1) *generalisation* versus *specialisation,* which is termed *classification* and 2) *aggregation* versus *separation.* The two different ways of relating entity types to each other are shown in figure 4.

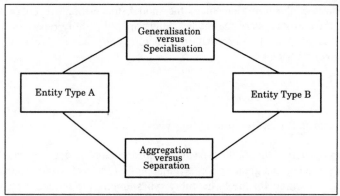

Figure 4   Fundamental relationships between entity types

In classification, the entity types are organised in a *class tree,* a *taxonomy,* by *inheritance* of attributes from *super-types* to *sub-types.* By means of *aggregation* versus *separation,* the relationships between *component types* are defined. The taxonomy defines a *static structure* of the information model whereas component relationships define *dynamic structures* of the model.

When the modelling is being carried out, it is important to remember that a major objective for making decisions about structures is to avoid *redundancy,* i.e. to ensure that every piece of entity information occurs only once.

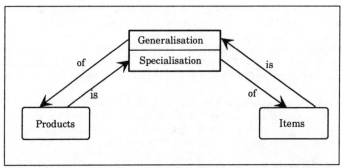

Figure 5   An example of generalisation versus specialisation

In figure 5 it is shown that the PRODUCTS type is a specialisation of the ITEMS type, and, correspondingly, that the ITEMS type is a generalisation og the PRODUCTS type.

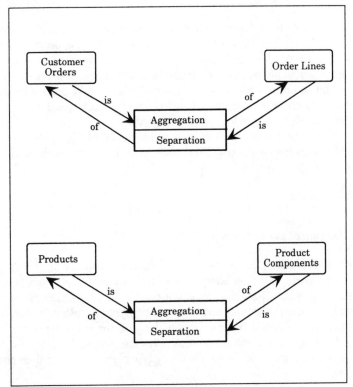

Figure 6   Two examples of aggregation versus separation

In figure 6 it is shown that the CUSTOMERORDERS type is an aggregation of the ORDERLINES type and, correspondingly, that the ORDERLINES type is a separation of the CUSTOMERORDERS type. Likewise, the relationship between PRODUCTS and PRODUCTCOMPONENTS is shown.

As the taxonomy defines a static structure, the first investigation will be to perform classification. This is done by evaluating the generalisation-specialisation relationships.

3.4 Preliminary definition of entity types

In order to create the taxonomy, a preliminary set of *entity types* must be selected from the initial list of names. An entity type is defined by a name and a set of attributes with which a set of entities, the instances of the type, can be described. Thus, entity types will usually be selected among the names representing sets of entities. *Individual entities* may sometimes imply entity types, also. This is often the situation where only one single entity of this type will ever exist.

Such names as PRODUCTS, CUSTOMERS, SUPPLIERS, EMPLOYEES and MACHINES are obviously representing entity types, whereas such names as PROJECTS, ACCOUNTS, DELIVERYDATES and INVOICES could be representing both entity types and attributes.

PRODUCTINVENTORY could be regarded as the name of an individual entity. If such an entity is to be described, an entity type, e.g. INVENTORIES, could be defined. It must be a type even though only one single instance is present.

A necessary step, in defining the entity types, is to list a set of *attributes* for description of object properties. From these attributes, it is important to define the *identifiers* of the type. Except for certain entity types, where only one or a few entities exist, it would be necessary to define such identifiers. In the daily work, with information and an information system, the users have to make precise identification of real world objects. Furthermore, identifiers make it secure and easy to retrieve certain individual objects from the information system.

The attributes which should be included in the model are highly dependent of the purpose of the information system. Thus, some evaluations and decisions have to be made in the modelling process.

3.5 Classification of entity types

Because classification is the process in which the taxonomy of entity types is constructed, the focus is set on entity types which are related to each other by the static relationship generalisation - specialisation (see figure 4). General entity types are placed higher in the taxonomy and special entity types are placed lower.

Sometimes, classification is quite obvious because the entity types are based on well defined concepts for which a thesaurus may exist. Often, the concepts require to be organised by classification so that the entity types are defined by their position in the taxonomy.

To find these relationships, the attributes of all entity types have to be examined for *resemblance* and *differences*. Two entity types, which have identical attributes, are related to each other in the taxonomy. If the attributes of one entity type are completely contained in the other, it is a super-type to the other. If two entity types share common attributes but each have their own attributes, both of them will have the same ancestor-type, possibly the parent.

When the taxonomy has been arranged, it can be verified, in order to ensure that all possible entities can be described by the attributes of the entity types and the inherited attributes. Also, it can be verified that no attributes are identical.

Normally, the initial attributes of the entity types are *data attributes*. Thus, an important step, in the development of the entity types, is to add other attributes, primarily methods, to the types. This may possibly lead to a rearrangement of the taxonomy. The methods are defined in order to provide operations to be performed on the data attributes, some of which may be encapsulated.

In general, an entity type corresponding to a real world object should have attributes, both data attributes and methods, to represent all the properties of the object. This principle is termed the *completeness principle* or *responsible principle* (Ref. [34]). Thus, attributes are defined in order to provide a complete set of services according to the purpose of the type.

Entity types, which have been developed according to this principle, are well prepared for reuse in other information models and information systems. If the services, which have been provided by such entity types, are sufficient, they will be reliable to use because they are developed for a general purpose and not to any specific application. Reuse of well equipped entity types, in future development projects, will increase the productivity, significantly.

4 Object-Oriented Information Models

In the following, the characteristics of object-oriented information models are presented in detail. This includes the fundamental concepts classification, inheritance, encapsulation, collection and complex entities. The object-oriented view of entity relationships, by means of references and associations, is also described.

4.1 The Taxonomy

Classification is used in the early phases when the entity types are to be defined in a model. As a result, the entity types will be related to each other in a classification structure *(taxonomy)* where entities are grouped by means of the similarities of attributes. Because of this classification technique, entity types are also termed *classes,* and the taxonomy is a *class tree.* The class tree defines the levels of abstraction in the model. (Ref. [8] and [18])

In the initial steps of building the example information model, some obvious real world object types and their corresponding entity types are suggested. Four of these are: CUSTOMERS, SUPPLIERS, ITEMS and ORDERS.

Examining some examples of attributes of the entity type ITEMS, it appears that many of them depend on what category of item is in question. Therefore, it is possible to define sub-types of ITEMS. In the exemplified model, three sub-types, RAWMATERIALS, PRODUCEDITEMS and PRODUCTS, are defined by the following arrangement of attributes:

ITEMS	PRODUCTS
- ITEMNAME	- SALESPRICE
- WEIGHT	- COLOUR
PRODUCEDITEMS	RAWMATERIALS
- MATERIALPRICE	- MATERIAL
- PROCESSINGPRICE	- UNITOFMEASURE
- PROCESSINGTIME	- PURCHASEPRICE

The ORDERS type is also changed in this step. It is divided into CUSTOMERORDERS and INTERNALORDERS. Furthermore, the type COMPANIES is added, containing common information about CUSTOMERS and SUPPLIERS, i.e. COMPANYNAME, ADDRESS, etc.

The result of this classification is illustrated in figure 7.

A class tree is a structure of *generic relationships*, which is arranged by means of the *inheritance* concept. Inheritance means that an entity type can be defined, on the basis of another (generic) entity type. As a result, each type inherits the attributes from its super-type. (Ref. [9], [22] and [30])

Consequently, each individual entity of a certain type is created as a *compound entity,* consisting of a set of concatenated entity-parts, one for each of the inherited types.

58

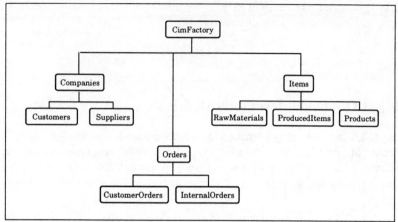

Figure 7   The result of the first refinement of entity types.

In figure 7, the type ITEM is the super-type of RAWMATERIALS, PRODUCEDITEMS and PRODUCTS. Therefore, all of these types will inherit the attributes ItemName and Weight of ITEMS.

The root type CIM FACTORY is the super-type of all other types, and any attribute of this type will be inherited by all of these types.

In accordance with the characteristics of tree structures, each type in a taxonomy (except the root type) has one, and only one, super-type and it may have one or more disjoint sub-types.

The top level entity types are the most general types, and the types at the lower levels are the most special types. Every added attribute is positioned up or down in the class tree by considering how general or special it is.

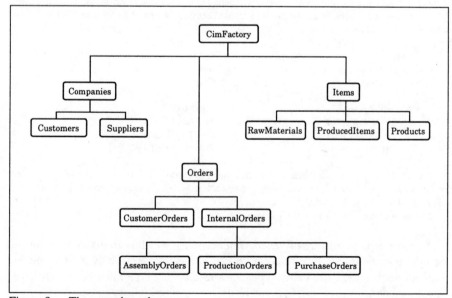

Figure 8   The complete class tree.

Two new attributes, INVENTORYQUANTITY and MINIMUMINVQTY, are considered general for all item types and they are placed in the ITEMS type. An attribute STANDARDBATCHQUANTITY is relevant only to PRODUCEDITEMS and, therefore, it is placed in this type. A new attribute, TOTALCOST, is added to PRODUCTS.

Figure 8 shows the complete class tree where INTERNALORDERS are further refined.

A new type COMPANIES is placed in the taxonomy as a super-type of CUSTOMERS and SUPPLIERS. COMPANIES contains the attributes which are needed in both sub-types: NAME, ADDRESS, PHONENUMBER, FAXNUMBER, etc. The sub-types, on the contrary, are defined by attributes which are different from each other e.g. a DEBITBALANCE attribute in CUSTOMERS and a CREDITBALANCE attribute in SUPPLIERS.

Sometimes, it is convenient to allow an entity to inherit attributes from more than one other entity type. This is termed *multiple inheritance,* instead of the singular inheritance which has been described above.

An entity type MIXCOMPANIES could be added, if it is necessary to describe companies which are both customers and suppliers. This means that the DEBITBALANCE attribute in CUSTOMERS and the CREDITBALANCE attribute in SUPPLIERS will both be inherited by the new type.

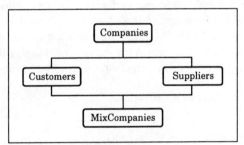

Figure 9   Multiple inheritance in
MIXCOMPANIES

Multiple inheritance may introduce some conflicts about names of attributes. It might be possible, from two different entity types, to inherit attributes with the same name. Therefore, when specifying a multiple inheritance, it is possible to rename attributes which are inherited from super-types. Multiple inheritance will often be used in connection with pre-developed or public libraries of entity types, available for many different purposes.

When the information system is implemented, a number of application programs need to be developed. Depending on the user interface, of the computer, a number of library entity types are often available in order to handle the basic functions for creating screen forms and reports. Thus, multiple inheritance may be applied in order to add such methods to the entity types of the information model.

Multiple inheritance implies that the classification structure is not actually a tree, but rather a network. Therefore, in order to preserve a good overview of the entity types, it is practical to preserve the tree structure by indicating primary and secondary relationships (Ref. [26]).

4.2 Attributes of Entities

Corresponding to the categorisation of the properties of objects, attributes can be divided into *factual attributes* or *behavioural attributes*. In the information system, this will correspond to *data* and *operations* respectively.

Factual attributes contain *state information,* describing, at any time the *state* of the system. They are represented as *simple* or *structured data attributes.* Simple data attributes or *data elements*, e.g. length, width, material, colour or age, each have three components: name, data type and data value (e.g. age:number:21). Structured data attributes are compositions of simple attributes or other structured data attributes. *Constant attributes* are special data attributes, where the values are defined at the time of entity creation and remain unchanged, thereafter.

The main data types are *number* and *text strings,* and other data types are date, *time* and *boolean* (true/false, on/off, yes/no). Special data types may be defined for types such as diagrams, fotographs, etc.

Behavioural attributes contain procedural information and describe all relevant operations (services) which may be performed in connection with the state information, e.g. routines for changing the information or routines for presenting the present state information. In OOIM, these attributes are termed *methods*. Methods consist of a *head* and a *body*. The head is the name of the method and a description of the input and/or output of the method. The body contains the method operations.

In the example, the following methods are defined:

In ITEMS:
- UPDATEINVENTORYQTY(INPUT: ADDEDQTY)
  Add the parameter value to the present inventory quantity.

In PRODUCEDITEMS:
- ENTERPROCESSINGTIME(INPUT: NEWTIME)
  Store the parameter value as the new processing time.
- COMPUTEPROCESSINGPRICE(INPUT: TIMEFACTOR)
  Calculate and store a new price as PROCESSINGTIME*TIMEFACTOR.
- PRICES(OUTPUT: MATERIALPRICE,PROCESSPRICE)
  Fetch the two price attributes.

Entity types, which contain both data structures and methods, are sometimes termed *abstract data types.* The idea is to stress that every data structure should be provided with descriptions of all relevant operations in the same way as standard data types, i.e. integer and real, where the operations addition, subtraction, multiplication, division, etc. are defined.

A queue is a well-known data structure and, if it is described as an abstract data type, it will include the definitions of the operations for inserting elements, removing elements, determining the number of contained elements, etc.

4.3 Encapsulation

Attributes (data structures and methods) in entity types may be specified as *visible* or *hidden* attributes. Visible attributes are accessible from outside the entity whereas hidden attributes are not directly accessible. They are only accessible from the methods of the entity. Some *visible data attributes* may be *read-only* data attributes which can be read but not updated. All bodies of methods are hidden. Hidden attributes are said to be *encapsulated* in the entity type (figure 10). Visible attributes are sometimes termed *public* attributes while hidden attributes are termed *private*.

In an entity type, the visible attributes, as a whole, is termed the *form* or the *interface*. It defines the *services* which are provided by the type. In the constructed information system, these services of the *server object* can be invoked by one or more *client objects* as often as necessary, throughout the life time of the system. Seen from the clients point of view, the call of an interface method is like sending a *message*. Therefore, the attributes of an interface constitute a *protocol* between the server and its clients.

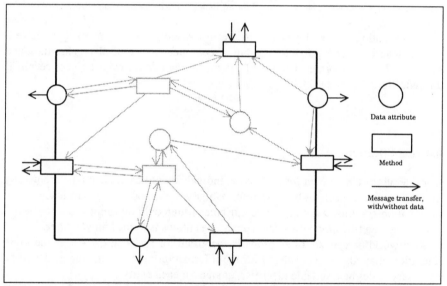

Figure 10   Attributes of entity types - visible and hidden attributes.

The purpose of encapsulation is to provide a clear distinction between what the entity type offers to the environment and what are the internal details. (Ref. [30]) Updating of information within entities will be performed only by means of methods in the interface, and this restriction assures that all constraints for the operations are safely checked. As a special consequence, an entity type is occasionally created because of a need for methods rather than data, and hidden attributes will sometimes be used only in connection with the methods (for storing temporary or intermediate information).

Encapsulation also means that it is not possible, from the outside of an entity, to see the difference between a read-only stored data attribute and a method without parameters (for example an arithmetic function). Instead of consuming space for storing data permanently, a method can be defined. It will compute values, only when it is called upon. Such a method is also termed a *virtual data attribute*.

Attributes, which are defined in sub-types with the same name, will *overwrite* the definitions from super-types, but the methods with the same head and different bodies can be specified as *virtual* or *polymorphic methods*. Such methods may be invoked in the same way in all the concerned entities, but the functionality will be defined according to the actual entity type in which the method is invoked.

This means that a common interface of an entity type and its descendants can be defined once, and in the relevant sub-types, the virtual methods can be accommodated to the individual needs. The way in which a virtual method should be invoked is defined at one level in the class tree, and the content of the method may be modelled differently at lower levels.

In the example, a SHOW method may be defined in the root type. The purpose of this method could be to present the data content of an entity. By defining this method in all entity types as a virtual method, it may be invoked identically in all entities, but it will be implemented differently in the entity types (This will indeed be necessary if the attributes in the entity types are different).

Encapsulation makes it possible to create tool-boxes where entity types have a simple interface, and possibly contain a complicated set of data attributes and methods. These important details will be hidden and could even be kept secret, if desired. (Ref. [10] and [32])

4.4 Identification

Considerations about *identification* of the individual entities of each type are usually very important, especially because information retrieval operations are essential for information systems. A set of visible attributes (often only one simple data attribute) containing identification information about the entities is termed an *identifier*.

Real world objects are often properly identified for practical daily use, and those identifiers may also be considered for the information model. Sometimes, it will be necessary to define a suitable identifier, unique for each entity.

It is assumed that the values of identifiers never change. Therefore, it is recommended not to include any attribute describing properties of the corresponding real world object. Often, a serial number will be ideal as identifier.

Occasionally, entities are created for temporary reasons only, and for such entities it may not be necessary to define identifiers in the information model. Furthermore, it is assumed that every entity is automatically equipped with a unique *entity identifier* which, in the information system, will be a system generated unique identifier - *object identifier* (OID). (Ref. [1] and [17])

According to the class tree, identifiers may be placed anywhere in the tree. An identifier in the root class, however, will define a universal identifier for all entities in the system. Often, none or only one entity of the root type will exist, and, in such cases, an identifier is not necessary. Even if a universal identifier is defined it may be necessary to have separate identifiers in the sub-types.

In the example in figure 8, it may be practical to have an identifier (ITEMNUMBER) for all ITEMS, regardless of the sub-types, and it may even be appropriate to have an additional identifier (PRODUCTID) for PRODUCTS. This identifier would be used by sales staff, customers etc.

4.5 Collections

All entities, of the same entity type, will jointly form a collection. Therefore, all entities of a taxonomy are assumed accessible via their own type-collection. Thus, in the information system, it will always be possible to scan the objects of each type in order to retrieve specific information.

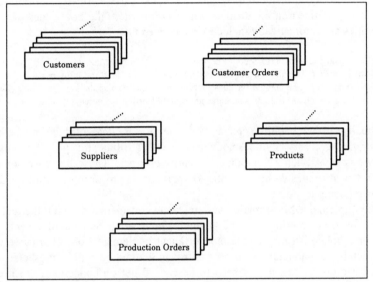

Figure 11   Entities of each type are collections

In the example, entities of the types CUSTOMERS, SUPPLIERS, PRODUCTS, etc. are regarded as collections. One for each type.

By default, a collection is not organised in any sequential order. This may even mean that two scannings of a collection at different times may result in two different sequences. This is first of all defined for the purpose of preserving a flexibility for the physical storing mechanism.

Sequencing of collections can sometimes be desirable. This means that, when the members of a sequenced collection are accessed, it will always be in the defined order. For such collections, a selected attribute (or set of attributes) must be specified as the *key,* and the members will be sequenced in ascending or descending order of the key values.

Normally, the collection of ORDERS will grow during the time, while new orders will be accepted. This means that most operations will be performed on the new orders while old orders gradually will lose their interest. Therefore, it would be appropriate to sequence the collection with the ORDERNUMBER as the key.

Sequential ordered collections can contain sub-collections. If a collection is defined with a key attribute, which is inherited, such a collection will automatically include the entities of the sub-types.

However, it will often be necessary to access a collection in more than one order. Because it is not possible to organise in more than one sequential order, it must be possible to create *indexes* on the basis of a selected key. An index is a sequenced collection of the key values, which has been created separately, but with references to the members of the base collection. When a query is carried out, based on the index key, the search is performed in the index. From the matched elements in the index, the corresponding entities are found via the references. It is possible to define more than one index on each type if queries on different keys are necessary.

In object-oriented information models, it is arranged so that an entity of any type can be replaced by an entity of one of its sub-types. This is termed *substitution*. Thus, if an index is defined on an attribute which is inherited, the key values of the sub-type entities are automatically included in the index.

The key of an index on ITEMS could be the attribute ITEMNUMBER, in order to find particular items by the identifier. This index would contain all the entities of all three sub-types.

Another index could be created on PRODUCTS with PRODUCTID as the key. This index would be appropriate for a search of products where the additional identifier is known.

Normally, an index will be created with one or more *data* attributes as the key. If, however, the necessary key does not exist as a data attribute, it is possible to use *data functions* as keys. Such a function of an entity returns a data value, based on values of other data attributes within the entity or, perhaps, in one or more entities which the entity is referring to.

In the implemented information system, some considerations about the realisation of the collection ordering may be applied. For a small number of objects in the collection a linked list may be the most efficient implementation. Alternatively, a B-tree, which is a dynamically balanced multi-way search tree, will be most efficient.

The basic operations, which need to be carried out on collections, are normally provided automatically. If not, they can be defined as methods of a separate entity type where the implementation details would be hidden. Typical methods are provided for the addition and removal of entities, for a search of a particular entity and for displaying the number of entities in the collection.

4.6 Complex Entities

The taxonomy defines a world of entity types, and, in the modelling process, it is decided which types to include. Often, the defined entity types may reveal new worlds because it can be the origin of several other entity types. Each entity type which contain internal entity types, is termed a *complex entity type* (Ref. [27] and [31]).

Each entity of CUSTOMERORDERS will usually relate to a number of order line entities, one for each ordered product. The order lines will define the ordered quantity, wanted delivery date, price agreement, etc.

The CIM Factory entity type in figure 8 is the root of the taxonomy, but it may also be regarded as one type among others in a larger world where several different types of factories and organisations exist. If the taxonomy is seen in the local view then, normally, only one entity will be present, but, when seen in the wider perspective, many entities may exist.

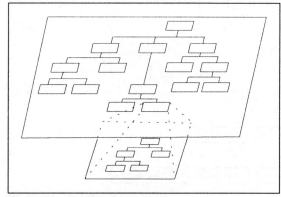

Figure 12   Levels of taxonomies

For each complex entity type, an internal taxonomy can be defined in exactly the same way as described above. This means that the overall structure of an information model will be a hierarchy of taxonomies, and in the modelling process, it is necessary to decide where to put the entity types. Simple information models may only contain one taxonomy.

In the example, an internal taxonomy of CUSTOMERORDERS is defined in order to describe a classification of order lines. In this, the different types of lines are defined, as illustrated in figure 13, where two sub-types are shown: STANDARDPRODUCT lines and CUSTOMERSPECIFIEDPRODUCT lines. When a customer specify a modified product together with the CIM Factory, the ProductId and Quantity will not exist at order entry time, and the order line will appear otherwise than in standard products.

ORDERLINES
- ORDERLINENB
- DELIVERYTIME

STANDARDPRODUCT
- PRODUCT
- QUANTITY
- PRICEAGREEMENT

CUSTOMERSPECIFIEDPRODUCT
- BASEPRODUCT
- MINIMUMQUANTITY
- INITIALPAYMENT

The attributes PRODUCT and BASEPRODUCT will be defined as references (see section 4.7).

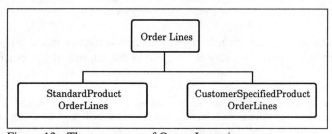

Figure 13   The taxonomy of ORDERLINES in
CUSTOMERORDERS

4.7 References

Identifiers of entities can be used to establish connections between entities. Such a connection is termed a *reference* and is a special attribute which can contain the value of the identifier of an entity. By means of references, it is possible to provide direct accesses from one entity to another. Often a reference is used by an entity in order to obtain services from another entity. Thus, the connection is established as a *client-server relationship*.

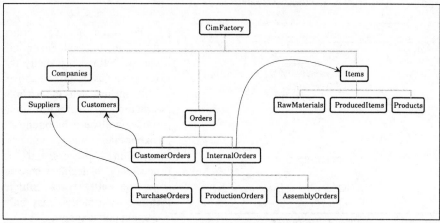

Figure 14   References indicated on the basis of the taxonomy

In the information system, a reference can be implemented by using the object identifier. This makes it possible to construct a flexible and efficient mechanism for reaching other entities.

The use of references is also a means of avoiding information redundancy because it eliminates the need for having multiple copies of data in different entities.

From each entity of CUSTOMERORDERS, a connection will be made to the entity of CUSTOMERS which placed the order in the company. Thus, from each order, it will also be easy to access information about the customer. Likewise, connections from PURCHASEORDERS to SUPPLIERS can be established.

As indicated in the previous section, a reference with the name PRODUCT is defined in STANDARDPRODUCT ORDERLINES internally in the CUSTOMERORDERS, pointing towards entities of PRODUCTS. Likewise, in CUSTOMERSPECIFIEDPRODUCT ORDERLINES a reference with the name BASEPRODUCT is defined, also pointing towards entities of PRODUCTS.

Substitution also applies for references. This means that the actual reference may also be pointing towards any entity of the types in the sub-tree of the specified entity type.

In the example, it is defined that each entity of INTERNALORDERS should specify a demand on only one item (product, produced item or raw material) to be handled. (In AssemblyOrders, the item will be the assembled item and the internal structure will specify which components are to be assembled.) Thus, from INTERNALORDERS a reference may be defined pointing towards ITEMS allowing all entities of the sub-types to substitute.

All the defined references are shown in figure 14 as arrows between pairs of entity types.

Although being an attribute, a reference is not exactly like other attributes which are describing a property of the corresponding real world object. It is only doing it indirectly. It is not the reference itself which is describing properties, but, the entity it is referring to. Each value of a reference represents an extention of the description by including the content of another entity.

4.8 Associations

An *association* is an oriented relationship between two entity types. One entity type is the *anchor type* of the association and the other entity type is termed the *body type*. In each instance of an association, one entity of the anchor type will exist while zero, one or more entities of the body type will exist. From each anchor entity, the members of the body can be accessed as a whole, or individually, in an undefined order. An association is a substantiation of the *aggregation/separation* relationship.

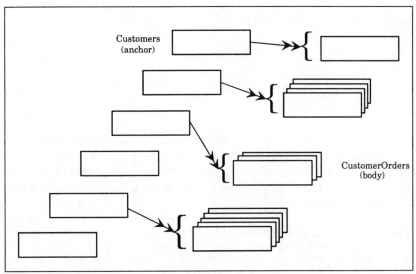

Figure 15   Instances of an association between two entity types

Often, references are indicating the existence of associations. When references are defined from one entity type to another, an association may normally be defined in the opposite direction.

As already stated, it is necessary to specify a reference from each entity of CUSTOMERORDERS to an entity of CUSTOMERS. Seen from the opposite side, it might be convenient from each customer to define an association of customer orders which are submitted by this customer. In such an association, each entity of CUSTOMERS is an anchor and all the entities in CUSTOMERORDERS which are referring to this anchor will be the body of the association.

An assembly structure is an association where each entity of the PRODUCTS entity is an anchor entity and PRODUCEDITEMS and RAWMATERIALS are body entities.

Usually, an association is representing a *master-detail* relationship where the master type is the anchor type and the detail type is the body type.

The internal association of ORDERLINES in ORDERS may be characterised as a master-detail relationship. Each order is an anchor and the pertaining order lines form the body.

As indicated, the internal organisation of an association is undefined and sometimes, it is important to establish an exact *information structure* because certain physical relations have to be modelled.

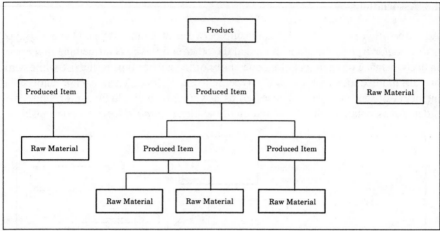

Figure 16   The assembly structure

Information structures may be defined on the basis of associations. An information structure is an organisation of the association, so that a structure is build between each anchor entity and the corresponding body. Examples of information structures are *sets, lists* (including *stacks* and *queues*), *trees* (including *balanced search trees*) and *networks*.

A product is often regarded as an assembly of items which, again, may be assembled by items. An assembly structure is usually formed as a tree, but if some items occur more than once, the structure will be a network. In assembly trees, all consisting items of a product will be nodes in the tree. The product is the root node and raw materials are leaf nodes.

Because two entity types are required in the assembly association, the type ITEMS can be specified as the type of the body, so that the concerned sub-types can be the substitutes. Alternatively, a new type could be added to the taxonomy. This should be placed as a sister type to PRODUCTS and super-type to PRODUCEDITEMS and RAWMATERIALS. It could be named PRODUCTCOMPONENTS.

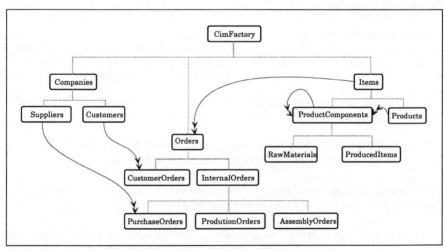

Figure 17   Associations indicated on the basis of the taxonomy

The order lines, which belong to each order, will often be accessed in the order, in which they arrive. Therefore, it could be defined that the association of order lines for each order should be structured as a linked queue structure. The anchor will refer to both the first and the last order line. The first order line will refer to the second and the second to the third, etc.

Occasionally, the definition of associations appear to be impossible because no anchor type exists. In such a case, a new entity type must be defined in the taxonomy.

If an association is to be created for a particular group of entities of PRODUCTS, e.g. standard products in stock, a product inventory entity could be defined as the anchor entity. In the taxonomy, an entity type for this entity will be placed as a sub-type of the root type.

Often, when orders are scheduled, a priority is assigned to each of them. In such a case, it is relevant to specify that the orders are structured in a priority queue, defining that when an order is to be carried out, it will always be the order with the highest priority. A separate anchor entity type should also be defined, in this case.

As in the case of collections, a number of basic operations, to be performed on associations, havee to be available. At least, methods must be provided for the addition and removal of entities, for a search for a particular entity and for determining the number of entities in the association. These can be defined as methods of a separate entity type. Usually, some *auxiliary entities* will have to be generated in order to hold variant number of references in the association.

As indicated above, associations may internally be organised with, or without, information structures. This distinction leads to different implementation aspects for the information system. If no information structure is defined, it is not necessary to maintain the structure when new members are added and existing ones are removed. Thus, if an association is implemented without an information structure, and a specific set of connections in the association is required, later, then the information system objects must be examined by scanning.

It is possible from each item of ITEMS to define an information structure of INTERNALORDERS, in which the item is referred to. If it is decided not to form such a structure, there will be no connections in the direction from ITEMS to ORDERS. Therefore, if it is necessary to find orders with demands for a particular item, then all orders will have to be examined. If so, it would be efficient, to have an index on INTERNALORDERS with ITEMNUMBER as the key.

Another association, without an information structure, could be defined from PRODUCTS to CUSTOMERORDERS indicating that, for each product, a number of customer orders may exist, containing at least one request for the product.

5 Representation of Information Models

Information models may be represented in different ways depending on the mode of communication of the model. A graphic representation is excellent for the purpose of providing the reader with a good overview of the structures of the model. A text representation is suitable for simple transmission of the model.

5.1 Graphic representations

Object-oriented information models, to a great extent, may be represented graphically. It provides the opportunity to work in a visual form and to communicate easily with others. (Ref. [2] and [3])

The taxonomy of entity types serves as a starting point, as the *back-bone* of the information model. The taxonomy is very convenient, because it can be viewed in many ways, and many further structural considerations may be linked to the taxonomy.

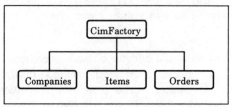

Figure 18   The top levels of the taxonomy.

It is possible to look at different sub-trees, by selecting any tree-node as a root node. This is useful if the tree has many nodes, and when it is inconvenient to display the total structure in one diagram. A viewing and zooming technique allows the designer to select a certain local view of the model and to concentrate on the necessary types there.

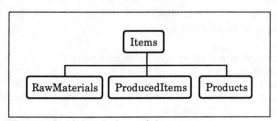

Figure 19   A local view of the taxonomy

The diagram in figure 8 indicates the full taxonomy, but, as already indicated, the model could be further refined and the tree could develop wider and deeper. Moreover, it is possible to zoom in and concentrate on a particular sub-tree or branch. In the figure, local discussion could be held, e.g. on ITEMS, RAWMATERIALS, PRODUCEDITEMS and PRODUCTS.

This viewing of sub-trees could be further extended by selecting only certain parts of the branches and concealing some irrelevant parts on certain levels. Tree structures are particularly suitable for this kind of operations, because they are easy to remember and therefore more easy to navigate without losing orientation.

Because the sequence in which the branches are connected to a parent node is arbitrary, the tree can easily be rearranged so that entity types, which have the closest relationships, are placed near each other. Such rearrangements may be conducted differently in different displays and in different types of entity relationship diagrams.

A special zoom operation could open up each entity type box and reveal the name and type of all the attributes. Furthermore, structured attributes could be shown together with the internal structure. When opening up entity types, it is important for the modelling approach to be able to distinguish between the visible and the hidden attributes depending on the user. E.g. only the visible attributes could be displayed.

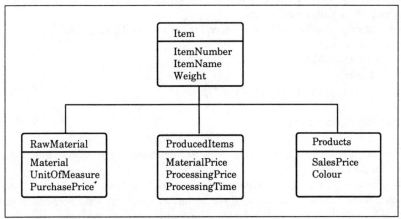

Figure 20   A view of the attributes of entity types

All Entity Relationships may be specified as one or more layers on top of the taxonomy, and the individual relationships in the layers could be shown with different graphic symbols. E.g. associations are shown as arrows between pairs of entity types. See figure 17.

5.2 Representation of information models in text form

Information model designers may desire to communicate with each other and exchange models in a simple text representation. Moreover, it would be necessary, from a graphically represented model, to generate a text representation, or the opposite, to generate the graphs of the model from a source text. In order to do this, an *information model language* must be available.

Different forms of text representation may exist, depending on what it will be used for. One form could be developed for expressing the entity types, including data attributes, methods, constraints and triggers. Another form may include the details of the graphic representations. A special language could exist for producing back-up files of information models.

A language defines syntax and semantics of a set of declarations and statements for describing the elements of information models. Many languages have been proposed and most of the proposals are originating from database schema languages (Data Definition Languages) and from object-oriented programming languages.

However, the Structured Query Language (SQL) for definition and manipulation of relational databases has become a widely accepted database language. Originally, SQL was created as the database language for IBM System R, published in 1976 (Ref. [4], [15], [16] and [19]). Since then, the language has been changed several times. The current version was standardised in 1992, termed SQL-92.

The standardisation committee has been working on proposals for the further development of the language (SQL-3) for more than three years. These proposals contain an extensible type system including certain elements of object-orientation. This standard is expected to be completed in 1995-96. In the following, an information model language is presented which is inspired by various proposals for SQL-3 and by various object-oriented programming languages. This language is presented, informally, step by step.

An *entity type*, in the most simple form, is defined by the statement

```
type NAMEOFTYPE
    begin
        DATAELEMENTNAME DataType;

        . . . .

    end NAMEOFTYPE;
```

This simple form is compatible with the definition of *database tables* in the relational data model.

In addition to the simple data elements, data structures of any kind may be defined. The available datatypes will not be further specified here. A *constant* can be specified implicitly by replacing the DataType with a *data expression*.

In the CIMFACTORY example, the type Companies could be defined as
```
type COMPANIES
    begin
        COMPANYID number(4);
        COMPANYNAME char(50);
        ADDRESS char(80);
        PHONENUMBER char(20);
        FAXNUMBER char(20)
    end COMPANIES;
```

Methods are also defined within the type declaration. They are termed *procedures* and *functions*. Procedures may have both input and output parameters. Functions only have input parameters and return one value.

```
type NAMEOFTYPE
    begin
        . . .
        procedure NAME(ListOfParameters)
        begin
            LocalAttributes;
            Statements;
        end NAME;
        function NAME(ListOfInputParameters)ReturnType
        begin
            LocalAttributes;
            Statements;
            return Expression
        end NAME;
        . . .
    end NAMEOFTYPE;
```

The entity type ITEMS could be defined as

```
type ITEMS
    begin
        ITEMNUMBER number(5);
        ITEMNAME char(30);
        WEIGHT number(10,2);
        INVENTORYQTY number(5);
        CREATETIME date;
        REVISIONDATE date;
        function Age time
        begin
            return SYSDATE - CREATED
        end Age;
        procedure UPDATEINVENTORYQTY(ADDEDQTY number(5))
        begin
            INVENTORYQTY := INVENTORYQTY + ADDEDQTY
        end UPDATEINVENTORYQTY
    end ITEMS;
```

Encapsulation can be specified by a number of constructions. First of all, the specification of an entity type can be separated into two definitions with the same name: an *interface type* and an *implementation type*.

```
interface type NAMEOFTYPE
begin
    . . .
end NAMEOFTYPE;

implementation type NAMEOFTYPE
begin
    . . .
end NAMEOFTYPE;
```

In the interface type, all *public attributes* are specified, and in the implementation type, all *private attributes* and *bodies of methods* are specified. This means that the interface type may be published while the implementation type may be kept secret by the implementor.

Moreover, the interface and implementation versions of the type declaration make it easy to specify *read-only data elements*. In the interface type, they will simply be defined as functions without parameters and in the implementation type they will be specified as stored data elements. *Read-only* data attributes may be understood as functions of which the purpose is only to return the stored data value.

The bodies of functions and procedures are only specified in the implementation type.

The entity type Items may now be revised in the following way:

```
interface type ITEMS
    begin
        ITEMNUMBER number(5);
        ITEMNAME char(30);
        WEIGHT number(10,2);
        function INVENTORYQTY number(5);
        function CREATETIME date;
        function INVENTORYUPDATETIME date;
        function AGE time,
        procedure UPDATEINVENTORYQTY(ADDEDQTY number(5))
    end ITEMS;
implementation type ITEMS
    begin
        INVENTORYQTY number(5);
        CREATETIME constant SYSDATE;
        INVENTORYUPDATETIME date;
        function AGE time
        begin
            return SYSDATE - CREATED
        end AGE;
        procedure UPDATEINVENTORYQTY(ADDEDQTY number(5))
        begin
            INVENTORYQTY := INVENTORYQTY + ADDEDQTY;
            INVENTORYUPDATETIME := SYSDATE
        end UPDATEINVENTORYQTY
    end ITEMS;
```

Because INVENTORYQTY and INVENTORYUPDATETIME are specified as functions in the interface type and stored data attributes in the implementation type, they are read-only data attributes. The attribute CREATETIME is implemented as a constant, specifying the time when the entity is created. The function AGE is an ordinary function.

Inheritage to a sub-type is specified by indicating one or more super-types when the type is declared.

```
type NAMEOFTYPE inherits from NAMEOFTYPE , NAMEOFTYPE, . . .
    begin
        . . . .
    end NAMEOFTYPE;
```

In this way, a complete taxonomy can be described. *Multiple Inheritage* to a sub-type is simply specified by indicating more than one super-type. *Virtual methods* are declared with the leading word virtual.

In the example, the entity type CIMFACTORY should be declared first, and the types COMPANIES and ITEMS, in their declaration, would inherit the attributes from CIMFACTORY.

```
interface type CIMFACTORY
begin
    virtual procedure SHOW;
    . . . .
end CIMFACTORY;
```

```
interface type COMPANIES inherits from CIMFACTORY
begin
    virtual procedure SHOW;
    . . . .
end COMPANIES
```

The virtual procedure Show will also be declared in all other types.
The types CUSTOMERS and SUPPLIERS would both inherit from COMPANIES.

```
interface type CUSTOMERS inherits from COMPANIES
    begin
        DEBITBALANCE number(15,2);

        . . . .
    end CUSTOMERS
interface type SUPPLIERS inherits from COMPANIES
    begin
        CREDITBALANCE number(15,2);

        . . . .
    end CUSTOMERS
```

The type MIXCOMPANIES is declared to inherit from both CUSTOMERS and SUPPLIERS in this way
```
interface type MIXCOMPANIES inherits from CUSTOMERS, SUPPLIERS
    begin

        . . . .
    end MIXCOMPANIES
```

Inherited attributes can be listed in order to facilitate the reading af the type.

```
type NAMEOFTYPE
    inherits from NAMEOFTYPE(ListOfAttributes)
    begin

        . . . .
    end NAMEOFTYPE;
```

Inherited attributes can be *redefined* by declaring a local attribute with the same name. Otherwise, inherited attributes can be *renamed* by adding new names in the attribute list.

Furthermore, renaming makes it possible to use a name of an inherited attribute without affecting this.

If in CUSTOMERS, the attribute COMPANYNAME from COMPANIES should be renamed to CUSTOMERNAME, it is declared in this way

```
interface type CUSTOMERS
    inherits from COMPANIES
        (rename CUSTOMERNAME for COMPANYNAME)
    begin

        . . . .
    end CUSTOMERS
```

Each entity has a unique *entity identifier* which is automatically generated when the entity is created. An entity can be created by a predefined function which will take values for all public data attributes as arguments and returns the value of the entity identifier. This value may be stored in a *reference*. Such an attribute is declared with a specification of the entity type to which it will refer.

```
type NAMEOFTYPE
begin

    . . . .

    REFERENCENAME reference NAMEOFTYPE;

    . . . .

end NAMEOFTYPE;
```

The type CUSTOMERORDERS will have a reference attribute to CUSTOMERS.

```
type ORDERS inherits from CIMFACTORY
    begin
        DELIVERYTIME date;

        . . . .

    end ORDERS;
type CUSTOMERORDERS inherits from ORDERS
    begin
        CUSTOMER reference CUSTOMERS;
        TOTALSALES number(15,2);
        DISCOUNTLIMIT number(15,2);
        DISCOUNT number(15,2);

        . . .

    end CUSTOMERORDERS
```

When a customer order entity is created, the specified customer will be selected by scanning the collection of all customers and the value of the entity identifier will be copied to the CUSTOMER attribute. From this point, a *client-server* relationship is established between the customer order entity and the customer entity. Therefore, the *public services* of the customer entity can be used from the customer order entity by sending *messages*.

6 Application Development

Information models are intended to be used for constructing information systems, and the entity types will form the information system object types. Thus, when an information system is implemented, persistent objects of these types can be created in one or more *objectbases*.

In order to manipulate these objects of the information system, some applications have to be available. *Applications* are program modules, such as screen form applications, diagrams, printed reports, etc. with which data of the information system objects can be entered, updated and retrieved. The most often used type of application is screen forms.

An application can be regarded as an interface to one or more objects, prepared for the users of the system. Through the interface, the user can communicate with the objects. The interface provide a set of services which can be executed on the request of the user. In this respect, the object will act as a server for the user. Some of the services will be the public methods which are defined in the entity types.

In the object-oriented approach, the focus is on the objects and not on the applications. This is only natural, because, when working with an information system, the user will have the attention mainly on the data. The operations will be attached to the data and the manipulation of the data. The applications are shells around the data which implement the limitations of the data and control the data manipulations. It is assumed that the interface to objectbases is provided by the *objectbase management system.*

The components of applications can be attached to the information model as other kinds of information. This means that the application will be activated from the object and not the other way around.

The applications must be matched with the tasks which the users have to perform. Therefore, some function modelling will have to be performed in advance.

Based on the information model, some applications can be generated automatically by use of a development system. The better developed the information model is, the better the generated applications will be. A great number of functions can be performed by means of such applications. But, in practice, some applications will have to be modified or specifically designed. (See figure 21.)

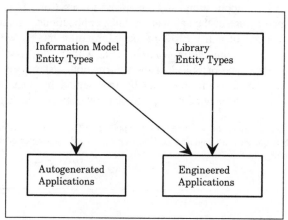

Figure 21   Building information system object types
and applications

The basic information for generating applications is an information model with its entity types. Furthermore, the specifications of structural relationships will be important. These are the references and the associations, which establish links between objects. In order to control the manipulations of the objects, the semantics is used. The constraints will set limits and the triggers will add some functionality to the user manipulations. The simplest applications are those which only perform retrieval operations. For such applications, the semantics is not necessary. It will only be used for applications with which it is possible to update the objects of the information system.

When developing applications, which are based on an information model, it is important as to how well the entity types of the model are equipped with available services. Most of the basic operations, which have to be performed on the objects, can be obtained from the specifications of the entity types in the information model. But, in general, some additional functionality has to be attached.

A number of additional features can be attached to applications if *library entity types* are available (see figure 21). These library types may be combined with the model types by inheritance, possibly by multiple inheritance, or they could be used dynamically on a client-server basis (Ref. [25] and [32]). The library entity types can often be selected from commercially available *tool packages* for object-oriented programming languages.

In the following, some aspects of developing screen form applications will be described. It is assumed that a comprehensive development tool is available for modelling of the application on the basis of a stored information model. Some of the required functionality of such a tool will be indicated. Two extremes are considered, automatic generation of applications and specific design of applications by use of an object-oriented programming language. Only screen forms aimed at a windows user interface are examined.

Autogeneration of Applications

A window screen form application can be prepared for a set of different tasks which are to be performed by use of the application. Such applications, which are generated from the specifications of entity types in the information model, will, in a simple version, consist of an *application window* and, within this, one *data window* corresponding to each type of object of the information system (see figure 22). In data windows, all the data attributes can be represented by data fields. When a development tool generates a window, a particular style can be selected by the user.

In figure 22, an example of a screen form application is showed. A data window for CUSTOMERORDERS is opened with a number of fields representing the attributes of the corresponding information system objects.

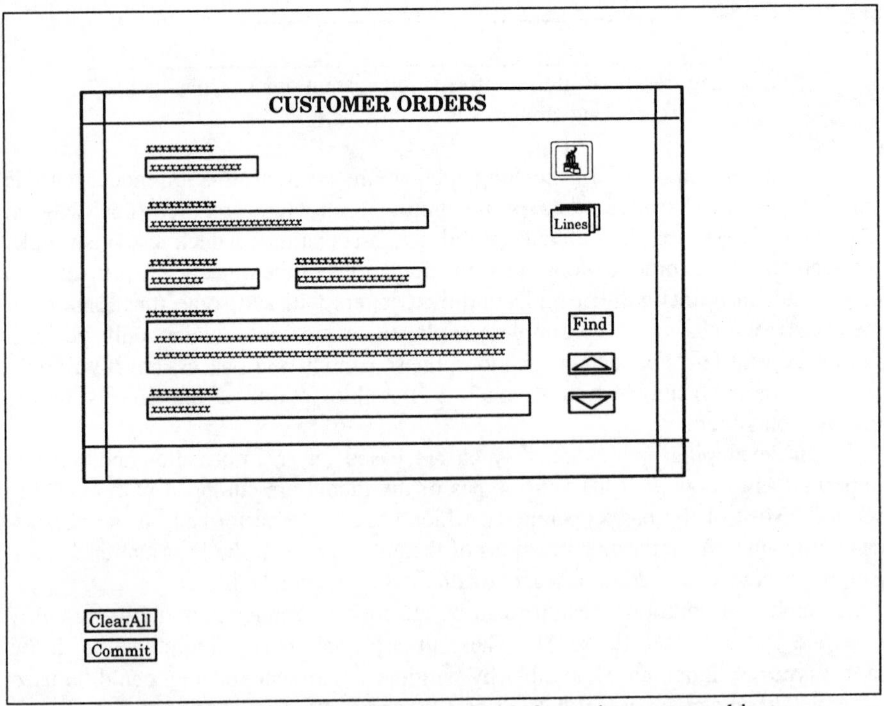

Figure 22   Window form presenting fields of an information system object

The specifications of each attribute of the entity types makes it possible to generate an appropriate data field for each data type of the attributes. Number and text string fields can be displayed in the traditional character based fields. Binary valued data can be specified as on/off fields or switch fields of different kinds. Diagrams, photographs and other graph data types can be presented directly.

The functionality, which will be offered to the user, can be represented by menu items or graphical symbols on the screen. These functions can be activated by a pointer device, e.g. a mouse or a pen. The graphical symbols can be regarded as push buttons. By means of these symbols, the user is able to enter new objects into the information system or retrieve objects from the system. A number of functions will automatically be included in all windows for standard window manipulations. These manipulations include sizing, scrolling, moving, etc. A window can normally be reduced, minimised, to a small graphic symbol termed an *icon*.

Other functions, related to data manipulations, can be generated on the basis of the methods which are defined in the entity types of the information model. Standard functions can be added automatically by the development tool, eventually, selected by the application engineer from a list of possible functions.

In the lower left corner of the application window and in the lower right corner of the data window in figure 22, some function symbols are indicated. The Find function can initiate a retrieval operation in the object base. The Commit function will commit changes, which are made by the application, in the object base.

The references between objects can be represented as links between data windows. Therefore, it will be appropriate to represent these as icons. When such an icon is activated, the corresponding window form will be opened and display the attribute values of the referred object.

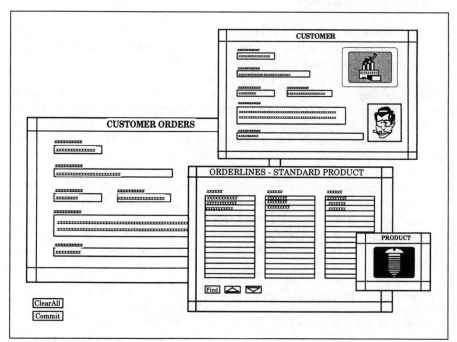

Figure 23   Window forms opened to display data of referred objects

In the data window in figure 22, an icon in the upper right corner is indicating a reference by which a data window for the CUSTOMERS object type can be created on the screen. This is illustrated in figure 23. Furthermore, as customer orders are complex objects, data windows for internal order line objects can also be opened by activating an icon. Figure 23 shows an opened window for ORDERLINES objects. In this case, it is of STANDARDPRODUCT ORDERLINES. The icon will appear when the window is minimised.

Besides the simple data windows, other windows can be made on the basis of the collections and associations which are defined in the information model. Normally, these windows will limited to retrieval operations, only.

Screen forms can also be generated for collections. As previously defined, a collection can include objects of a given type and all its sub-types. In such collections, sub-types can substitute the super-type. This means that, when objects are presented in a particular sequence, the data window will change automatically according to the object type.

For an association, two linked data windows will be generated. One for the anchor type and one for the body type. As substitution also applies to associations, the body window will change, automatically, if different sub-type objects are retrieved.

When different operations on the information system objects are performed with an application, the defined semantics will be used to control the operations. When updating the data values, eventual constraints will be checked and attached triggers will be executed. Additional constraints and triggers, attached to the application, can me developed by the application engineer. Here, the defined methods in the information model can be used.

Engineering of Applications

If applications are specifically engineered, a development tool will also be useful. Often, the first version of the application can be generated by specifying some development criteria to the tool system. Afterwards, this system can be modified by the application engineer assisted by the development tool.

When the applications are generated, the development tool may be able to formulate the applications in different programming languages. First of all, it will be suitable to focus on object-oriented languages. In such languages, it will be possible to take full advantage of the object-orientation of the information model. Furthermore, it will be easier to attach library entities.

If library entity types are not included in the information model taxonomy, such types can be attached to object types when the application is developed. E.g. object types for creation and manipulation of windows, displaying different kinds of information, can easily be included in the applications.

Library entity types should also be available for creating and maintaining the various information structures which can be specified according to their defined associations.

Library entity types can be attached to the type ITEMS in order to construct the assembly structure of PRODUCTS. Likewise, if a priority queue is defined for the orders, it can be organised and maintained by inheriting the necessary services from one or more library entity types.

Modification of an application will often include changes of the properties of data fields. Furthermore, the functionality will usually be modified. This is done by adding new application-specific methods. The methods can be attached to particular fields and windows or they can be attached to functions which are made available to the user.

Use of library entity types will increase the reliability of application development because these types are supposed to be tested in advance by the supplier. Furthermore, if new entity types are constructed, it could be considered to include such types in the library in order to use them in other applications.

Applications, which are developed according to these principles, are easy to extend. If new attributes and entity types are added to the information model, the applications can easily be changed in order to include the extensions. New fields can be added to data windows and new data windows can be generated.

Because the applications, to a great extent, are developed on the basis of an object-oriented information model, it is also very safe to reuse applications or data windows of applications in other development projects, where parts of the same information model is used.

7 Conclusion

The object-oriented paradigm is well suited for the design of information systems. The components of such systems are regarded as active, living objects which, to some degree, are prepared for network distribution. The information system objects are communicating with each other as clients and servers.

For information modelling, object-orientation is an excellent approach, and models of this kind are well prepared for rapid development of information systems. Object-oriented information modelling is data driven instead of process driven and, consequently, the result, to a greater extent, is based on invariant real world structures.

The object-oriented methodology for information modelling, which is presented here, is developed according to two fundamental abstraction mechanisms; generalisation and aggregation. Specifically, two orthogonal evaluations are defined: generalisation versus specialisation (classification) and aggregation versus separation. Classification focuses on the static structures of the model as a foundation for the design of dynamic structures.

Therefore, the design of information models is, primarily, performed by means of classification, whereas the entity types (classes) of the model are placed and defined in a hierarchy, the taxonomy. The arrangement is performed in respect to the resemblance and differences of the attributes of the entity types. This leads to the use of inheritance of attributes from super-types to sub-types in the taxonomy. Variations of the taxonomy appear when multiple inheritance requires to be introduced in the model, and when complex entities are described.

Compared with traditional entity relationship modelling, the object-oriented approach focuses on information structures which are designed by references and associations between the individual entities. These structures are defined on the type level and can be envisaged as separate layers on top of the taxonomy.

In order to facilitate information models for the automatic generation of information system applications, it is important to define methods within the entity types. In general, they should be developed according to the completeness principle, which states that all entity types should provide a complete set of services, relevant to the included data structure.

The object-oriented information modelling methodology is not difficult to use. The main technique, classification, is directly understandable, even for the non experienced participants in development projects.

This implies that information systems can be developed with a higher fulfilment of the requirements and with a better quality. The systems are more reliable, reusable and extendable. Furthermore, the productivity will be improved.

References

[1]   S. Abiteboul, P. Kanellakis: Object identity as a query language primitive. Proceedings of the 1989 ACM SIGMOD, Portland, Oregon, June 1986. 1989.

[2]   Michael Ackroyd, Dana Daum: Graphical notation for object-oriented design and programming. Journal of object-oriented programming. New York 1991.

[3]   Marc Andries, Marc Gemis, Jan Paredaens, Inge Thyssens, Jan Van den Bussche: Concepts for Graph-Oriented Object Manipulation. EDBT '92. Springer-Verlag 1992.

[4]   M. M. Astrahan et al. System R: Relational Approach to Database Management. ACM Transactions on Database Systems Vol 1, no. 2, June 1976.

[5]   M. Atkinson, Francois Bancilhon, Klaus Dittrich, David Maier, David DeWitt, S.B. Zdonik: The Object-Oriented Database System Manifesto. Elsevier Science Publishers 1990.

[6]   Francois Bancilhon, F. Velez, P. Richard, P. Pfeffer, C. Lecluse, S. Gamerman, C. Delobel, V. Benzaken, G. Barbedette: The design and implementation of O2, an object-oriented database system. Advances in object-oriented database systems, 2nd international workshop on object-oriented database systems. 1988.

[7]   Francois Bancilhon: Understanding Object-Oriented Database Systems. EDBT '92. Springer-Verlag 1992.

[8]   J. Banerjee, H.T. Chou, J. F. Garza, D. Woelk, N. Ballou, H. Kim, Won Kim: Data Model Issues for Object-Oriented Applications. ACM TOOIS v5 n1. 1987.

[9]   R. Brachman: What IS-A is and isn't: An analysis of taxonomic links in semantic networks. IEEE Computer sept. 1983. 1983.

[10] Luca Cardelli, James Donahue, Lucille Glassman, Bill Kalsow, Greg Nelson, Mick Jordan: Modula-3 Language Definition. ACM Sigplan Notices Vol 27, no. 8. ACM Press 1992.

[11] Michael Carey, David DeWitt, Goetz Graefe, Joel E. Richardson, Eugene J. Shekita, Scott L. Vandenberg, Daniel T. Schuh, David M Haight: The EXODUS Extensible DBMS Project: An Overview. Readings in Object-Oriented Database Systems. Madison Computer Science Dept. 1988.

[12] P. Chen: The Entity-Relationchip Model - Toward a Unified View of Data. ACM Transactions on Database Systems v1 1976.

[13] P. Coad, E. Yourdon: Object-Oriented Analysis. Yourdon Press, 1991.

[14] P. Coad, E. Yourdon: Object-Oriented Design. Yourdon Press, 1991.

[15] E.F. Codd: A Relational Model of Data for Large Shared Data Banks. Communications of the ACM v13 no6. 1970.

[16] E.F. Codd: Extending the Database Relational Model to Capture More Meaning. ACM Transactions on Database Systems v4 no4. 1979.

[17] G. Copeland, S. Khoshafian: Object Identity. Proceedings on OOPSLA, Portland, Oregon USA. 1986.

[18] Ole-Johan Dahl, C.A.R. Hoare: Hierarchical Program Structures. APIC Studies in Data Processing no8: Structured Programming. Academic Press, N.Y. 1972.

[19] S. Dar, N.H. Gehani, H. V. Jagadish: CQL++: A SQL for the Ode Object-Oriented DBMS. EDBT '92. Springer-Verlag 1992.

[20] Brian Henderson-Sellers, Julian M. Edwards: The Object-Oriented Systems Life Cycle. Communications of the ACM, 1990, vol.33 no.9. ACM, New York 1990.

[21] C.A.R. Hoare: Notes on Data Structuring. APIC Studies in Data Processing no8: Structured Programming. Academic Press, N.Y. 1972.

[22 Alfons Kemper, Guido Moerkotte: A framework for strong typing and type inference in (persistent) object models. DEXA'90. Universität Karlsruhe, Germany.

[23] J. M. Keer: The information engineering paradigm. Journal of systems management, vol.42, no.4, USA 1991.

[24] Leo Mark, N. Roussopoulos: Metadata Management. IEEE Computer. 1986.

[25] Bertrand Meyer: Lessons from the Design of the Eiffel Libraries. Communications of the ACM, 1990, vol.33 no.9. ACM, New York 1990.

[26] Guido Moerkotte, Andreas Zachmann: Multiple Substitutability without Affecting the Taxonomy. EDBT '92. Springer-Verlag 1992.

[27] Larry Rowe, Michael Stonebraker: The POSTGRES Data Model. Proceedings of the XIII Int. Conf. on Very Large Databases. Morgan Kaufmann Publ. 1987.

[28] J. M. Smith, D. C. P. Smith: Database Abstractions: Aggregation. Communications of the ACM v20, no6. 1977.

[29] J. M. Smith, D. C. P. Smith: Database Abstractions: Aggregation and Generalization. ACM transactions on Data Base Systems, v2, no.2. 1977.

[30] Alan Snyder: Encapsulation and Inheritance in Object-Oriented Programming Languages. Reading in Object-Oriented Database Systems. ACM, New York 1986.

[31] Michael Stonebraker, Lawrence A. Rowe: The design of Postgres. SIGMOD'86. New York 1986.

[32] Clemens A. Szyperski: Import is Not Inheritance - Why We Need Both: Modules and Classes. ECOOP'92. Springer-Verlag 1992.

[33] Gio Wiederhold: Knowledge versus Data. On Knowledge Base Systems. Springer-Verlag, N.Y. USA 1986.

[34] Wirfs-Brock: Designing Object-Oriented Software. Prentice Hall, 1990.

4 Databases for CIMS and IMS

F.B. Vernadat

1. Introduction

One of the trends of modern manufacturing is to progress towards Computer-Integrated Manufacturing (CIM) for better communication, co-ordination, and efficiency throughout the enterprise. Communications systems and information systems are key components for achieving true integration of distributed manufacturing systems (horizontal integration). Furthermore, information is the basic ingredient used for decision-making at the various decision levels of any manufacturing enterprise (vertical integration). Information is the "glue" which ties together all the modules of a manufacturing system while the underlying integrating infrastructure (i.e., a set of computer services connecting modules and ensuring transparent information exchange) is the backbone of its information system [2].

Information systems are usually computerized in the form of (distributed) databases, which tend to replace conventional computer files as widely used in the past. Compared to file systems, database systems provide decisive advantages which include [6, 34]:

- *Data independence*: Physical data storage aspects are independent from logical data description and manipulation, and are unknown from users and applications.
- *Data sharing*: Data can be shared and accessed concurrently by different users and/or applications. Data access conflicts are handled by the database system.
- *Controled data redundancy*: The same information must be stored once, therefore reducing the risk for inconsistent updates of redundant data as commonly found in file systems.
- *Data integrity and security*: Rules can be expressed and verified to ensure that the state of the database is semantically correct, i.e., it complies with the reality of the enterprise, and that data can be protected from unauthorized accesses, misuses, or from system failures.
- *Ad hoc query facility*: Queries on data can be expressed and executed to provide answers to questions formulated by applications or on-line users.

The aim of this section is to discuss database technology as it applies to modern manufacturing systems. Manufacturing information systems and basic database terminology are first presented. Database design techniques are then discussed on the basis of the entity-relationship model. Next, relational databases and the SQL language followed by object-oriented databases are both considered for information system implementation. Finally, distributed database technology is discussed before concluding.

2. Manufacturing Information Systems

Manufacturing systems are concerned with designing and engineering new products and their components, producing necessary information for their manufacturing, planning and controling their manufacturing, performing assembly and shipping operations, and managing related stocks (raw-materials, components and sub-assemblies, finished products). To support operations of these functions, the following information systems and associated (distributed) databases can be found in most modern manufacturing enterprises (Fig. 1) [28, 29, 35]:

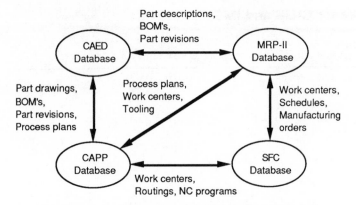

Fig. 1. Manufacturing information systems and their relationships

- *Computer-Aided Engineering and Design* (CAED) information system: It concerns information related to product design and engineering. This includes product specifications, part descriptions, group technology codes, part and assembly drawings, revisions, bills-of-materials (BOMs), finite element analysis data, etc.
- *Computer-Aided Process Planning* (CAPP) information system: It concerns manufacturing processes and manufacturability data such as process plans, metal cutting data, tooling data, numerical control (NC) and robot programs, etc.
- *Manufacturing Resource Planning* (MRP II) information system: It concerns information used to plan and control production. This includes inventory control data, purchasing and accounts payable, suppliers data, master schedules, production schedules, work centers, resource capacities, manufacturing orders, etc.
- *Shop-Floor Control* (SFC) information system: It concerns information about shop-floor operations such as part routings, work-in-process, resource status, manufacturing order status, etc.

Administrative databases are also used frequently to store information about personnel, suppliers, customers, invoices, accounting, sales, financial data, etc.

3. Basics of Database Technology

Data and information:
Database systems are computer systems used to store, manage, and retrieve data and information.
- A *datum* (or *data item*) is a piece of information made of two things: a fact and a meaning. For instance, "my_height = 1.65 m" is a datum. 1.65 is the fact and the meaning is that it represents a person's height in meters.
- *Information* is the basis for communication and decision-making. It is made of data and subject to rules. For instance, if I tell you that my hight is 1.65 m, then I give you an information. Furthermore, to be plausible, this information must comply with the rule which says that "any human being's height is comprised between 0.30 m and 2.20 m".

Attributes, entities, and relationships:
In database technology, it is assumed that the real-world is made of entities (e.g., persons, things, facts, events, ...) linked by relationships (e.g., this car belongs to that person), and described by attributes, where attributes are defined as data.
- *Attributes*. An attribute describes a characteristic or property of an entity or a relationship (e.g., the color of a car, the name of a person, the date of purchase of the car by the person). An attribute can be total (its value is always defined), or partial (its value

can be left unknown). It is defined by its name and takes its values from a set called a *domain* defined over a data type. An attribute can be atomic (i.e., made of one fact), or aggregated. A domain defines the set of admissible values that an attribute can take. If a_i is the name of an attribute and if D_i is the name of its domain, we note $a_i : D_i$.

- *Entities*. An entity E contains a set of concrete or abstract real-world objects (i.e., entity occurrences) described by means of a common set of attributes. If a1, a2, ... are the attributes of entity E, we note E (a1, a2, ...).

- *Relationships*. A relationship R among entities El, E2, ... describes the set of relations (i.e., relationship occurrences or links or connections) relating objects of El, E2, ..., and (possibly) sharing some common attributes. The functionality of a relationship between a pair of entities can be defined either as 1:1 (each occurrence of an entity is linked to at most one occurrence of the other entity), 1: n (each occurrence of an entity can be linked to more than one occurrence of the other entity), n:1 (symmetric link to 1:n), or m: n (pure relation or subset of the Cartesian product of the two entities). The arity of a relationship is the number of entities linked by the relationship. It must be a strictly positive integer (unitary relationships, binary relationships, n-ary relationships). In practice, it is recommended to mostly define binary relationships.

- *Identifiers* or keys. An identifier is an attribute or a collection of attributes, the values of which are never null, and which universally identify each occurrence of an entity or a relationship. In practice, the identifier of a relationship contains at least the identifiers of the entities involved in the relationship (as long as there is no functional dependency among these attributes).

Integrity constraints:
Attributes are subject to integrity constraints [13]. An *integrity constraint* is a Boolean expression defined as a first-order logic formula used to ensure validity, correctness, or plausibility of values of an attribute, or consistency of these values with values of other attributes. If some data violate the integrity constraints, they are rejected by the system.
Example 1: To ensure that any person's age is a positive integer smaller than 125, we can write, if age is an attribute of the entity person defined over the domain of integers:

$$0 \leq person.age \leq 125$$

Example 2: To ensure that the volume of a room is the product of its length by its width by its height, using the same notation system where room is the entity, we can write:

room.volume = room.length * room.width * room.hight

Databases and database management systems:
- A *database* (DB) is a large collection of (possibly distributed) shared operational data stored for the purpose of various applications and users of some enterprise [6,34].
Users are on-line operators on the database while applications are commercial software packages or special purpose computer programs written in conventional programming languages (Pascal, Fortran, Cobol, C, ...). Usually, databases are stored on mass storage devices since their size may vary from a few K-bytes up to several G-bytes. However, main-memory database systems have also been recently investigated for special applications requiring very fast data access [19].
- A *database management system* (DBMS) is a complex piece of software which supports the creation of one or more databases and allows one or more users and/or applications to access and/or modify data of these databases. Functions of a DBMS include data access methods, transaction and concurrency control, data integrity and security management, and recovery management in case of system failures [6,10,34]. All accesses to data in a database are performed via the DBMS. Database management systems are commercially available for mainframe, mini- and micro-computers. They are usually dependent on a given technology imposed by their underlying data model.
- A *data model* is a formalism used to represent the syntax, and possibly part of the semantics, of data in a database. It usually provides a formal notation to describe the generic structure (or conceptual schema) of a database. In most cases, it is supported by a graphical notation to produce a graphics representation of the database schema for better

understanding by business users. Classical data models include the hierarchical model, the network model, the relational model, and the entity-relationship model [6,34].

- A *database transaction* is set of operations or a routine to be executed on a database, and to be considered as a whole, i.e., either the full transaction is executed or not at all. It must have a beginning and an end, and at its end it is either committed (correct execution) or rolled back (discarded). A database transaction transforms a database from a consistent state to another consistent state. However, consistency may be violated during transaction execution. Database transactions are always serialized and cannot be nested.

ANSI/X3/SPARC structure:

The ANSI/X3/SPARC structure [33] has been proposed to harmonize the global structure of database systems. Its aim is to separate the users' conceptual view of data from the burdens of physical implementation details. The ANSI/X3/SPARC structure is made of three types of schema, one for each level of the structure (Fig. 2).

- A *conceptual schema* is a canonical description of a database structure (called intention) expressed by means of a data model. It provides a unified view of the whole database structure at the conceptual level. It is usually defined and maintained by the database administrator, and it takes into account all user and application needs. The *database administrator* is the person responsible for defining, implementing, and maintaining the conceptual schema and the internal schema of the database system, as well as most of the external schemata.

- An *external schema* (or external view) is a restriction on a database tailored to specific application needs. It is used at the external level, i.e., the level of users and applications. It acts like a filter on the database and only allows a group of users or applications to access data from a well-defined portion of the database. This is one way to guarantee data security (a user cannot access or "see" data not covered by his/her external schema). This is also a way to reduce system complexity in complex databases, since users do not have to be concerned by the full complexity of the whole database schema, but can concentrate on the entities and relationships of interest to them.

- An *internal schema* provides a description of the physical implementation of a database in terms of internal storage structures (e.g., indexed sequential files, direct access files, B-trees, etc.), indexing mechanisms, and data access authorization (definition of passwords, granting data access permission to users in read and/or write mode). It is used at the internal level or physical level, i.e., at the level of mass storage on disks and operating system.

The languages and formats used at these three levels can be quite different, ranging from user-friendly to system-oriented ones.

The most important feature of the 3-schema ANSI/SPARC structure is that it provides a sound framework for achieving true data independence. Indeed, the internal storage structures of data specified by the internal schema can be changed by the database administrator with no impact on the application programs and without being known by the end-users. Conversely, application programs and even external schemata can be modified with no change implied on the data structures on disk.

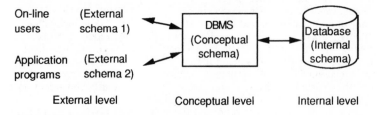

Fig. 2: ANSI/X3/SPARC structure

Data languages:
Three types of languages are used in database technology [6,34]:
- The *data definition language* (DDL) is used to declare the conceptual schema of the database to the database management system and to define external schemata. It is used by the database administrator.
- The *data manipulation language* (DML) is used to insert, delete, and update data of a database. It can be used by on-line users and is extensively used in application programs.
- The *query language* (QL) is mostly used by on-line users to query the database. However, in some cases, it can also be used in application programs.

These languages are sometimes integrated into one language as is the case for SQL, a standardized data language for relational databases described in sub-section 6. Application programs are written using conventional programming languages (Pascal, C, Fortran, Cobol, ...). However, in this case, a programming language interface must be provided, with the accompanying pre-compiler, so that commands of the DML and QL can be used in the program code, therefore simplifying data access and manipulation procedures, and reducing program development and maintenance life cycles.

4. Flexible Manufacturing Cell Example

This example will be used throughout the remainder of this section. Let us consider a Flexible Manufacturing Cell (FMC) made of machine-tools, measuring machines and handling devices such as robots (machines, for short), pallets, cutting and measuring tools (tools, for short). The FMC can produce several parts of different types according to predefined work schedules. A part must be mounted on a pallet to be processed. The sequence of machining operations to be applied to a part is given by a process plan. Alternative process plans can exist for a given type of parts.

5. Database Design

The goal is to design and prepare the implementation of a database conceptual schema and its related external schemata, in compliance with user requirements, in order to optimize information systems operations.

Database design process:
The traditional database design process commonly used for business and administrative database applications involves four major steps:
- *Database requirements definition and data collection*: The aim is to collect informal user requirements concerning the application, and to define the so-called universe of discourse. *Non-formatted data requirements* (e.g., task descriptions given by users, written procedures, business policies, etc.), and *formatted data requirements* (e.g., forms, screen layouts, report layouts, file formats, etc.) are collected from business users.
- *Conceptual design*: The aim is to model static and dynamic aspects of data from data requirements collected in the previous step. Static properties are expressed in the conceptual schema of the database obtained by stepwise refinements using a semantic data model [17] such as the semantic binary model [27] or the extended entity-relationship approach [32], and by the relevant external schemata defined for each group of users or application programs. Dynamic properties concerning data access and manipulation are specified by means of database transactions or routines on the database. Implementation considerations are not yet taken into account at this stage.
- *Logical design*: The aim is to translate the conceptual schema developed during the conceptual design step into a logical schema using the concepts of a target implementation model (e.g., relations for the relational model and objects, classes, and methods for an object-oriented model).

- *Physical design*: The aim of the last step is to produce the physical or implementation schema of the database (i.e., the one which is going to be implemented with the DDL of the selected DBMS). It consists in tailoring and optimizing the logical schema according to physical constraints on the database imposed by its operational environment, or by the DBMS itself (e.g., need for indexes, data security protections, data access permissions, data distribution, data replication, etc.).

Specific data models and database design methodologies have been designed for CIM, either to capture more of the semantics of design and manufacturing data, or to better adhere to the overall CIM system design process. Among these, we can mention SAM* [31], SHOOD [25], TSER [15], and M* [36]. In this paper, we use models and the database design process as defined in the M* methodology for CIM databases. This process comprises three main phases and is illustrated by Fig. 3. Each phase is discussed in turn in more details.

Organization Analysis:
Objectives: The aim of the organization analysis phase is to model as broadly as possible, but not in full detail, functional areas of the enterprise to define a CIM strategy (i.e., to model the current business processes of the enterprise and to engineer desired new ones), to assess the impact of CIM on the enterprise, and to define the CIM information system requirements and environment. The goal is to produce an integrated organizational model (not necessarly executable) stating the requirements of well-defined part(s) of the enterprise, and which contains:
- a functional model (describing the enterprise functionality, i.e., the WHAT),
- an information model (describing the information flow and enterprise objects), and
- a behavior model (describing the flow of control of business processes, i.e., the HOW)

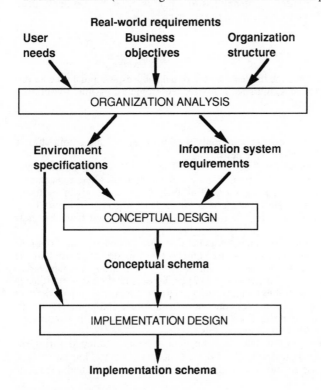

Fig. 3: CIM database design process

Approach: In the M* methodology, organization analysis is supported by an *organization model* in which subsets of the enterprise, called *environments*, are modeled in terms of organization nets connecting object views to functions. The structure of high-level Petri nets, more specifically predicate-transition nets (PrT-nets) [12], are used for the definition of organization nets.

More formally, an *organization net* is a bipartite graph N = (O, F; A), where O and F are two disjoint sets of places and transitions, called object views and functions, respectively, and A is a set of directed arcs connecting object views to functions and vice versa (A ⊆ O X F ∪ F X O, O ∩ F = ∅, O ∪ F ≠ ∅).

An organization net is thus a descriptive causal model of a part of the enterprise. It is made of object views (represented by circles) and functions (represented by bars) in which the functions indicate how object views are consumed, used, or produced. It is used for AS-IS analysis of the enterprise (i.e., describing the current state), and for TO-BE analysis (i.e., describing the desired state). A CIM strategy is then defined as a migration path from the AS-IS state to progress to the TO-BE state.

Object views represent external appearances or states of objects of the enterprise (concrete or abstract) as they are perceived (and described) by users and applications. Enterprise objects represent any physical objects, entities, or resources as well as pieces of information, data, messages, events, etc. of the enterprise environments. CIM object views in M* may belong to four classes for which different graphical place symbols can be defined:

- *messages*: They represent real-world events or requests to execute a function as well as pre-conditions (input messages) and post-conditions (output messages) to the execution of functions. Messages can be used to represent timed-events such as starting scheduled actions (e.g., start at 5:00 p.m.).
- *information*: They represent information objects made of data used/produced by functions. Four sub-classes have been defined: *data* for information items, *files* for structured sets of data items, *forms* for structured documents or computer screens, and *text* for free-format text.
- *materials*: They represent physical matters used/produced by functions (e.g., raw materials, parts, products, components, etc.).
- *resources*: They represent physical means used to carry out functions (e.g., tools, fixtures, machines, people, application programs, etc.).

Functions represent tasks (or functionalities), i.e., things of the enterprise to be done with their inputs and outputs (defined as object views). A task description, obtained from users, describes the function operations (pure text or pseudo-code). Constraints (either defined as input rules and called guards, i.e., predicates defined on attributes of input object views and free variables, or as inscriptions attached to the input and output arcs) can be imposed on functions. A function can be executed if its input object views contain enough occurrences, if input object view occurrences can satisfy the input arc inscriptions, and if the guard is true, if any. Execution of a function means removing object view occurrences selected from their respective input places, waiting a certain time corresponding to the average execution time of the function, and adding object view occurrences to the output places according to output arc inscriptions.

A function at one level can be further described as an organization net at a more detailed level, therefore creating a hierarchical functional structure. Thus, organization nets are used in the organization analysis phase of M* to define environments to be analyzed, to identify their functional structure (flow of control and hierarchy of functions), to define the information and material flow (flow of object views), and to collect formatted and non-formatted requirements for the information system of the CIM system to be designed.

Example: In the FMC example, shop floor control activities may consist of two stand-alone functions: a planning function (which generates the schedule), and an activity control function (which controls execution of the schedule) with their related inputs and outputs (Fig. 4). The activity control function has a guard stating that it only works on the last validated schedule and can only be performed between 8 a.m. and 6 p.m. "plan-req" (respect., "start-ac") is a message representing a planning request (respect., a start activity control request) and used to trigger the planning (respect., the activity control) function. <p-p>, <s> and <rslt> are free variables for arc inscriptions indicating how many occurrences of object views are consumed or produced. ¢ is used for control tokens (black dots). The planning function is in fact a process consisting of a scheduling activity followed by a schedule analysis activity (Fig. 5). Additional activities are also involved to update the shop status (shop floor data acquisition), to update the shop capacity (addition of manpower or machines) and optionally for rescheduling. Input and output places of a decomposable function must be input and output places of its organization subnet.

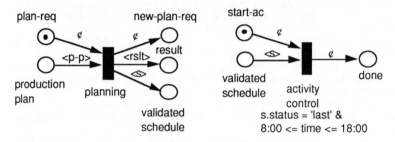

Fig. 4. Planning and activity control functions

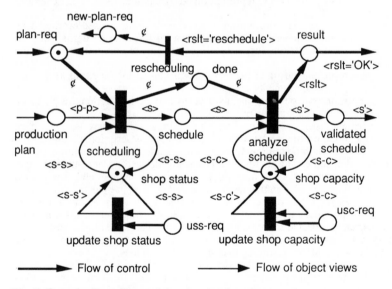

Fig. 5. Organization sub-net of the planning function

Conceptual Design:
Objectives: The goal is to produce the database conceptual schema and its related external schemata as an implementation independent specification of the CIM information system from the data requirements. Static and dynamic properties of data must be covered.

Approach: The full specification of a CIM information system is made of three parts: a semantic data model, a transaction model, and a process model. The *semantic data model* is used to specify the static properties of data (derived from the definition of object views and from formatted and non-formatted requirements). The *transaction model* is used to define database queries and database transactions. The *process model* is used to specify dynamic properties of data, i.e., application behavior from user requirements and function descriptions as defined in the organization model. Semantic data models used in practice in M* are extended entity-relationship models and object-oriented models.

Extended entity-relationship model: Extended entity-relationship (EER) models [8,32] are extensions of the basic entity-relationship model originally proposed by Chen. The EER model used in M* is defined as a 5-tuple $(\mathbf{D}, \mathbf{E}, \mathbf{A}, \mathbf{R}, \mathbf{X}, \mathbf{Q})$ where:

1) $\mathbf{D}$ is a finite set of not necessarily disjoint *domains*. Basic domains D_i of $\mathbf{D}$, $i = 1, 2,$ ..., are sets of atomic values (e.g., integers, reals, characters, strings, ...). Compound domains can be constructed from basic domains by means of two mechanisms:
- the Cartesian product (noted X) of domains: e.g., Date = Day X Month X Year

- the power set $\Pi(S)$ of a set S: $\Pi(S) = \{S' / S' \subseteq S\}$;

2) $\mathbf{E}$ is a finite set of entity classes (or simply *entities*). Each entity class $E_j \in \mathbf{E}$, $j = 1, 2,$ ... describes a set of identifiable objects of the real-world (or entity occurrences) which have an identical structure defined by a common set of attributes. Two types of abstraction hierarchies are defined on entities:
- The *subset hierarchy*. An entity S is a subset of an entity E, or E is a generalization of an entity S, denoted by $S \Rightarrow E$, if each occurrence of S is an occurrence of E.
- The *partition hierarchy*. Entities $P_1, P_2, ..., P_n$ form a partition of entity E, denoted by $E \leq \{P_1, P_2, ..., P_n\}$, if each occurrence of E is also one and only one occurrence of either P_1 or P_2 or ... or P_n.

3) $\mathbf{A}$ is a finite set of *attributes* defined on domains of $\mathbf{D}$ and used to define characteristics of entities or relationships and their identifiers.

4) $\mathbf{R}$ is a finite set of relationship classes (or simply *relationships*). A relationship $R \in \mathbf{R}$ between entities $E_1, E_2, ..., E_n$ describes the set of relationship occurrences (or links) linking objects $e_1 \in E_1, e_2 \in E_2, ..., e_n \in E_n$, i.e., $R = \{<e_1, e_2, ..., e_n>\} \subseteq E_1 X E_2 X ... X E_n$.

5) $\mathbf{X}$ is a set of *variables* on entities or relationships. A variable $x \in \mathbf{X}$ relatively to entity E or to relationship R will be denoted by x|E or x|R, respectively.

6) $\mathbf{Q}$ is a finite set of *operators* on values of attributes, entities, and relationships. Operators on values of domains are usual operators (such as addition, multiplication, division, comparision, ... for integers and reals; concatenation, comparision, sub-string extraction,... for character strings; etc.). Operators on entities include *insert, update,* and *delete* to add, modify, and remove occurrences of an entity class, respectively. Operators on relationships include *connect* and *disconnect* to add and remove occurrences in a relationship, respectively.

The EER model allows the definition of a wide range of integrity constraints. Among them, we can mention relationship cardinalities, and existence and identification dependencies.
- A *relationship cardinality* is a pair (min, max) defined for each entity implied in a relationship where "min" indicates the minimum number of times that each occurrence of the entity can be involved in the relationship occurrences (can be 0, 1, constant, or n if unbounded), and "max" indicates the maximum number of times that each occurrence of the entity can be involved in the relationship occurrences (can be 1, constant, or n).
- An entity A (called a weak entity) presents an *existence dependency* on an entity B (called a regular entity), if the existence of every occurrence of A depends on the existence of an occurrence of B. An entity A presents an *identification dependency* on an entity B, if it has an existence dependency on B, and each occurrence of A must be identified by the identifier of an occurrence of B. A weak entity has always cardinality (1,1) in a relationship with its regular entity.

Entities are represented graphically by rectangular boxes with their name inside, relationships by named diamonds, attributes by small circles, and hierarchies by thick arrows (Fig. 6). Weak entities are represented by double rectangles. Identifiers are indicated by dashed lines pointing from the attributes of the identifier to the entity that they identify. Conceptual database schemata can be represented graphically using this formalism as diagrams, called entity-relationship diagrams (ERD).

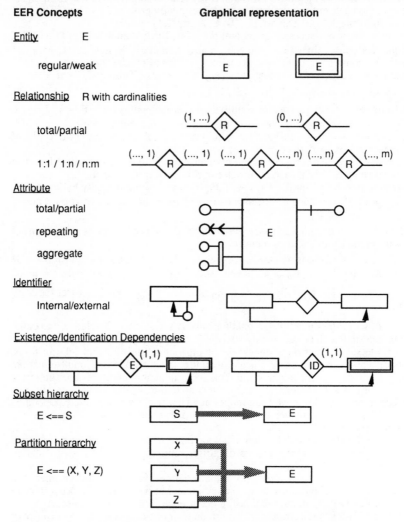

| **EER Concepts** | **Graphical representation** |

Entity   E

    regular/weak

Relationship   R with cardinalities

    total/partial

    1:1 / 1:n / n:m

Attribute

    total/partial

    repeating

    aggregate

Identifier

    Internal/external

Existence/Identification Dependencies

Subset hierarchy

    E <== S

Partition hierarchy

    E <== (X, Y, Z)

Fig. 6. Graphical formalism of the extended entity-relationship model

An ERD for the flexible manufacturing cell (FMC) example is given by Fig. 7. The manufacturing cell is made of machines, tools, and robots. It produces parts according to predefined schedules (entities PART and SCHEDULE) and for given due dates (relationship LP). Schedules are produced by a planning function from a production plan and are described by their identifier (s#), status (stat), and a date. Each part has an identifier (p#), a type (reference to its PART FAMILY), a status whose values are elements of the set {'raw', 'ready', 'mounted', 'in-process', 'worked', 'completed'}, and a location (loc) whose values will be the input buffer IN, the output buffer OUT, or the in-process buffers of the cell. Buffers have not been defined as entities (this could have been done if more information on them had to be used). Each part must be mounted on a pallet by means of fixtures for processing (entity PALLET and relationship PP). Notice that at some points in time, this relationship can be empty (minimum cardinalities equal to 0 on both legs). Pallets are defined by their identifier (f#), type (partial attribute), and condition (cond) whose values are elements of {'free', 'held'}. Any part belongs to a part family (entity PART FAMILY and relationship PFP). The entity PART is a weak entity of the entity PART FAMILY (existence dependency). The same pallet can be used for different part families and a part of each family can be mounted on different possible pallets (relationship FP). The sequence of operations to be executed for a part of a given part family is given by a process plan (entity PROCESS PLAN). This is also a weak entity of PART FAMILY (identification dependency). The primary process plan is linked to the entity PART FAMILY by PPP. Alternate process plans for the same part family are recursively given by the relationship SS if there are any. SS.succ (succ stands for successor) and SS.pred (pred stands for predecessor) indicates the role played by the process plan in SS. Each process plan is identified by its part family identifier (pf#) and a sequential number (s#). The name of the process plan designer is also recorded (planner). Each operation (entity OPERATION) has an identifier (o#), a description (descr) whose values are elements of {'mfg' for machining, 'asy' for assembly, 'cnt' for quality control}, and a processing time (run_time). The sequence of operations in a process plan is given by relationship PO and the order in the sequence by the attribute seq#. Each operation of a process plan is linked to a tool (entity TOOL) and to a machining machine (entity M_MAC) by the relationship OMT if it is a machining operation. It is linked to a measuring machine (C_MAC) by the relationship OCM if it is a quality control or measuring operation. Machining machines and measuring machines form a partition of the set of machines (entity MACHINE from which they inherit attributes). Each machine is described by a machine number (m#), a type, and a condition (cond) whose values are elements of {'available', 'working', 'down'} depending if they are idle, working, or unavailable. Similarly, tools are described by a tool number (t#), a type, and a condition (cond) whose values are in {'available', 'in-use', 'worn-out'}. Relationships MTP and MCP keep track of the links between parts being machined in the cell and machines depending on the machine type. Finally, parts on pallets are handled in the cell by robots (entity ROBOT). Relationship PR between PART and ROBOT indicates parts being handled by robots at a given time. A part in the cell is therefore either in a buffer, or an a machine, or being moved by a robot.

Transaction model: It provides a way: (1) to express queries on the database defined by means of the semantic data model, and (2) to specify transactions on this database. It is based on the concepts of views, messages, and routines.

- The *view* mechanism allows users to express complex queries on data of the database described by its conceptual schema. A view W on a database DB, defined by its schema S, is a restriction (or a subset) of the database defined by a subschema S' defined over S. It can be expressed by[*]:

```
VIEW W (<db_parameters>)
[RANGE <data_segments> [, <data_segments>]]
[WHERE <selection>]
```

1) W is the name of the view.
2) Parameters <db_parameters> are defined as:
   <variable> [, <variable>]|<db_element>

---

[*] Reserved words are capital letters. <.> denotes a non-terminal symbol and [.] means zero or many occurrences of the clause.

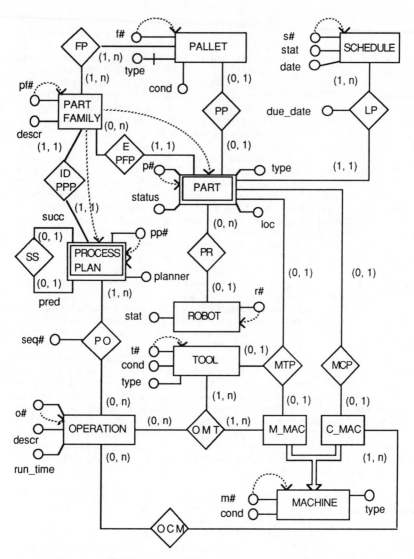

Fig. 7. Entity-relationship diagram for the FMC

where <db_element> represents an entity or a relationship of the database schema S and <variable> is the name of a variable whose values are occurrences of this entity or relationship.

3) The optional RANGE clause allows for the description of relationships involved in the view and is made of data segments (<data_segments>) which define access paths to data in the database schema. A data segment has the following form:

$$\pi_h<\mu_i>\pi_k \ldots <\mu_m>\pi_n$$

in which π_h, π_k, ..., π_n are entity terms, and μ_i, ..., μ_m are relationship terms. An entity term (respectively, a relationship term) can be a variable specified in <db_parameters>, or can be a variable z|Z defined on-the-fly on element Z of the database schema S. A sequence of terms of the form $\pi_h<\mu_i>\pi_k$, called a *link predicate*, means that an occurrence $e_h \in E_h$ (represented by the variable corresponding to the term π_h) must be connected by a relationship $r_i \in R_i$ (represented by μ_i) to a valid occurrence $e_k \in E_k$ (represented by π_k). Using a tuple notation, we have:

$$r_i = <e_h, e_k> \in R_i \subseteq E_h \times E_k$$

Occurrences of a data segment are obtained by concatenating entity occurrences which satisfy the link predicate chain. Different data segments can be connected by means of common variables, i.e., variables defined in a term of a data segment and used in another data segment. For instance, in the flexible manufacturing cell example, the following RANGE clause:

RANGE mIM_MAC<olOMT>ylOPERATION, <o>tITOOL

describes the full set of combinations of machines with associated tools for all operations in the database. This example makes use of a ternary relationship.
4) The selection predicate (<selection>) in the WHERE clause specifies a logical formula on variables introduced in the view parameters or in the RANGE clause. Occurrences of the view are obtained by retaining only occurrences of the RANGE clause which satisfy the selection predicate. For instance, let us consider the following views defined on the data schema of Fig. 7:

W1: "Select all parts which are stored in the input buffer IN"
    VIEW W1 (plPART) WHERE p.loc = 'IN'

W2: "Select all available tools and machines for all operations of type machining"
    VIEW W2 (mIM_MAC, ylOPERATION, tITOOL)
    RANGE m<olOMT>y, <o>t
    WHERE m.cond = 'available' AND t.cond = 'available'
    AND y.descr = 'mfg'
Remark: The RANGE clause could also have been written: y<olOMT>m,<o>t

W3: "Retrieve the main process plan for all parts in the input buffer IN"
    VIEW W3 (plPART, pflPART_FAMILY, pplPROCESS_PLAN)
    RANGE p<r1lPFP>pf<r2lPPP>pp
    WHERE p.loc = 'IN' AND pf.pf# = p.type AND pp.p# = 1

B1: "Select all parts which have been processed on machines and are stored in buffer B1"
    VIEW B1 (mIM_MAC, plPART, tITOOL)
    RANGE m<xlMTP>, <x>p
    WHERE p.loc = 'B1'

- A *message* can be defined as a view M with a list of <db_parameters> reduced to external variables, no RANGE clause and no WHERE clause, i.e., as a n-argument predicate. It is used to model events, i.e., requests to trigger the processing of something, to provide indications that something has happened (e.g., signals), or to pass external run-time data. Example, to initialize a part number part# to the value 'PX105':

MESSAGE IN (part# = 'PX105')

- A *routine* T on a database (or database transaction) can be defined by:

ROUTINE   T (ACCEPT <i_messages>, <i_views>;
               RETURN <o_messages>, <o_views>):
[RANGE   <data_segments>; [<local_views>]]
[WHERE   <input_rule>]
{<transaction_block}

1) T is the routine name.

2) The ACCEPT clause specifies input messages <i_messages> and input views <i_views> of T. The RETURN clause specifies output messages <o_messages> and output views <o_views> of T. A view is specified by a term of the form P $(p_1, ..., p_n)$ in which P is the name of the view and $p_1, ..., p_n$ is a list of variables corresponding to parameters of P. Variables specified in the input/output views are used in the routine as well as those which might have been introduced in the RANGE clause or in the definition of local views (local_views).

3) The optional RANGE clause specifies the relationships on data which are involved in the routine. The syntax is the same as for views. Use of local views (local_views) may be required. They are defined as:

Wi (<db_parameters>) <= { <data_segments> WHERE ... };

in which Wi is the name of the view and <db_parameters>, <data_segments> and the WHERE clause have the same meaning as defined for global views.

4) In the WHERE clause of the routine, the input rule <input_rule> specifies a formula that must be verified by the data which are involved in the routine. Such a rule can be used to express conditions on local views using set-theory operators (such as contained_in, contains, difference, member_of, equals). The routine body <transaction_block> consists of a sequence of operations on data variables specified in input/output views. It describes the processing to be done on database data using the view definitions. Control structures which can be used include sequential, conditional, iterative, and case structures. They are expressed using operators on entities, relationships, and sets. For example, let us consider the following routine using views defined previously and operations on entities and relationships (update and connect):

T1 : "Perform the first operation on part PX105 (indicated by input message IN) using the required machine and tool and put the part in buffer B1"

```
    ROUTINE T1 (ACCEPT IN, W3 (p, pf, pp), W2 (m, y, t);
        RETURN B1 (m, p, t)):
    RANGE pp|PROCESS_PLAN<PO>y
        /* connects process plan to operation */
    WHERE p.p# = part# AND PO.seq# = 1
        /* filters data for part PX105 and first operation in its main process plan */
    {update m (m.cond ← 'working'); /* changes machine condition */
    update t (t.cond ← 'working'); /* changes tool condition */
    update p (p.loc ← 'B1'); /* changes part location (after processing) */
    connect MTP (m, t, p)} /* adds new occurrence <m, t, p> to MTP */
```

The process model: The process model of M* is used to specify data processing aspects of functions of the CIM system identified in the organization nets during organization analysis. Functions and processes are specified in terms of elementary routines as process nets to be finally implemented as application programs.

A *process net* defines causal relationships between database routines to specify a function behavior or a CIM business process. It is defined as a PrT-net [12] by:

- a directed net $(P, T; F)$ that represents the control structure of the system. P is a set of places such that $P = V \cup M$, where V is a set of views and M is a set of messages, T is a set of routines on the information system, and F is a set of arcs connecting each routine to its input and output places ($F \subseteq P \times T \cup T \times P, P \cap T = \emptyset, P \cup T \neq \emptyset$);

- a database schema S to specify system objects (entities and relationships);

- the marking D of the process net consists of the collection of populations of its views and the message conditions (active or inactive);

- a routine $T \in \mathbf{T}$ models how the system reacts to a given marking D; we will say that D enables T if all the input messages of T are active, all the output messages of T are inactive, and there exists an input substitution on the parameters of the input and local views which satisfies the guard predicate;

- a firing rule: if the routine T is enabled, it can be executed. The execution of operations in the routine-block must guarantee that input tuples (which take part in the admissible substitution) are removed from the input views, output tuples are added to the output views. Moreover, input messages become inactive (i.e., they no longer hold) and output messages become active (i.e., they begin to hold).

As an example let us consider the dynamic description of the FMC system as illustrated by the process net of Fig. 8. Parts stored in the input buffer and ready to be machined (view R) are mounted on pallets with proper fixtures (view F) by routine Mount and get status 'mounted' (view M). A part from M is processed by routine Work according to a sequence of operations performed by an available machine and tool (view U). At the end of each operation, the machine and tool are released (view Q) and the part must be prepared for the next operation. The routine Prep returns the machine used in the previous operation and selects the next operation (view W). If all the operations on the part have been done, the part is dismounted and is stored in an output buffer (view C) by routine Dism, otherwise, a routine Send sends the part back into the cell for further processing (view M). Views R, M, T, F, Q, U, W, and C must be specified using the view mechanism and transitions Mount, Work, Prep, Send, and Dism must be specified using the routine mechanism.

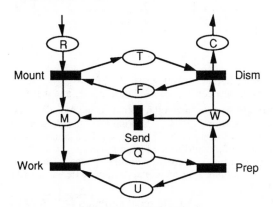

Fig. 8. Process net for the FMC example

Implementation Design:
Objectives: The aim is to perform the logical and physical design of the database. The goal is (1) to produce the physical database implementation schema, (2) to implement it using the DDL of the DBMS to be used, and (3) to code routines and database transactions of application programs as well as on-line queries using statements of the DML and QL of the DBMS.

Approach: At the implementation design level, the conceptual schema, the views, and the routines specified during the conceptual design phase are coded either directly using an object-oriented database system or are derived in terms of a relational database. Relational databases and object-oriented databases are discussed in turn in the next sections.

6. Relational Databases

The relational database technology has developed from the seminal work done at IBM and reported by Ted Codd in 1970 [5]. The original objectives of the relational model were simplicity, data independence, and rigor. The relational model has its mathematical foundations in predicate calculus and set theory.

The relational model:

Simply stated, the relational model structures data in two-dimensional tables called relations. A relation can represent an entity or a relationship among entities. Columns of the table represent attributes of the relation, and rows are tuples representing occurrences of the relation.

For instance, Fig. 9 presents a relational database made of three relations: PARTS, PALLETS, and MOUNTED describing parts, pallets and their links, respectively.

PARTS	p#	status	location
	P1	ready	IN
	P2	mounted	B1
	P3	ready	IN
	P4	mounted	B2

PALLETS	f#	cond	type
	F1	free	T1
	F2	used	T1
	F3	used	T2

MOUNTED	p#	f#	fixture
	P2	F3	yes
	P3	F2	no

Fig. 9. A relational database

More formally, the relational model can be defined as follows [6,10,23]:

- A *relation schema* R (or relation intention) consists of a relation name, and a list of attributes $\langle A1, A2, ..., An \rangle$. Each attribute Ai is defined on a domain dom(Ai). The type of R is defined as dom(R) = dom(A1) X dom(A2) X ... X dom(An). A *relation* (or relation extension) of R is a set of n-tuples in dom(R). In other words, a relation is a subset of the Cartesian product of domains of the attributes of the relation.

- A *relational database schema* D (or relational database intention) is a set of relation schemata {R1, R2, ..., Rm}. A *relational database* (or relational database extension) of D is set of relations of R1, R2, ..., Rm.

- A *key* of a relation schema R $\langle A1, A2, ..., An \rangle$ is a finite set of attributes $\langle A1', A2', ..., Al \rangle$ $(1 \leq n)$ of R defined on domains having no null values, and such that l-tuples of the key have different values for all different n-tuples of relations of R.

The sound foundations of the relational model have led to the theory of data normalization for database design, and the relational algebra for data manipulation and query languages.

Data normalization:

The theory of data normalization [6,21,23] provides a rational way to verify if a relational database schema is well-formed in order to avoid unecessary data redundancies and potential inconsistencies due to data insertion, update, or deletion anomalies. It is based on the analysis of functional dependencies among attributes. In practice, a relational database schema must be at least in 3rd normal form, but preferably in Boyce-Codd normal form.

- *Functional dependencies*: An attribute A is functionally dependent on another attribute B (or B functionally determines A), if for every occurrence of B there exists one and only one possible occurrence of A. We write, B $\rightarrow$ A. Functional dependencies also apply to subsets of attributes of a relation schema R. Multi-valued dependencies are defined in the case where for every occurrence of an attribute B there is a set of dependent values for attribute A. These are special forms of integrity constraints on attributes.

- *1st normal form* (1NF): A relation schema R is in 1st normal form if every attribute of R is defined on an atomic domain (i.e., there is no aggregate attribute).

- *2nd normal form* (2NF): A relation schema R is in 2nd normal form if it is in 1st normal form, and if every non-key attribute is functionally dependent on each attribute of the key of R and not on a subset of key attributes.

- *3rd normal form* (3NF): A relation schema R is in 3rd normal form if it is in 2nd normal form, and if every non-key attribute is not functionally dependent on another non-key attribute of R (i.e., there is no transitive dependencies).

- *Boyce-Codd normal form* (BCNF): A relation schema R is in BCNF if whenever $X \rightarrow$ A holds in R, and A is not in X, then X is a superkey of R, i.e., X is or contains a key. The Boyce-Codd normal form is the strongest form of the four types.

For instance, the following relational schema, with the following functional dependencies, is not in normal form, even not in 1st normal form.

PARTS (P#, Price, (Buffer#, Location, Quantity)) Key: P#
P# $\rightarrow$ Price; Buffer# $\rightarrow$ Location; (P#, Buffer#) $\rightarrow$ Quantity

To normalize this data schema, the relation PARTS must be broken down into three simpler, non decomposable, relational schemata as follows:

PARTS (P#, Price) Key: P# BUFFERS (Buffer#, Location) Key: Buffer#
STORAGE (P#, Buffer#, Quantity) Key: (P#, Buffer#)

The relational algebra:
It defines operations on relations [5,10,23]. Among these the most important ones are:
- The *projection,* which is a unary operation used to decompose a relation vertically, i.e., to project the relation over a subset of its attributes.
- The *selection,* which is a unary operation used to decompose a relation horizontally, i.e., to retrieve only tuples verifying a specified condition (called qualification).
- The *join,* which is n-ary operation used to combine two or more relations using a join condition.

These operations can be combined to perform complex manipulations and data queries on relations. Other operations on relations can be defined from set theory such as union, intersection, and difference of relations having the same schema, or Cartesian product of n relations. These operations will be illustrated using the SQL language, which is based on the relation algebra.

The SQL data language:
SQL (Structured Query Language) is a ISO-standardized data language for relational database structures [7]. It embeds features of a DDL, a DML, and of a query language.

The power of the database query capabilities of the SQL language essentially relies on its SELECT statement. Its general form is:

SELECT [DISTINCT] <fields>
FROM <tables>
[WHERE <predicate>]
[GROUP BY <fields> [HAVING <predicate>]]
[ORDER BY <fields>];

where <fields> represents one or more attribute names separated by comma and <tables> represents one or more relation names separated by comma. The optional clause DISTINCT is used to eliminate redundant tuples in the resulting relation, if any.

It can be used for pure selection operations. For instance, if we consider the relational database defined in Fig. 9, to retrieve all parts ready to be mounted on a pallet, we write:
SELECT *
FROM PARTS
WHERE status = 'ready';

* replaces all attribute names of the relation specified in the FROM clause. The resulting relation has the same structure as PARTS and will contain two tuples: <P1, ready, IN> and <P3, ready, IN>.

The SELECT statement can be used for pure projection operations. For instance, to retrieve only the status and location of all parts, we write:
```
SELECT status, location
FROM PARTS;
```

The result will be a relation with two columns (status and location) and containing four tuples: <ready, IN>, <mounted, B1>, <ready, IN>, and <mounted, B2>.

The SELECT statement can be used to perform join operations. For instance, to retrieve all the information on parts mounted on pallets, we write:
```
SELECT *
FROM PARTS P, MOUNTED M
WHERE P.p# = M.p#;
```

The result will be a relation made of all attributes of PARTS and MOUNTED and containing only two occurrences (<P2, mounted, B1, P2, F3, yes> and <P3, ready, IN, P3, F2, no>). Note the use of short names in the FROM clause and the use of the dotted notation to remove ambiguities.

The SELECT statement can be used as a combination of these operations. For instance, to use a selection and a projection on PARTS to retrieve the number and location of mounted parts ordered in the descending order of their location, we write:
```
SELECT p#, location
FROM PARTS
WHERE status = 'mounted'
ORDER BY location (DESC);
```

ORDER BY is used to classify the resulting occurrences in ascending order (ASC) or descending order (DESC) of attributes indicated. Ordering is first done according to the first attribute, then according to the second one to break ties, and so on. GROUP BY is an optional clause to group occurrences into sub-categories.

The SELECT statement is a powerful SQL command to query data of the database. Important aspects of the SELECT statement include:
- It specifies what has to be retreived without explicitly stating how the retreival must be done. It is therefore a non-procedural statement. It can however be embedded in procedural languages (such as Pascal, Fortran, C, ...) to simplify data access code for relational database application programs. A pre-compiler is needed to convert SQL statements into relational DBMS calls in corresponding program statements.
- It filters data of the database returning only attributes of interest and tuples satisfying the predicates in the WHERE clause. The result is a new relation. The predicates can be combined using the usual Boolean operators AND, OR, and NOT with parentheses whenever required.
- Built-in functions can be used in the SELECT clause. They include COUNT to return the number of values in a column, SUM to sum numeric values in a column, AVG to compute the average value of numeric values in a column, MAX to get the largest value in a column, and MIN to get the smallest value in a column.
- SELECT statements can be nested. The results of one selection can be the operand for another. For instance, to retrieve part number and location of parts mounted on pallets of type T1, we write:
```
SELECT p#, location
FROM PARTS P, MOUNTED M
WHERE P.p# = M.p# AND f# IN
    (SELECT f#
    FROM PALLETS
    WHERE type = 'T1');
```

The reserved word IN must be used if the nested SELECT returns a set of values. If there is always one value returned by the SELECT, the symbol "=" can be used. SELECT statements can be nested up to 256 levels.

A data definition language (DDL) of a relational database management system is used to declare, modify, or destroy relational schemata of the database schema. DDL statements in SQL allow the definition of base tables (relations) and views. They include:

- CREATE TABLE to declare a relational schema. For instance, to define the relation PARTS, we write:

```
CREATE TABLE PARTS
    (p#: CHAR (5)  NOT NULL,
    status: CHAR (10),
    location CHAR (3));
```

Basic domains (or data types) provided by SQL are INTEGER for signed integers, SMALLINT for signed short integers, DECIMAL (p, [,q]) for signed reals (p defines the digit precision and q defines the decimal part), FLOAT for signed doubleword floating point numbers, CHAR (n) for character strings of length n, VARCHAR (n) for varying length character strings, DATE, and MONEY. NOT NULL indicates that null values are not allowed for the specified attribute. NOT NULL is mandatory for key attributes.

- CREATE VIEW to define a view on relations of the database. A view in SQL is a virtual relation which is not stored but can be manipulated as any other relation. For instance, to create a view for mounted parts from relation PARTS, we write:

```
CREATE VIEW P_MOUNTED
AS SELECT p#, location
    FROM PARTS
    WHERE status = 'mounted';
```

- DROP TABLE (respect., DROP VIEW) to destroy a relation (respect., a view).

A data manipulation language (DML) is used to modify the state of a database. DML statements in SQL include:

- INSERT to add a tuple to a relation. For instance, to add a new part P5 in the relation PART, we write:

```
INSERT INTO PARTS
VALUES ('P5  ', 'ready    ', 'IN ');
```

- DELETE to remove tuples from a relation. For instance, to delete information about part P5, we write:

```
DELETE FROM PARTS
WHERE p# = 'P5';
```

- UPDATE to modify the value of stored data of a relation. For instance, to change the status of part P2 from 'mounted' to 'completed', we write:

```
UPDATE PARTS
SET  status = 'completed'
WHERE status = 'mounted';
```

EER schemata translation into a relational schemata:

A conceptual schema of an information system modeled by the EER model can be automatically translated into a logical schema expressed in terms of the relational model during the implementation design phase. Conversion rules are given by Fig. 10 and 11. Fig. 10 provides rules to convert an EER structure into a basic entity-relationship model. Fig. 11 provides rules to convert basic entity-relationship structures into relational structures. Note, however, that for some EER structures, several equivalent relational structures are possible. These rules provide a relational schema in 3rd normal form.

An equivalent relational schema of the ERD for the FMC example of Fig. 7 is:

SCHEDULE (s#, sta, date), key: s#
PART (p#, type, status, loc), key: p# /* type equals pf# in PART_FAMILY */
PART_FAMILY (pf#, descr), key: pf#   FP (pf#, f#), key: (pf#, f#)
ROBOT (r#, stat), key: r#   PR (p#, r#), key: (p#, r#)
PALLET (f#, type, cond), key: f#   PP (p#, f#), key: (p#, f#)
LP (s#, p#, due_date), key: (s#, p#)
PROCESS_PLAN (pf#, pp#, planner), key: (pf#, pp#)
SS (pf#, pred, succ), key (pf#, pred) /* pred and succ have pp# values */
OPERATION (o#, descr, run_time), key: o# PO (pf#, pp#, seq#, o#), key: (pf#, pp#, seq#)
TOOL (t#, cond, type), key: t#   MACHINE (m#, cond, type), key: m#
OMT (o#, t#, m#), key: (o#, t#, m#)   OCM (o#, m#), key: (o#, m#)
MTP (p#, m#, t#), key: (p#, m#, t#)   MCP (p#, m#), key: (p#, m#)

This relational schema could have been directly produced from the data requirements (object views and user descriptions) to serve as a conceptual schema, then normalized, and then optimized to become the physical data schema. However, the M* methodology recommends to perform the conceptual design phase first using the extended entity-relationship model and then to convert it into a relational schema if relational database technology is to be used. Indeed, the models produced during conceptual design are implementation independent, can be used and shared by a large community of users within the enterprise (as part of a consistent enterprise documentation), and can be converted into implementation models. The implementation models can thus be modified or changed if the implementation technology changes or evolves.

Database transactions or routines of process nets are either implemented as sequences of SQL commands started by a BEGIN TRANSACTION statement and finished by the statement COMMIT (to validate operations on data) or ROLLBACK (to undo operations on data), or as procedures written within an application program and making use of DML statements of SQL for data manipulations [6,7].

Remark: In the relational model, key attributes of entities can be modified. Therefore, they are surrogates, and not true identifiers in the strict sense. A true identifier of an object should remain the same for the whole life cycle of the object. This is a basic principle of object-oriented models.

Well-known relational database systems available on the market and supporting the SQL language are ORACLE, INGRES, INFORMIX, and DB2.

Partition Hierarchy

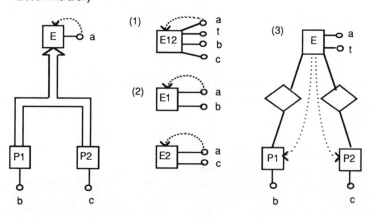

Fig. 10. Conversion rules to translate EER structures into entity-relationship structures

EER structures   **Relational structures**

<u>Entities and attributes</u>

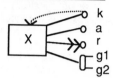

$X(\underline{k},a,g1,g2)$ $X'(\underline{k},r)$

Underlined attributes denote key attributes

<u>Case a: 1:1 type</u>

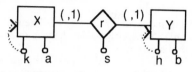

1) $X(\underline{k},a)$ $Y'(\underline{h},k,b,s)$

or

2) $X'(\underline{k},h,a,s)$ $Y(\underline{h},b)$

<u>Case b: 1:n type</u>

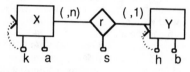

$X(\underline{k},a)$ $Y'(\underline{h},k,b,s)$

<u>Case c: m:n type</u>

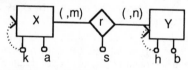

$X(\underline{k},a)$ $Y(\underline{h},b)$ $r(\underline{k},\underline{h},s)$

<u>Case d: ring type</u>

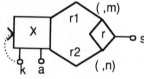

$X(\underline{k},a)$ $r(\underline{kr1,kr2},s)$

r1 and r2 are the roles of the entity X in relationship r
kr1 and kr2 are values of k in r1 and r2

<u>Case e: n-ary type</u>

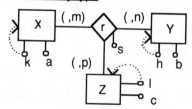

$X(\underline{k},a)$ $Y(\underline{h},b)$ $Z(\underline{l},c)$

$r(\underline{k,h,l},s)$

Fig. 11. Translation rules to convert EER schemata into relational schemata

7. Object-Oriented Databases

As opposed to the relational model, for which SQL has become a standard, no definitive paper has been published which defines the mathematical basis of object-oriented (OO) models. However, several papers have discussed fundamental aspects of object-oriented models from the database perspective [1,3,20]. A complete bibliography on object-oriented database management is provided by Vossen [37].

Simply stated, in addition to the entity-relationship or relational models, which both define static entities and relationships on atomic attributes, object-oriented models enforce object uniqueness, allow an attribute to be itself an entity (i.e., accept non-first normal form (N1NF) relations), and associate operations to entities to access and modify attributes (i.e., provide behavior to objects).

Object-oriented data models:
All object-oriented data models are based on the following fundamental concepts:
- Each occurrence of a real-world entity is modeled as an object. Each object is characterized by a unique identifier (called OID), which cannot be changed over the full life cycle of the object, even if the object state is changed.
- Each object is an instance of an object class. A class is defined by its name, a set of attributes (defining the object state), and a set of methods (defining the object behavior). An attribute of an object can be defined as a simple data item, another object, or a set of objects. Methods are named procedures which can be activated by sending a message to the object. A message is a call to a method specified by the name of the object class, the name of the method, and the list of parameters for the method.
- A class can be defined as a specialization of one or more classes. A class defined as a specialization of another class inherits attributes and methods from its superclass.

More formally, an object-oriented data model can be defined by a 5-tuple (**D, C, A, X, O**) as in the M*-Object methodology [9], where:
1) **D** is a set of basic *domains* (including integers, reals, characters, strings as well as record structures, enumerated data types, and range data types).

2) **C** is a set of *classes* of objects. Each class $C_i \in C$ describes a set of like objects having the same structure (i.e., the same set of attributes), and the same behavior (i.e., the same set of operators or methods). Objects are occurrences of classes. Each object is identified by a unique *object identifier* (OID), and belongs to one class only. Object classes can be organized into class hierarchies. There are three types of class hierarchies (Fig. 12):
- the *generalization hierarchy* [30], which allows a class (the subclass) to be defined as a specialization of another existing class (the superclass). The subclass inherits attributes and methods from the superclass and may add or overwrite specific attributes and methods. It is also called the "is-a" semantic link. Multiple inheritance is not allowed in this model;
- the *aggregation hierarchy* [30], which allows a class (the composite class) to be specified as a composition (or Cartesian product) of existing classes (the component classes). It is also called the "part-of" semantic link;
- the *association hierarchy*, which allows to define a class (the grouping class) as a set of elements belonging to existing classes (the basic classes). It is also called the "member-of" semantic link.
3) **A** is a finite set of *attributes* A_{ki}, i.e., characteristics of classes of **C**, defined by:

$A_{ki}: C_i \rightarrow D_k$ where C_i is a class and D_k is a domain or a class.
An attribute is either
- a member attribute, i.e., a structural property of the object class (e.g., the price of a machine, operation time, etc.),
- a set attribute, i.e., a property of the whole class C_i (e.g., the number of machines). It does not appear in objects of the class,

- a relationship attribute, i.e., a link property between the object class and another class (e.g., a machine "uses" a tool). Reverse links can be defined as inverse mappings.

4) **X** is a set of *variables* on the classes of **C** declared as follows:
    use p | P

meaning that variable p ∈ **X** will represent any object of class P ∈ **C**.

5) **O** is a set of *methods* applicable to objects of the classes of **C**. Methods are procedures which express object behavior. They are invoked by sending a message (i.e., a request) to an object.

Furthermore, built-in methods are included in the model to access data or manipulate objects. These are:

- Select, to select an occurrence of a class
- Create, to create new occurrences of a class
- Update, to update values of an object
- Delete, to destroy occurrences of a class
- Specialize (respect., Generalize), to specialize (respect., generalize) an occurrence of a class into a sub-class (respect., super-class)
- Add (respect., Remove), to add (respect., remove) an object to (respect., from) a composite or grouping object

Figure 13 gives an illustration of an OO data model for the FMC example. The graphical formalism is close to the one used for the entity-relationship model. Boxes represent object classes, arrows represent class hierarchies, and diamonds represent relationships. Only binary relationships are defined since a relationship is a mapping from one class to another (semantic binary model). Relationships have no name in the schema, but their legs are identified by the names of relationship attributes of object classes linked. A leg of a relationship can be left undefined, which means that reverse relationship has not been defined. In the schema of Fig. 13, the class MACHINE represents machine-tools, part handling machines including robots, and control machines (no sub-classes have been defined like in the EER model). Since there is one part per manufacturing order in the schedule, the attribute due_date has been defined in the class PART. Finally, the actual association between a part and a pallet is represented as an aggregation between elements of classes PART and PALLET. The aggregation WORK has been defined in a similar way to keep track of actual association between parts and machines.

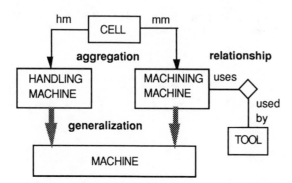

Fig. 12. Examples of class hierarchies

108

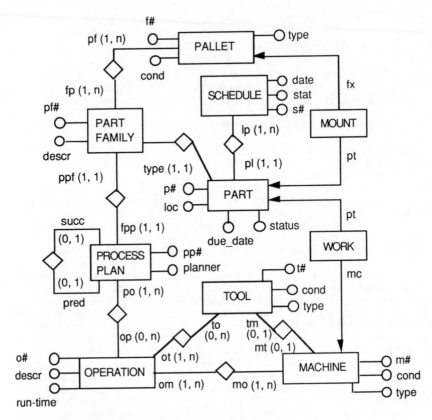

Fig. 13. Object-oriented data model for the FMC example

The class MACHINE with its relationships to classes TOOL and OPERATION can be described as follows (Underlined text denotes reserved words):

```
class MACHINE
identifier m#;
member attributes
    m#: integer;
    cond: (available, working, down);
    type: (machining, control);
relationship attributes
    mt (0,1): TOOL inverse-of tm;
    mo (1,n): OPERATION inverse-of om;
methods
    insert (num: integer; T: setof op-code): string
```

In such a specification, names of relationship attributes must be unique in the data schema (because of inverse links). The methods are defined by their signature. The code of the method insert (which is used to add a new machine that can perform a set T = {t1, t2, ...} of operations of OPERATION) is hidden from object class usage (i.e., only the object interface is visible to the outside world). The code is specified in the implementation part of the class specification as the object behavior. The implementation part of the class MACHINE follows:

```
class MACHINE implementation
procedure insert (mn: integer; T: setof op-code): string
var t: op-code;
use wIOPERATION, mIMACHINE;
begin
m.Select (m#=mn);
if m≠nil then return ('error') end
else begin m.Create (m#: mn; ...; mo: nil);
    foreach t in T do
            begin w.Select (o#=t);
            w.Update (om: om+m);
            m.Update (mo: mo+w);
            end; return ('ok'); end;
end.
```

The object orientation provides a number of advantages which can significantly impact future CIM system development from a computer science point of view as well as from a systems engineering point of view. These include: information hiding of internal structures (data structures and data operations), data abstraction capabilities using semantic links (such as generalization, aggregation, and association), inheritance of object properties, and late and dynamic binding of an operator to its code. However, the most important features of the object-oriented approach for CIM systems engineering rely on the close relation between functions and data of an object of the system (object encapsulation), system modularity since each object is a module with its own characteristics and behavior, reusability of existing objects (as reference or partial models) leading to economy of design effort and coding, system expandability since new objects can be easily added to the existing system, and system evolution for the management of change during the system life cycle (object evolution).

Views, messages, and routines:
Views, messages, and routines as defined for the extended entity-relationship model can be defined for the object-oriented model for the conceptual design phase.
   A *view* is defined by a statement of the form:

```
VIEW <view-name> (<target-list>)
USE <variable-declaration-list>
WHERE <predicate>
```

1) <view-name> is the name of the view.
2) <target-list> specifies what objects must be retrieved from the database as a set of tuples $<o_h, o_k, ...>$ where $o_h \in C_h$, $o_k \in C_k$, ...
3) <variable-declaration-list> in the USE clause is used to introduce a set of local variables defined on object classes in the form $v_m|C_m$, $v_n|C_n$, ...
4) The WHERE clause is used to select occurrences of classes C_h, C_k, ... by means of logical predicates connected by AND, OR or NOT operators.
   For instance, the following views retrieve, at a given instant in time, all parts ready for processing, all available pallets, parts with status 'mounted', and pallets used by mounted parts of the FMC database:

```
VIEW R (p|PART) WHERE p.status = 'ready'
VIEW F (f|PALLET) WHERE f.cond = 'free'
VIEW M (p|PART) WHERE p.status = 'mounted'
VIEW T (t|MOUNT)
```

A *message* is used to define input and output conditions on routine triggering, or to pass input/output data at run-time to/from a routine. It can be declared as:

MESSAGE <message-name> (<argument-list>)

1) <message-name> is the name of the message.
2) <argument-list> is made of predicates (free variables or character strings) to pass run-time data to routines or to display messages.

For instance, the message Process providing a part number part# (previously initialized) is written as follows:

MESSAGE Process (part#)

The *routine* mechanism defines pieces of functionalities to be executed on objects of the system under the request of an event (such as an order). An event is described by means of a *message* which is issued at a given instant in time. A routine can receive and send messages, and views as inputs and outputs and perform some data processing. It is declared in the following form:

```
ROUTINE <routine-name>
    [ACCEPT <input-place-list>;
    RETURN <output-place-list>]
USE <local-variable-list>; <local-view-list>
GUARD <predicate>
EXEC <routine-block> END
```

1) <routine-name> is the name of the routine (or database transaction).
2) <input-place-list> provides the list of input messages (i.e., triggering events or external run-time data), and input views (i.e., stored data) required by the routine.
3) <output-place-list> indicates the list of ouput messages and output views which can be generated as result of the routine execution.
4) The USE clause allows the definition of local variables on elements of input and output views as well as of local views.
5) The GUARD clause defines a predicate which specifies conditions on input and local data which must be satisfied to execute the routine.
6) The <routine-block> defines the sequence of operations to be executed (using sequential, conditional, parallel, rendez-vous, and iterative control structures).

For instance, the mount operation in the FMC application, i.e., mounting a part on a pallet, can be written as follows using the previous message and views:

```
ROUTINE Mount [ACCEPT Process (part#), R(p), F(f); RETURN M(p), T(t)]
USE fam|PART_FAMILY
GUARD p.p# = part# AND fam.pf# = p.type AND f IN fam.fp
EXEC p.Update (status := 'mounted');
    f.Update (cond := 'held');
    t.Create (pt: p, fx: f) END
```

Process models can be defined in the same way as the extended entity-relationship model and use the same graphical formalism as the one used in Fig. 9.

At the implementation level, object-oriented database management systems (OODBMSs) such as O2, ORION or IRIS can be used [3,20,37]. The data language offered by OODBMSs is usually an extension of an object-oriented programming language (C++, LISP extensions, Eiffel, etc.). Thus, the implementation part and especially methods of object classes are usually coded using these languages, and are therefore system-dependent. The same applies to application programs. Thus, views and routines specified at the conceptual design level must be coded with the object-oriented languages selected for system implementation.

OODBMSs lack the flexibility, data independence, and ease of use of relational database management systems (RDBMSs). Conversely, the high descriptive power of the object-oriented model allows the description of complex objects which provides a more faithful image of the real object structure (natural modeling reducing the semantic gap). Furthermore, OODBMSs allow the direct handling of complex objects (such as design artifacts, signals, images, etc.). While RDBMSs suit very well the data management needs of administrative and production management activities, OODBMSs are more suited to more specific applications dealing with complex objects such as CAED and CAPP activities. However, the use of object-oriented databases will continuously increase as their response times improve and with the growing use of multi-media data, networks, and devices. Finally, extensions of SQL to object-orientation are under investigation [11].

8. Distributed Databases

Large scale manufacturing systems are distributed systems. A manufacturing enterprise is usually made of several plants and departments located on different sites, sometimes in different countries. Even on one site, there may be several buildings. Each department or service has its own computers and databases, creating a need for intra-enterprise integration and transparency while preserving local autonomy and diversity of implementations. Furthermore, strategic alliances, just-in-time, and agile manufacturing are creating the need for inter-enterprise integration. Thanks to the advancement of the computer communications, networking, and VLSI technologies, it is now technically and economically possible to interconnect remote terminals, workstations, and all kinds of computers via wide area or local area networks, and to share data files and databases. However, distributed systems are much more complex to control, coordinate, and use than centralized systems.

A *distributed database* (DDB) is a collection of multiple, logically interrelated, databases distributed over a computer network (Fig. 14) [14,26].

A *distributed database management system* (D-DBMS) is the software which manages the distributed database and provides an access mechanism that makes the distribution of data transparent to the users or applications [18]. The software is itself distributed on the system nodes.

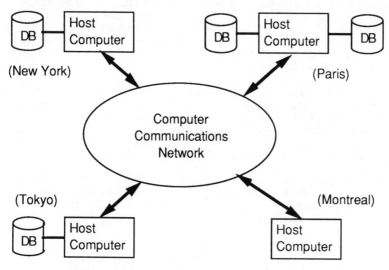

Fig. 14. Distributed database environment

Two types of transparency can be provided by distributed database systems: full transparency, and database federation.
- Full transparency requires full agreement among all sites concerning the data model, the schema interpretation, the data representation, the available functionalities, and where the data is located. All the data at the remote sites are accessible, just if they were local. Use of a global conceptual schema is required. Semantic data unification is achieved at data schema build-time. Implementation is made easier if the same commercial software (such as ORACLE/SQL* or INGRES/STAR) is used on each node, even under different operating systems.
- Federation concerns integration of heterogeneous distributed databases, i.e., databases stored on different database systems, possibly using different DBMS, and possibly under different operating systems. In federated architecture, there is only agreement on the data exchange formats and on the functions provided by each site. This is an open and flexible architecture in the sense that databases can be easily added to or removed from the system. The federated architecture is easier to implement than full transparency.

Another classification distinguishes among distributed databases (DDB) and multi-databases (MDB): In a distributed database, the database is divided into several (possibly redundant) local databases. In a multi-database, several existing databases are integrated using a meta-database to form a larger information system [16].

Distributed and multi-database systems require extensions to the SQL language to access and manipulate data separate autonomous databases. A proposal for an extension is the MDSL language [22] using the following generic statement:

```
OPEN name1 [mode1] name2 [mode2] ...
DB (abbrev1 name1) (abbrev2 name2) ...
    <auxilary clauses>
RANGE (tuple-variable relation) ...
SELECT <target list>
WHERE <predicate>
VALUE value-list
    <query commands>
CLOSE name1 name2 ...
```

where name1, name2, ... are names of databases or multi-databases to be openned (and later closed) for data access or modification (indicated by the mode mode1, mode2, ...), the DB clause allows to define nick names, the RANGE clause indicates data paths, and the SELECT clause indicates data to be retrieved. The WHERE and VALUE clauses for data selection and manipulation require substantial extensions compared to the SQL language to deal with data distribution [22].

Although distributed database systems are definitely required to support modern manufacturing system integration and operations, there are still problems. These include:
- Database design problems: How to distribute the database? Which alternative is best: full data replication, partial data replication, or data partition?
- Recovery and reliability problems: Recovery is made more difficult since there may be multiple copies of data to bring up-to-date. Making the system resilient to failures is another challenge.
- Concurrency and transaction control problems: The two-phase commit procedure is used but implies an overload of message exchange on the network for synchronization.
- Problems of scale: When the number of nodes increases, the size of problems increases and the system becomes more vulnerable.

9. Conclusion

Database technology remains one of the essential ingredients for the implementation and exploitation of modern manufacturing systems, and especially for computer-integrated manufacturing systems.

Careful database design remains a critical task in designing and implementing CIM environments. Indeed, the database system provides data integration and semantic unification for CIM applications. Furthermore, the quality of the design and implementation of the CIM information system using database technology can have serious a impact on the overall enterprise performance.

The next generation of database technology will be concerned with multi-media database systems for which object-oriented database systems in distributed environments are a pre-requisite. This will provide the manufacturing world with the desired flexibility and interoperability in terms of data management and exchange.

Another major challenge still to be addressed concerns the evolution of information systems to miror the enterprise (and market) evolution and support the management of change. The information stored in enterprise databases represent a tremendous capital and intellectual investments. However, it is not only data. It also structures a large part of the knowledge base of the enterprise. It is therefore important that enteprises capitalize on this kind of resource and protect this investment for their future operations.

However, one must always keep in mind that no matter how good the design and implementation of the database system is, and how sophisticated the technology utilized can be, the efficiency of the system also strongly depends on the reliability of stored data. Therefore, timely data acquisition and validation are also key issues in building a consistent environment.

10. References

[1] Atkinson M. et al. The object-oriented database system manifesto. Proc. Int. Conf. Deductive and Object-Oriented Databases, Elsevier, Amsterdam, 1989, pp. 40-57.

[2] Beeby W.D. The heart of integration: A sound database. *IEEE Spectrum*; 20(5):44-48; 1983.

[3 Bertino E., Martino L. Object-oriented database management systems: Concepts and issues. *IEEE Computer*; 24(4):33-47; 1991.

[4] Date C.J. *An Introduction to Database Systems, 4th Edition*, Addison-Wesley, Reading, MA, 1986.

[5] Codd E.F. A relational model for large shared data bases. *Communications of the ACM*; 13(6):377-387; 1970.

[6] Date C.J. *An Introduction to Database Systems* (5th Ed.), Vol. I. Addison-Wesley, Reading, MA, 1989.

[7] Date C.J. *A Guide to the SQL Standard*. Addison-Wesley, Reading, MA, 1987.

[8] Di Leva A., Giolito P., Vernadat F. Executable models for the representation of production systems. Proc. IMACS-IFAC Symposium MCTS-91, Lille, France, 7-10 May 1991, pp. 561-566.

[9] Di Leva A., Giolito P., Vernadat F. M*-Object: An object-oriented database design methodology for CIM information systems. *Control Engineering Practice*; 1(1):183-187; 1993.

[10] Gardarin G. and Vaduriez P. *Relational Databases and Knowledge Bases*. Addison-Wesley, Reading, MA, 1989.

[11] Gardarin G., Valduriez P. ESQL: An extended SQL with object and deductive capabilities. Research Report, INRIA, No. 1185, March 1990.

[12] Genrich H.J. Predicate/Transition nets, In W. Brauer, W. Reisig and G. Rozenberg (eds.), *Petri Nets: Central Models and Their Properties*. Springer-Verlag, Berlin, 1987. pp. 208-247.

[13] Grefen P., Apers P. Integrity control in relational database systems - An overview. *Data & Knowledge Engineering*, 10: 187-223; 1993.

[14] Gupta A. (Ed.). *Integration of Information Systems: Bridging Heterogeneous Databases*. IEEE Press, New York, NY, 1989.

[15] Hsu Cheng, Rattner L. Information modeling for computerized manufacturing, *IEEE Trans. on Systems, Man, and Cybernetics*; 20(4):758-776; 1990.

[16] Hsu Cheng et al. The metadatabase for manufacturing system integration. Proc. INCOM'92 7th IFAC/IFIP/IFORS/IMACS/ISPE Symposium on Information Control Problems in Manufacturing Technology, Toronto, Canada, May 25-28, 1992. pp. 663-668.

[17] Hull R., King R. Semantic database modeling: Survey, applications, and research issues. *ACM Computing Surveys*; 19(3):201-260; 1987.

[18] IEEE. Special issue on distributed databases systems. *Proceedings of the IEEE*; 75(5); 1987.

[19] IEEE. Special issue on main memory databases. *IEEE Trans. on Knowledge and Data Engineering*, 4(6); 1992.

[20] Joseph J.V.et al. Object-oriented databases: Design and implementation. *Proceedings of the IEEE*; 79(1):42-64; 1991.

[21] Kent .W. A simple guide to five normal forms in relational database theory. *Communications of the ACM*; 26(2):120-125; 1983.

[22] Litwin W., Abdellatif A. An overview of the multi-database manipulation language MDSL. *Proceedings of the IEEE*; 75(5):621-632; 1987.

[23] Maier D. *The Theory of Relational Databases*, Computer Science Press, 1983.

[24] Meyer B. *Object-Oriented Software Construction*. Prentice-Hall, Englewood Cliffs, NJ, 1988.

[25] Nguyen G.T., Rieu D. An object model for engineering design. Proc. ECOOP'92, Utrecht, NL, June 1992.

[26] Özsu M.T., Valduriez P. *Principles of Distributed Database Systems*, Prentice-Hall International, Englewood Cliffs, NJ, 1991.

[27] Rishe N. A methodology and tool for top-down relational database design. *Data & Knowledge Engineering*; 10:259-291; 1993.

[28] Sartori L.G. *Manufacturing Information Systems*. Addison-Wesley, Reading, MA, 1988.

[29] Scheer A.W. *Enterprise-Wide Data Modelling*. Springer-Verlag, Berlin, 1989.

[30] Smith J.M., Smith D.C.P. Database abstractions: Aggregation and generalization, *ACM Database Transactions on Database Systems*; 2(2); 1977.

[31] Su S.Y.W. Modeling integrated manufacturing data with SAM*. *IEEE Computer*; 19(1):34-49; 1986.

[32] Teorey T.J., Yang D., Fry J.P. A logical design methodology for relational databases using the extended entity-relationship model. *ACM Computing Surveys*; 18(2):197-222; 1986.

[33] Tsichritzis D.C., Klug A. (eds). The ANSI/X3/SPARC DBMS Framework: Report of the Study Group on Database Management Systems. *Information Systems*; 3; 1978.

[34] Ullman J.D. *Principles of Database Systems, 2nd Edition*. Computer Science Press, Rockville, MA, 1982.

[35] Vernadat F. A conceptual schema for a CIM database. In: *CAD/CAM Integration and Innovation*, SME, Dearborn, MI, 1985. (Also in Autofact 6 Conf. Proc., Anaheim, CA. October 1-4, 1984. pp. 11.24-11.41).

[36] Vernadat F., Di Leva A., and Giolito P. Organization and Information System Design of Manufacturing Environments: The New M* Approach. *Computer-Integrated Manufacturing Systems*; 2(2):69-81; 1989.

[37] Vossen G. Bibliography on object-oriented database management. *ACM SIGMOD Record*; 20(1):24-46; 1991.

5   An Object-Oriented Approach to Design of Process Parameters

Yu-To Chen and Andrew Kusiak

1   Abstract

This paper describes a framework of an object-oriented system for design of process parameters. The shortcomings of models for design of process parameters are described, followed by the introduction of a general architecture of the proposed system. The system has a star topology, where the central node is a blackboard and all the surrounding nodes are design objects representing different design perspectives. Conflicts among different design objectives are resolved by numerical methods, fuzzy inference, or neural networks. In addition, the parameter design problem has been reduced to the inversion function problem, i.e., given a set of design objectives which can be represented by a set of simultaneous equations, it is desired to produce a set of design variables that satisfy all the constraints imposed by the equations. The paper concludes by summarizing the current research activities, and points out directions for future research to fully realize the proposed object-oriented framework for parameter design.

2   Literature Review

A cooperative product development environment has been introduced to realize the concurrent engineering concept [1]. Although general ideas about how to utilize computer-based design tools have been discussed, step-by-step engineering procedures have not been developed. Any mechanical design can be viewed as a combination of physical entities or building blocks [2]. In their model, the building blocks of a design are expressed as objects; however, they fail to fully utilize the representation power of object-oriented modeling in that they did not consider methods associated with the design objects. An associative memory model for design [3] has been proposed to argue that, given a set of functional requirements, a human designer can identify a set of structures. In this way, neural networks (NN) can be applied to implement the above idea. Unfortunately, a detailed implementation of the concept does not exist. Brandon and Huang [4] introduced an agent-based system for mechanical design; however, they did not address the issue of conflict resolution. Kannapan and Marshek [5] proposed the concept of a design diagram and applied game theory to resolve conflicts between different design objectives. Although the design diagram concept depicts parameters and their relationships involved in the design of a component, it fails to visualize the function inversion in parameter design. Nor does it provide insights into how a parameter design process is modeled. Moreover, game theory is not capable of coping with complex, nonlinear analytical design models.

3   Parametric Design

Thinking backwards is a major feature of parameter design. Traditionally, design elements are synthesized first to form a coarse design. Next, life-cycle issues are considered to generate a detailed design. This process requires a number of iterations. The idea behind parameter design is to reduce the number of iterations in the design by bring up life-cycle issues to the early design stage, i.e., life-cycle issues act as constraints imposed upon the design process. In other words, if move backwards, i.e., starting from life-cycle issues to the coarse design, the parameter design process is analogous to constraint propagation in a reverse order.

We argue that parameter design problem can be reduced to the generalized function inversion problem [6]. For example, consider the design of a cantilever beam:

$$\Delta = \frac{F\,L^3}{3EI} \tag{1}$$

where: Δ is the deflection,
 F is the applied force,
 L is the length of the beam ,
 E is the modulus of elasticity of the beam,
 I is the second moment of area bout the neutral axis.

The objective of parameter design problem is to generate a set of independent variables (design variables), given the desired dependent variables (design objects). In this case, the design variables are the applied force, the length of the beam, and so on, while the design object is the deflection. The relationships between design objects and design variables may be represented by equations, e.g., equation (1), production rules, heuristics, and so on. The approach to determine the reverse relationship from a given model, such as equations or production rules, is called a generalized inversion function method. As will be described in this paper, the generalized inversion function can be accomplished by numerical methods, fuzzy inference, or neural networks.

4   System Architecture

The system for parameter design consists of a blackboard and several design objects (see Figure 1).

4.1   Design object

Design objects represent the different viewpoints of the candidate design held by the corresponding experts who are responsible for the design at different levels or with different interests. For example, in the design of a cantilever beam, the deflection can be represented as a design object, while the maximum bending stress is represented by another design object.

In general, design knowledge is viewed in terms of a set of objects which encapsulate both state and behavior. The state of an object is the set of values for the attributes of the object, and the behavior of an object is the set of methods which operate on the state of the object. Methods are invoked through message sent to the object by the user or by other objects.

In the proposed system, each design object has its own methods that explicitly express the associated design knowledge in the form of equations, production rules, or heuristic strategies. As an example, consider a cantilever beam with a constant rectangular cross-section. The design object of the maximum bending stress has its associated method expressed as the following equation:

$$\sigma = \frac{MY}{I}$$

where: σ is bending stress,
 M is momentum,

Y, I are geometric parameters.

One of the main advantages of an object-oriented system is in its functional message passing and method invocation mechanism, that allow for an easy extension of the program (method). Furthermore, individual objects can be easily added or removed without affecting the functionality of the system.

Blackboard

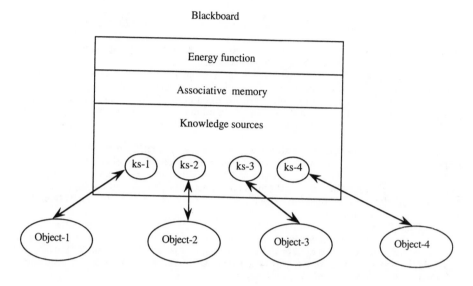

Design Objects

Figure 1. System architecture

4.2   Blackboard

The blackboard can also be viewed as an object. Its associated methods direct the flow of parametric design process and solve conflicts among design objects. It can receive a message from the outside world (the user) to perform a specific design step. Moreover, it can pass design message to change the state or a specific design object. The parametric design process is displayed on the blackboard for review by all design objects. Each design object can apply its knowledge, either critical or constructive, to the design process and express the corresponding results. Distributed and concurrent design activity is achieved since individual design object possess knowledge about different aspects of parametric design. Conflict resolution is required since the results of applying different aspects of parametric design knowledge may cause inconsistencies.

In general, the blackboard contains an energy function, a set of knowledge sources, and associative memory.

The energy function is similar to a performance index that measures the quality of the parameter design. For example, in the design of a cantilever beam, two design objectives are defined in advance: the deflection can not be excelled a certain value and the maximum bending stress should be within a certain range (when a force is applied to it.)

Two nominal values, Δ and σ are selected as target values for the deflection and the stress, respectively. Hence the energy function is defined follows:

$$E = \frac{1}{2}\left[\left(\frac{\Delta^*-\Delta}{\Delta^*}\right)^2 + \left(\frac{\sigma^*-\sigma}{\sigma^*}\right)^2\right] \qquad (2)$$

where: Δ^* and σ^* represent the actual values of a specific design for the deflection and stress, respectively.

In general, the overall design goal is to minimize the energy function so that the actual design values will reach the desired target values.

There is a one-to-one mapping from design objects to knowledge sources. Each knowledge source contains design specifications for its corresponding design object. In the above example, the knowledge source of the maximum bending stress could store the geometry parameters for the cantilever beam. In addition, knowledge sources provide a working space for their corresponding design objects: add, remove, or modify individual design knowledge.

"Associative memory is the ability to get from one internal representation to another or from one part of a complex representation to the remainder" [7]. Associative memory is basically of two types: pattern association and auto-association [3]. If an input pattern is associated with a different output pattern, then the association between the input and the output is called pattern association. If an input pattern is associated with itself, then the association is called auto-association.

The life-cycle issues of a parametric design are stored in associative memory. In other words, associative memory relates life-cycle design issues, such as working conditions, resource optimization, or life-cycle costs, to a parametric design. In the above example, prior experience may tell us how safe is the maximum bending stress and this knowledge can be stored in an associative memory.

5   Case Study - Parametric design of a surface grinding process

5.1   Background

The grinding process is considered as a crucial material removal operation with the increasing use of advanced materials [8]. However, grinding remains as one of the least-understood processes due to the following reasons [9]:
(1)   It is not possible to control the shape of hard particles which affects the grinding process.
(2)   Grinding processes have more process variables to be controlled than any other machining processes.
(3)   No mathematical model has been developed that describes the relationships among all grinding process variables.

This rises to a need to develop an intelligent system to assist in the design of grinding processes as to how operating parameters have to be adjusted in order to achieve the desired process conditions [10]. A grinding process advisory system with fuzzy logic has been proposed [9] and the implementation results have shown that the system can lead to the optimal design of a grinding process effectively. In addition, neural networks can serve as an alternative means of improving the design of grinding processes [6]. The goal of this paper is to show that the proposed object-oriented approach can be utilized as an effective tool in the automation of parametric design for complex processes such as the grinding process.

5.2   Model Building

As described earlier, the proposed system consists of a blackboard and several design objects. Each design object represents a specific viewpoint for the design of a grinding process. In addition, each design object has its associated methods describing how its design objective can be achieved by manipulating its corresponding design variables and knowledge source. The blackboard contains the energy function, a set of knowledge sources, and its associated method. The detailed methodologies will be presented in the following sections.

5.3   Design Objects

A number of grinding process models have been discussed in the literature. Unfortunately, most of the existing models describe only partial relationships between process variables and design variables. The objected-oriented approach proposed here is to view existing grinding process models as design objects, which represent the different viewpoints of the grinding process held by design engineers who are responsible for the design at different levels or with different interests. In addition, the proposed blackboard model provides a systematic way to synthesize these different design viewpoints and to resolve conflicts among them. In this paper, six design objects are identified as the most representative and important [10]. Note that as new theoretical advances are made and new grinding models are developed, they can be easily incorporated into the proposed system as new design objects without affecting the functionality of the system due to the representation power of the oriented-oriented approach.

5.3.1   Chip Object

In any metal removal process, the formation of proper chips is an important issue. Since grinding wheels consist of hard, randomly oriented grains acting as cutting edges during machining, it is difficult to predict exact chip lengths. The design objective here is to predict the grinding cutting profiles by using probabilistic techniques [11]. The following equations express the design knowledge associated with the chip object [10]:

$$a_n = 2\,L_i \frac{V_w}{V_s}\left[\frac{d}{De}\right]^{1/2} \tag{3}$$

$$l = \left[d\ De\right]^{1/2} \tag{4}$$

a_n: the maximum chip thickness (mm);
L_i: the circumferential distance between two active cutting grits (mm);
l: the chip length (mm);
V_w: the workpiece speed (m/s);
V_s: the grinding wheel speed (m/s);
d: the depth of cut (mm);
De: the equivalent diameter of wheel (mm);

Note that there are two subdesign objects: the chip thickness and the chip length, and three design variables: the depth of cut, the wheel speed, and the workpiece speed. Refer to Figure 2 for a schematic of the surface grinding process.

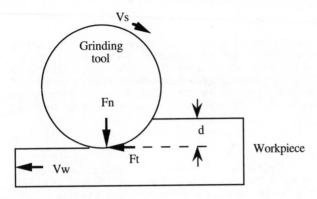

Figure 2. Surface grinding

5.3.2   Energy Object

Forces and energy play an essential role in all abrasive machining operations [12]. The design objective here is to calculate the energy and power consumed in a grinding process. For this energy object, four design variables are chosen: wheel speed, workpiece speed, width of cut, and depth of cut.

$$u = \frac{F_t V_w}{V_s b d} \tag{5}$$

u: the specific grinding energy (N-m/mm2-m);
F_t: the tangential force (N);
V_w: the workpiece speed (m/s);
V_s: the grinding wheel speed (m/s);
b: the axial width of cut (mm);
d: the depth of cut (mm);

5.3.3   Force Object

Fielding and Vickerstaff [13] developed empirical formulas to predict grinding forces based on several more easily accessible variables such as wheel speed, metal removal rate, dress lead, and metal being removed. The subdesign objects are the normal force and the coefficient of friction, while the design variables are the wheel speed, the metal removal rate, and the dress lead.

$$F_n = \left(\frac{1}{\mu}\right) F_t + C \tag{6}$$

$$\mu = 3.07 \, V_s^{-0.548} M^{-0.147} D_1^{-0.17} \tag{7}$$

$$C = -2.84 \, V_s^{0.608} M^{0.848} D_1^{0.077} \tag{8}$$

F_n: the normal force (N);
μ: the coefficient of friction;
F_t: the tangential force (N);
C: the intercept (N);
V_s: the wheel speed (m/s);

M: the metal removal rate (mm3/s);
D_l: the dress lead (mm/rev);

5.3.4 Surface Finish Object

For the surface finish, the root mean square (R_g) value is extracted as a design object, and the depth of cut is selected as a design variable. Based on the grinding wheel profiles and kinematic conditions, a simple formula to calculate the R_g value of the ground surface has been derived by Pandit and Sathyanarayanan [14].

$$R_g = \frac{A_c - 7.596 \left[\dfrac{F_n^2 A_c W_g^2}{b^2 D d} \left(K_w + K_g \right)^2 \right]^{1/3}}{\sqrt{2}} \tag{9}$$

R_g: the rms surface roughness value;
A_c: the amplitude of the secondary wave length of the wheel profile;
F_n: the normal thrust force on the wheel;
A_g: the amplitude of the primary wavelength of the wheel profile;
W_g: the wavelength of the primary wavelength of the wheel profile;
b: the width of cut;
D: the diameter of the wheel;

Note that K_w and K_g are constants, which depend on the workpiece material [8].

5.3.5 Stress Object

Brecker [15] reported that the effective stress required to fracture a variety of abrasive materials was measured using both static and roll crushing techniques. It was found that the grain strength strongly depends on the size of particles.

$$\sigma_e = 1001 \frac{P}{a^2} \tag{10}$$

σ_e: the effective uniaxial tensile strength (KP_a);
P: the load at fracture (N);
a: the dimension of abrasive grains (mm);

5.3.6 Temperature Object

Temperature is an important process condition that affects the wear of tool and surface burning [9]. Malkin [16] investigated the thermal aspect of grinding and calculated the surface temperature distribution. Based on the temperature model, he explained how surface burning can be caused.

$$Q = \left(9.0 \times 10^{-5}\right) d + \left(4.1 \times 10^4\right) D^{1/4} d^{1/4} V_w^{-1/2} \tag{11}$$

$$T_{max} = 1.595 \frac{R_3 q_3}{k} \left(\frac{Kl}{V_w} \right)^{1/2} \tag{12}$$

$$q_3 = \frac{F_h V_s}{(Dd)^{\frac{1}{2}} b} \tag{13}$$

Q: the energy flux, energy input per unit area ground;
d: the downfeed per pass;
D: the wheel diameter;
V_w: the workpiece velocity;
T_{max}:   the maximum grinding zone temperature;
R_3:   the fraction of grinding energy to workpiece;
q_3:   the grinding energy rate per unit area of grinding zone;
k: the thermal conductivity;
K: the thermal diffusivity;
l: the semilength of heat source;
F_h: the horizontal or power force component;
V_s: the wheel speed;
b: the width of workpiece;

From these equations, the energy flux and the temperature are selected as desired design objects, while the workpiece speed and the wheel speed are selected as design variables.

5.4   Blackboard

The blackboard is an object and its associated methods direct the flow of parametric design process and solves conflicts among design objects. While parametric design can be reduced to the generalized function inversion, the proposed methods embedded in the blackboard are responsible for the realization of the function inversion. There are three methods for the generalized function inversion: numerical methods, neural networks, and fuzzy inference. The detailed descriptions is presented next.

5.4.1   Numerical Methods

At the time a design object is built, it stores the design specifications in its corresponding knowledge source of the blackboard. Then the design object notifies the blackboard (by sending the blackboard a message) that it is ready (for processing a parametric design). After all the design objects are built, the blackboard invokes its associated method for solving the generalized inversion function problem. First, the blackboard aggregates all the design methods supplied by the design objects in the form of simultaneous equations:

Chip: $CT = C_0 \cdot (DOC)^{1/2} \cdot (WKS) \cdot (WLS)^{-1}$ (14)

$CL = C_1 \cdot (DOC)^{1/2}$ (15)

Energy: $SE = C_2 \cdot (WLS) \cdot (WKS)^{-1} \cdot (WOC)^{-1} \cdot (DOC)^{-1}$ (16)

Force: $COF = C_3 \cdot (WLS)^{-0.548} \cdot (DL)^{-0.17}$ (17)

$FN = C_4 \cdot (WLS)^{0.548} \cdot (DL)^{0.17} + C_5 \cdot (WLS)^{0.608} \cdot (DL)^{0.077}$ (18)

Surface finish: $RG = C_6 - C_7 \cdot (DOC)^{-1/3}$ (19)

Stress: $UTS = C_8 \cdot (GS)^{-2}$ (20)

Temperature: $EF = C_9 + C_{10} \cdot (WKS)^{-1/2}$ (21)

$T = C_{11} \cdot (WKS)^{-1/2} \cdot (WLS)$ (22)

where all the design objects and the design variables are defined in Table 1 and Table 2, respectively, and C_0 through C_{11} are constants.

Table 1.   The Design Objects of the Simultaneous Equations

Design Objects	Minimum Values	Maximum Values
CT: Chip Thickness	1.59	43.39
CL: Chip Length	14.14	39.33
SE: Specific Energy	66.30	3184.96
COF: Coefficient of Friction	0.35	1.22
FN: Normal Force	220.71	757.56
RG: Surface Finish Index	-0.72	-0.36
UTS: Tensile Strength	150.02	2205/70
EF: Energy Flux	0.0051	0.0091
T: Temperature	12.99	124.26

Table 2.   The Design Variables of the Simultaneous Equations

Design Variables	Minimum Values	Maximum Values
DOC: Depth of Cut	1.00	7.73
WKS: Workpiece Speed	0.50	1.57
WLS: Wheel Speed	5.00	44.97
WOC: Width of Cut	2.00	11.24
DL: Dress Lead	0.20	0.70
GS: Grain Size	1.00	3.83

In this paper, parametric design is informally defined as the initial determination of a set of optimal design variables for the purpose of achieving a set of design objectives. For example, to obtain a good surface finish in grinding, we have to limit the grinding force within some specific range and to keep the grinding temperature within some tolerances. In order to do that, one has to determine the initial values of wheel speed, dress lead, and workpiece speed that determine the desired grinding force and temperature. In this sense, parametric design of surface grinding can be reduced to solving an inversion of the above simultaneous equations, i.e., given a set of desired values of design objects, which are all on the left-hand side of the equations, we would like to get a set of design variables, which are all on the right-hand side, from the simultaneous equations.

This paper does not discuss numerical results of the generalized inversion function problem, but provides conceptual ideas on methodologies that can be applied to solve the problem at hand. Therefore, we will briefly survey numerical methods and software that are relevant to solve simultaneous nonlinear equations. Refer to Chen [17] for a comprehensive survey.

A system of simultaneous nonlinear equations can be expressed by

$$f_i(x_i, ..., x_n) = 0 \qquad for \qquad i = 1, 2, ..., n,$$

where the f_i's represent real-valued functions. In general, the multivariable Newton method can be used to solve a system of simultaneous nonlinear equations if their derivatives are available [18]. However, if the multivariable Newton method does not converge, there is no systematic way to generate a solution.

A strong link exists between nonlinear equations and optimization. Especially for a system of over-determined nonlinear equations, those redundant equations are generally treated as constraints so as to form a nonlinear optimization or a least squares problem. The classical techniques for the least squares problem includes a damped Newton iteration, a variation of Newton's methods, and so on [19].

A library of mathematical software is maintained by the International Mathematical and Statistical Libraries, better known as **IMSL**. **MINPACK**, one of the packages

124

distributed by **IMSL**, is capable of solving nonlinear systems of equations and nonlinear least squares problems [20]. **LANCELOT**, an acronym for **L**arge **A**nd **N**onlinear **C**onstrained **E**xtended **L**agrangian **O**ptimization **T**echniques, is a package of standard Fortran subroutines useful for solving large-scale nonlinearly constrained optimization problems [21]. It has been designed for problems where the objective function is a smooth function of many real variables and where the value of these variables may be restricted by a finite set of smooth constraints.

In summary, solving all classes of over-determined simultaneous nonlinear equations is a difficult problem [18]. Although the set of grinding equations described above seems to be relatively simple, in general, it could be complex and there is no guarantee it can be easily solved by numerical methods. Therefore, we propose neural networks as an alternate approach. Note that we do not make any claims regarding the quality of the neural solution versus other conventional methods.

5.4.2   Neural Networks

As reviewed by Chen [17], the inversion function problem can be solved by neural networks [22], [23]. However, neural networks have not been widely used for an inversion of simultaneous equations.

There are two inversion methods for feedforward neural nets: direct-inversion and indirect-inversion. The direct inversion inversion method works as follows:

(1)   Randomly generate values of independent variables, X_i within a specified range;
(2)   Feed these values to the set of equations to get values of dependent variables, Y_i;
(3)   Use (Y_i, X_i) as training pairs to construct a neural network;

Training mode

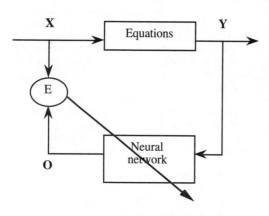

Operational mode

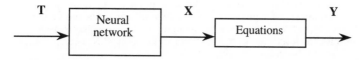

Figure 3. Direct-inversion

As shown in Figure 3, direct-inversion is composed of two modes. In the training mode, we provide **X** to the equations and obtain **Y** as output, then feed **Y** to the network as input and get **O** as actual output. In particular, we try to train the network by adapting its weights to minimize the error, $\mathbf{E} = \mathbf{X} - \mathbf{O}$. After training is complete, the network acts exactly as inversion of the equations, producing from the target response **T** a signal **X** that drives the output of the equations to $\mathbf{Y} \approx \mathbf{T}$.

For indirect-inversion, a neural network acting as an identifier of the equations is constructed first. As illustrated schematically in Figure 4, error signals, $\mathbf{E} = \mathbf{T} - \mathbf{O}$, are computed at the output layer and propagated back to the input layer to inform the input units in which direction and how much they should change their values in order to decrease the error. Again, the indirect-inversion is explained next:

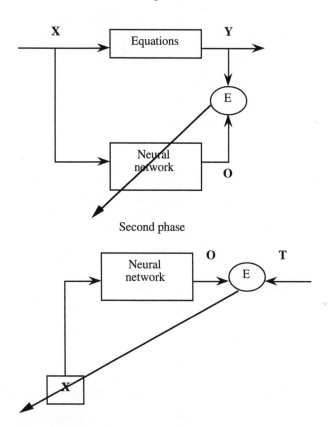

Figure 4. Indirect-inversion

(1)   Randomly generate values of independent variables, $\mathbf{X_i}$, within a specified range;
(2)   Feed these values to the set of equations to get values of dependent variables, $\mathbf{Y_i}$;
(3)   Use ($\mathbf{X_i}$, $\mathbf{Y_i}$) as training pairs to construct a neural network;

(4)   For each target value of dependent variables, T_j, initial guess each value of independent variables, X_j, compute error between actual output values, O_j and T_j;
(5)   Back propagate the error through the network to adjust X_j until the error is minimized;

5.4.3   Fuzzy Inference

An obvious drawback of the neural network method is that it needs numerical data for training. In other words, the neural network method is not capable of incorporating heuristic knowledge, e.g., production rules. In addition, the design methods of each design object in the parametric design of grinding processes are all highly nonlinear. In some cases, these methods are only useful for predicting approximate values due to the complexity of the underlying physical process. It implies that there will be a certain degree of uncertainty associated with each prediction. Using a conventional optimization methods will not only make it extremely difficult to obtain a numerical solution, but may also converge to local minima. In order to surmount these problems, fuzzy heuristic rules and fuzzy inferencing are employed to solve the generalized inversion function.

In everyday life, humans use fuzzy representation extensively in linguistic descriptions; for example, words like slow, hot, fast and cold are intuitively fuzzy. In this sense, the vagueness of some linguistic terms is an essential and useful characteristic of any natural representation for flexible systems. In a fuzzy rule based system, knowledge is represented by "if-then" rules associated with fuzzy variables, as well as through mathematical techniques of fuzzy parameters and variables [10].

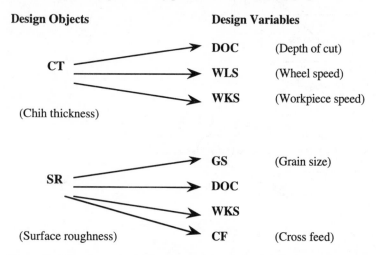

Design Objects

Design Variables

CT

(Chih thickness)

DOC   (Depth of cut)

WLS   (Wheel speed)

WKS   (Workpiece speed)

SR

(Surface roughness)

GS   (Grain size)

DOC

WKS

CF   (Cross feed)

Figure 5. Diagram representing the relationship between design objects and variables

As shown in the description of six design objectives to be accomplished in this paper, each design method describes a relationship between a few design objects and the corresponding design variables. Some examples of such pairs are shown in Figure 5. In most cases, these relationships are multi-modal, i.e., a design object may depend on several design variables as well as other design objects. Therefore, the rules to be generated must properly account for these multi-modal characteristics. However, considering all the design objects simultaneously can make the rule generation process difficult. For the effective generation of fuzzy rule sets, the blackboard considers only one pair of design object and design variable at a time while other objects and variables are treated as constants, i.e., each pair of design object and variable construct a fuzzy rule. For example, a grinding design object that contains two subdesign objects and two design variables will produce four fuzzy rules. Figure 6 illustrates this procedure more in detail. Then these generated rules are synthesized by the compositional rules of fuzzy sets and the multi-modal characteristics are fully recovered. This method makes it possible to use a set of simple fuzzy rules to represent multi variable relationships.

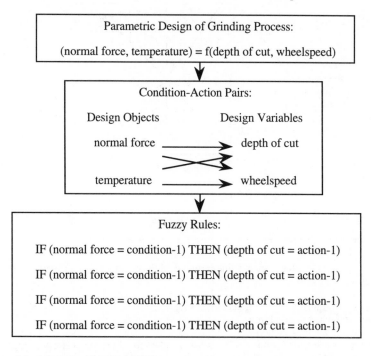

Figure 6. Matching condition-action pairs

Since the design methods for grinding processes are nonlinear equations as shown earlier, a single set of rules may not be uniformly applied to the entire domain of operating conditions in lieu of the equations. Instead, the entire domain is subdivided into small segments and labels are assigned to each segment depending on the rate of change in the design object within each segment. The labels are used to determine the type of rules to be used in that segment. However, this could easily become an overwhelming task, if performed manually with a large number of segments used. The blackboard adopts a new label generation method and a "grouping" technique that allow for the minimization of the number of rule sets to be generated. The detailed description on this automatic rule generation method has been presented in [9]. For example, the nonlinear relationship between residual stresses and temperature shown in Figure 7, can be divided into three regions and pertinent rules are generated for each subregion. Based on the maximum temperature value being estimated from another temperature object, a subregion under which the temperature value falls under is identified and the pertinent set of rules are extracted for the fuzzy logic inferencing. If a different temperature value belonging to

128

another subregion is to be used during the course of optimization, the corresponding rule set is retrieved.

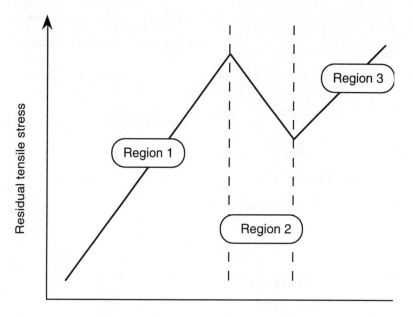

Figure 7. Subregioning to represent the non-linear relationship of temperature effects on a residual stresses [24]

The format of a linguistic fuzzy rule is as follows:
IF   Design Object = <u>Condition</u>
THEN   Design Variable = <u>Action</u>
If design object satisfies a specific condition, an action is taken to adjust the corresponding design variable. Therefore, a number of linguistic rules to be used must be determined to represent the conditions of design objects and the actions of design variables. The number of fuzzy linguistic rules to be used is generally based on a prior experience. While too many rules will make the subsequent fuzzy inferencing less efficient and increase the computational burden, too few rules will result in inaccuracy. The blackboard retrieves this kind of information from its associative memory.

Once all the rules are generated and synthesized, then tuning of design variables is performed by fuzzy logic. The tuning process uses the generated rule sets along with associated membership functions and involves fuzzy inference in order to get the elements of a fuzzy decision table. Note that the membership functions can be either provided by the blackboard's associative memory or by the message passing from the outside user. The elements generated are stored in the blackboard's private knowledge source are subsequently combined to form a complete decision table. Since it is very unlikely that the fuzzy tuning process will produce the optimal set of values in the first iteration based on the user-specified input, the entire process must be repeated, thus closing the loop, until the energy function is minimized.

The energy function includes a penalty which determines the relative importance of each design object to be regulated. If a grinding process involves **n** design objects, **y₁**,

y_2, ..., y_n, of which desired values are y_1^*, y_2^*, ..., y_n^*, respectively, the energy function can be defined as

$$E = \frac{1}{2}\sum_{i=1}^{n} P_i \left(\frac{y_i^* - y_i}{y_i^*} \right)^2 \qquad (23)$$

where P_1, P_2, ..., P_n are the penalties representing the relative importance of design objects. The goal here is to minimize this energy function. The best results would be obtained when the value of E equals to zero, i.e., all the design objects converge to the desired final values. In reality, however, it may not be feasible to get such a result due to the constraints imposed on the design variables and the conflicting effects of design variables on design objects. Instead, minimization of the objective function is pursued iteratively until there is no further improvement in the merit value.

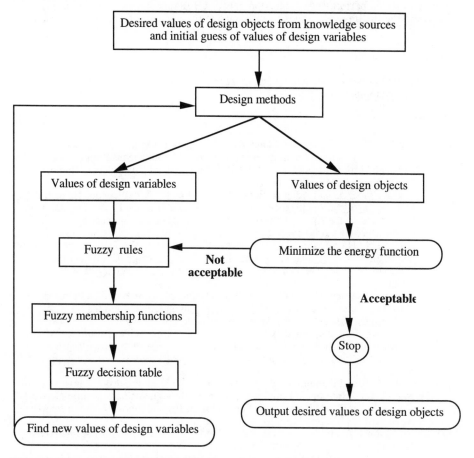

Figure 8. Flowchart of the blackboard's fuzzy design method

A set of design specifications which specify the limit values of design parameters is also used with the energy function. After the fuzzy decision table is generated, new values of design objects calculated by feeding the new values of the design variables into the design methods are used to evaluate the energy function. If the new process conditions do not meet the preset criterion, the entire process is repeated by generating new fuzzy rules with updated parameter values. The flowchart of the entire operation is schematically depicted in Figure 8. Detailed descriptions can be found in Chen [8].

6   Conclusion

In this paper, an object-oriented approach to parametric design was proposed. The main objective of this work was to demonstrate how an object-oriented system can be constructed to assist in parametric design by the introduction of a set of design objects and a blackboard. Since the parametric design problem has been reduced to a generalized inversion function problem, conflicts among different design objectives can be solved by numerical methods, neural networks, or fuzzy inference. Further developments of this research include performance evaluation between different function inversion methods, incorporation of life-cycle issues to fully realize the concept of concurrent parametric design.

Acknowledgment

This research has been sponsored by research contract No. DAAE07-C-R080 from the US Army Tank Automotive Command.

References

[1]   Lu S. C-Y., Subramanyam S. and Thompson J. B. (1989), "A Cooperative Product Development Environment to Realize the Simultaneous Engineering Concept," Proceedings of the 1989 ASME International Computers in Engineering Conference, ASME, New York, N.Y., pp. 9-18.

[2]   Ishii K., Goel A. and Adler R. E. (1989), "A Model of Simultaneous Engineering Design," Artificial Intelligence in Design, Proceedings of the Fourth International Conference on the Applications of Artificial Intelligence in Engineering, Cambridge, UK, July, 1989, Springer-Verlag, New York, N.Y., pp. 483-501.

[3]   Kumara S., Ham I. and Watson E. F. (1990), "AI Based Product Design - Review and Examples," Artificial Intelligence Based Product Design, CIRP International Working Seminar, The Pennsylvania State University, University Park, PA, May, 1990, pp. 21-52.

[4]   Brandon J. A. and Huang G. Q. (1993), "Use of an Agent-Based System for Concurrent mechanical Design," Concurrent Engineering: Automation, Tools, and Techniques, Kusiak A. (Ed), Wiley, New York, pp. 463-479.

[5]   Kannapan S. M. and Marshek K. M. (1993), "An Approach to Parametric Machine Design and Negotiation in Concurrent Engineering," Concurrent Engineering: Automation, Tools, and Techniques, Kusiak A. (Ed), Wiley, New York, pp. 509-533.

[6]   Chen Y. T. and Kumara S. (1993), "Design of Grinding Process via Inversion of Neural Nets," Intelligent Engineering Systems through Artificial Neural Networks, Vol. 3, Dagli C. H., Burke L. I., Fernandex B. R., and Ghosh J. (Eds), ASME, New York, N.Y., pp. 715-720.

[7]   Anderson J. A. and Hinton G. E. (eds) (1981), "Parallel Models of Associative Memory," Lawrence Erlbaum Associates, Inc., New York.

[8]   Chen Y. T. (1990), Design of Intelligent Grinding Process Optimizer via Fuzzy Logic, Master Thesis, The Pennsylvania State University, University Park, PA.

[9]   Shin Y. C., Chen Y. T., and Kumara S. (1992), "Framework of an Intelligent Grinding Process Advisor," Journal of Intelligent Manufacturing, Vol. 3, No. 3, pp. 135-148.

[10]   Chen Y. T. and Shin Y. C. (1991), "A Surface Grinding Process Advisory System with Fuzzy Logic," Control of Manufacturing Processes, Proceedings of the ASME Winter Annual Meeting, Atlanta, Georgia, DSC-Vol. 28/PED-Vol. 52, ASME, New York, N.Y., pp. 67-77.

[11]   Younis M., Sadek M. M., and EL-Wardani T. (1987), "A New Approach to Development of a Grinding Force Model," ASME Transactions: Journal of Engineering for Industry, Vol. 109, No. 4, pp. 306-313.

[12]   Brach K, Pai D. M., Ratterman E., and Shaw M. C. (1988), "Grinding Forces and Energy," ASME Transactions: Journal of Engineering for Industry, Vol. 110, No. 1, pp. 25-31.

[13]   Fielding E. R. and Vickerstaff T. J. (1986), "On the Relationship between the normal and tangential forces in cylindrical plunge-cut grinding," International Journal of Production Research, Vol. 24, No. 2, pp. 259-268.

[14]   Pandit S. M. and Sathyanarayanan G. (1982), "A Model for Surface Grinding Based on Abrasive Geometry and Elasticity," ASME Transactions: Journal of Engineering for Industry, Vol. 104, No. 4, pp. 349-357.

[15]   Brecker J. N. (1974), "The Fracture Strength of Abrasive Grains," ASME Transactions: Journal of Engineering for Industry, Vol. 96, No. 4, pp. 1253-1257.

[16]   Malkin S. (1974), "Thermal Aspect of Grinding," ASME Transactions: Journal of Engineering for Industry, Vol. 96, No. 4, pp. 1184-1191.

[17]   Chen Y. T. (1993), Process Design, Diagnostics, and Control in Manufacturing through Fuzzy Logic and Neural Networks, Ph.D. Dissertation, The Pennsylvania State University, University Park, PA.

[18]   Yakowitz S. and Szidarovszky F. (1986), An Introduction to Numerical Computations, Macmillan, New York, N.Y.

[19]   Orgega J. M. and Rheinboldt W. C. (1970), Iterative Solution of Nonlinear Equations in Several Variables, Academic Press, New York, N.Y.

[20]   More J. J., Garbow B. S., and Hillstrom K. E. (1980), User Guide for MINPACK-1, Report 80-74, Argonne National Laboratory, Argonne.

[21]   Conn A. R., Gould N. I. M. and Toint Ph. L. (1992), LANCELOT: A Fortran Package for Large-Scale Nonlinear Optimization (Release A), Springer-Verlag, New York, N.Y.

[22]   Psaltis D., Sideris A., and Yamamura A. A. (1988), "A Multilayered Neural Network Controller," IEEE Control Systems Magazine, Vol. 8, No. 2, pp. 17-21.

[23]   Linden A. and Kindermann J. (1989), "Inversion of Multilayer Nets," International Joint Conference on Neural Networks, Washington D. C., June, Vol. 2, IEEE Service Center, Piscataway, N.J., pp. 425-430.

[24]   Snoey R., Leuven K. U., Maris M., Wo N. F., and Peters J. (1978), "Thermally Induced Damage in Grinding," Annals of the CIRP, Vol. 27/2, pp. 571-581.

6   Hierarchical Production Management

Rakesh Nagi and Jean-Marie Proth
(In memory of George Harhalakis, for all he taught us in Production Management)

1. Introduction

Recent years have seen significant progress in automating and integrating Design, Process Planning, Machining, and Inspection, however, Production Management and Control seem to have received little attention. Decision making and optimal real-time control of a production system, subject to both endogenous (e.g., resource failures), as well as exogenous (e.g., unscheduled orders, delayed receipts of material) random events is a challenging problem. Maxwell *et al* [42] state: *"Billions of dollars are wasted in US each year by manufacturers of discrete parts because of inadequate procedures for controlling inventory and production."* The need for developing and implementing production planning systems for a wide variety of systems is obvious in light of competitiveness. Most currently available software systems, such as Manufacturing Resource Planning or MRP systems, fail to directly address some key aspects of the overall production control problem. This chapter defines an architecture for efficient production management, which can be well integrated in a CIM environment. A Hierarchical Production Management System (HPMS), organized in several hierarchical levels, is intended to address the complexity of global problems. The number of levels depends on the complexity of the manufacturing system. Progressive decision making centers are located at the strategic, tactical and operational levels. The hierarchy starts at the top-most corporate strategic center and flows through to the bottom-most execution module, where materials are transformed to finished products: the plant floor.

This chapter presents a methodology for decision making or problem solving in large scale production systems. A framework for building a Hierarchical System to model large scale production systems and the execution of such a system to effectuate optimal decision and control are presented.

1.1 Scope of Production Management

Production planning is a complex decision making process, the end result of which is to balance the demand and supply of part types, quantities, due-dates and resource levels over a relatively large horizon. Scheduling involves time phasing the production plan, to generate a Gantt-chart of activities over a relatively short horizon. Control function consists of monitoring production activities, and reacting to small random disturbances in real-time.

Two distinct approaches to production planning (long and medium-term) have been adopted in the past. The first is a monolithic approach, wherein the entire problem is formulated as a large mixed-integer linear programming type problem. The second is a hierarchical approach which partitions the global problem into a series of sub-problems that are solved sequentially, such that the solution at each level imposes constraints on the subsequent lower level. The fundamental advantages of the hierarchical approach are: (i) reduction of complexity, and (ii) gradual absorption of random events that may appear at the job-shop level.

This chapter primarily addresses the production planning function of decision making and proposes a hierarchical approach to it. In the future, these concepts can be extended to lower and higher levels of production management.

The basic goal of the hierarchical approach in production management is to decompose the global problem into a hierarchy of sub-problems. For example, figure 1.1 represents a possible decomposition of the planning function. The long term level plans production of an aggregate product entity on the entire facility over a long horizon. The medium term level disaggregates the long term plan over a shorter horizon, leading to the production plan of product families on manufacturing cells. Finally, the short term level disaggregates the medium term plan over an even shorter horizon, to provide the production plan of individual products, on individual machines.

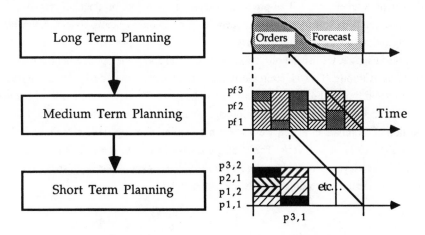

Figure 1.1: Hierarchical Production Planning

Planning decisions vary in terms of time-horizon, scope and focus, and have different impacts on the overall functioning of the manufacturing system. Time-horizon refers to the duration of time required for a strategy to have effect, or over which decisions are performed. In the context of production planning, a natural classification has been in place: short-, medium-, and long-term (or range) decisions. Short-term operations decisions may have an impact that can be measured in days or even hours. These include decisions regarding purchasing raw-materials, production and personnel scheduling, control of quality, equipment maintenance functions, short-term inventory control issues, etc. Medium-range decisions are those whose impacts can be measured with weeks and months. They include demand forecasting, employment planning decisions, decisions concerning distribution of goods, and setting company targets for inventory and service levels. Aggregate production planning is an important function accomplished at medium-term. This involves the structuring of a general plan for responding to forecasted demand through some combination of work force, output, and inventory loadings. Long-term decisions over a scale measured in years include capacity levels, timing, location and scale of construction of new manufacturing facilities. Introduction of new products or processes, and locating of facilities include long-term decisions. This time-horizon based decomposition of decisions does not assume a rigid structure in different applications. The terms short, medium and long term are merely relative in their time-scale. Depending on the context, they may refer to different functions or decisions as well as time horizons.

Another classification of decisions is proposed by Anthony [3]. This hierarchical taxonomy classifies decisions as: (i) strategic, (ii) tactical, and (iii) operational, based on their horizon and scope, as well as level of information detail, degree of uncertainty and level of management involvement, according to the most common practices in enterprises. Strategic decisions are defined as the decisions relating to long term marketing and financial policies as well as facilities design. Tactical decisions consist of deciding the work-force and over-time levels, as well as production rates of aggregate products. Typically, this involves aggregate capacity planning. Operational decisions typically concern short-term planning and detailed scheduling of parts.

In this chapter, we restrict ourselves to the tactical and some operational decisions as defined by Anthony. That is, strategic issues related to facilities design, and long-term marketing and financial policies, are assumed to be known or given. Furthermore, detailed operational decisions related to scheduling are not addressed. Alternatively, in view of the definition of Abraham et al [1], we address production planning (levels 2) through flow planning (level 3) in our work. The strategic decisions can be viewed as an additional adaptation type layer, that can be placed over the highest level of our suggested hierarchy, and that provide the constraints to apply to this hierarchy.

1.2   Issues in Hierarchical Production Management

The first step in this direction is to define the manufacturing system/environment to be addressed: (a) definition of the framework, (b) assumptions related to discrete parts manufacture, and (c) consideration of the most important random events. We restrict ourselves to the medium-term planning or tactical level decisions.

Some important issues involved in the **design** of such a hierarchy are:

(1) The appropriate number of levels, the related models, entities, definitions, planning horizons, domains etc.

(2) Aggregation schemes for determining higher level entities.

(3) Determination of horizon lengths for each decision making problem.

Some important issues involved in the **execution** of such a hierarchy are detailed below.

(1) Solution algorithms to resolve these optimization problems at each level have to be developed.

(2) Mechanism (or procedure) of top-down constraint propagation, i.e. how the solution of a higher level problem provides the constraints to subsequent lower level problem.

(3) Bottom-up feedback required for a closed loop real time control process. This consists of comparing the real inventory state with the solution of the planning problem, and providing the repair/failure state of the resources to the high levels of the hierarchy so that corrective action may take place if required.

In summary, in hierarchical architectures, the models at different levels are derived in a way that facilitates gradual planning decisions, with increased levels of detail. The computation of the control is a top-down process, which calls for the resolution of an optimization problem at each level of the hierarchy. The execution then involves the bottom-up feedbacks, and a revision of the plan is carried out if necessary.

This chapter is organized as follows. Section 2 presents the concepts of hierarchical decomposition of complex decision making problems, and a literature survey of related work in production management. Section 3 is devoted to the principles in hierarchical design, among which are: (i) the theory of aggregation of products and resources, (ii) temporal aggregation, and (iii) decomposition of controls. Section 4 presents the inputs to the hierarchical design process. The systematic design process is presented in section 5. Section 6 explains the operation of the hierarchy to perform production management. An application of this methodology to job-shop example is presented in section 7. Finally, in section 8 we draw our conclusions.

2. Hierarchical Decomposition: Basic Principles and Types

This section is devoted to the basic definitions and concepts involved in the hierarchical decomposition of complex decision making problems. The first part describes the principles of hierarchical decomposition, the major advantages, and classification of hierarchies. A literature survey of the major works in hierarchical production management, both from a historical perspective, as well as the current state-of-the-art constitutes the major contents of the second part of this section.

2.1 Motivation and Types of Hierarchical Decomposition Methods for Complex Decision Making Problems

A hierarchical decomposition can assume different meanings according to its context. However, the essential idea of all hierarchical decomposition schemes is the partition of a global problem into sub-problems. These sub-problems are either solved sequentially, such that the solution of a sub-problem imposes constraints on the subsequent sub-problem, or solved simultaneously in a coordinated fashion.

2.1.1 Advantages of Hierarchical Decomposition

The fundamental advantages of the hierarchical approach to complex problems are [14]:

(1) Reduction of complexity: Breaking a problem into sub-problems is a standard method of simplifying the solution process. The partitioning should be done in a way that either: (i) the interactions between sub-problems are acceptably weak (i.e., the sub-problems are as independent as possible), or (ii) the global problem is broken down into a series of sub-problems which are sequentially solved, in the sense that the solution of a problem at one level imposes constraints on the problem at the next lower level one; the global solution is obtained when all problems are solved. Thus, the initially intractable problem can be rendered pliable.

(2) Coping with uncertainty: Decision making in a system subject to random events and uncertainty could lead to frequent recomputations; monolithic models would require the entire problem to be resolved repeatedly, while the hierarchical approach can gradually absorb random events without the need to resolve any higher level problems. This results in large savings in computational burden, apart from added stability to the overall control. Decisions at various levels in the planning process are made at different points in time. Higher level decisions are more aggregate, and need not explicitly consider uncertain data at detailed levels. Random events with a lesser impact on the system can be absorbed at lower levels.

(3) Parallel with hierarchical organization of the physical systems: Hierarchical planning or decision making is often (though not always) performed parallel to the

organizational structure of the physical system at hand. This has important implications from the organizational aspects and personnel hierarchies.

(4) Reduced need for detailed information and better forecasting: Higher levels in hierarchies are more aggregate and do not require detailed information; this not only reduces dimensionality, but also allows a longer "look ahead" capability. Forecasting is usually easier and more accurate for aggregates than for detailed entities.

2.1.2 Basic Types of Hierarchies

Mesarovic *et al* [44], identify three types of hierarchical systems, which in a sense represent a classification of existing hierarchical systems.

Descriptive Hierarchies:

For the complete and detailed description of complex systems, while retaining simplicity in description and behavioral aspects, descriptive hierarchies are employed. A system is described by a family of models, each concerned with its behavior as viewed from different levels of abstraction. Models must be independent of the functioning for an effective description of a *"stratified system."* The levels of abstraction are referred to as *strata*. Each stratum in the hierarchy is associated with a different set of relevant variables, which allow the study to be confined to one stratum only. Understanding the system increases by crossing the strata: in moving down the hierarchy, one obtains a more detailed explanation, while in moving up the hierarchy, one obtains a deeper understanding of its significance. Such hierarchies are primarily employed for system description and detailing function behavior. They usually do not support decision making aspects or control, hence their applications are limited to system modeling and description.

Multilayer Hierarchies:

These hierarchies, also known as decisional hierarchies, appear in the context of a complex decision-making process. Complex decision-making is associated with a fundamental dilemma: on one hand, there is a need to take timely and prompt action, while on the other, there is an equally great need to understand the situation better and retain perspective of the consequences and relationships of actions in a complex situation. This problem may be resolved by multilayer hierarchies, where one defines a family of decision problems whose solution is attempted in a sequential manner, in the sense that the solution of any problem in the sequence imposed some constraints to the subsequent problem. This has to be done in a way that the latter is completely specified and its solution can be attempted. The solution of the original problem is achieved when all sub-problems are solved. Each decision-making unit receives information from the system. The output of a unit, say D_1, represents a solution, or the consequence of a solution, of a decision problem which depends upon a parameter fixed by the value of the input X_1, which in turn is the output of a unit on a higher level. Such a hierarchy of decision layers is termed as a multilayer hierarchy.

Multilevel Hierarchies:

These hierarchies are also known as organizational hierarchies. It is necessary that: (i) the system consists of a family of interacting sub-problems which are recognized explicitly, (ii) some of the subsystems be defined as decision (making) units, and (iii) the decision units are arranged hierarchically, in the sense that some of them are influenced or controlled by other decision units.

Designing such a system consists in the assignment of the tasks or roles which various levels or individual units have to perform. Decomposition of the overall systems task is performed, by assigning a sub-problem to each decision-making unit (infimal unit). Infimal units perform their decision-making independently, though *coordinated* through a supremal unit. Coordination is concerned with the existence of supremal control under which infimal units can solve their local control problems. The supremal-infimal (or master-slave) relationship exists between each consecutive echelon of the hierarchy.

2.2 Literature on Hierarchies in Production Management

Literature surveys in the field of hierarchical production management can be found in Gelders and VanWassenhove [23], Dempster, Fisher *et al* [14] and Libosvar [40]. The work surveyed here is classified as: (i) hierarchical production planning related literature, and (ii) hierarchical scheduling in FMS, AI approaches, and other conceptual architectures.

2.2.1 Hierarchical Production Planning

The work of Hax and Meal [29], developed at M.I.T. in the seventies has been considered a substantial contribution in the area of hierarchical management. Production planning is the major concern of this work. Hax and Meal [29], describe a hierarchical model designed for a particular implementation, but also provide a general direction to subsequent work. Four decision levels are considered for a multi-plant firm. The highest level model decomposes the problem into decoupled single plant problems, by determining the products to be produced in each plant. The subsequent hierarchy is intended for the management of a single plant. The highest level involves static (one time) decisions of a decoupling nature. Thus, it differs from the rest of the hierarchy, which is typically multi-layered for dynamic decision making. We restrict our attention to these lower levels. Bitran and Hax (1977) [8] discuss the hierarchical framework for production planning problems.

Based on the analysis of the production process, and a typical cost structure, three levels of aggregation are considered for products:

1. *Items* are final products to be delivered to customers.

2. *Families* are groups of items which share a common manufacturing set-up cost.

3. *Types* are groups of families that have similar production costs and similar seasonal patterns.

A production planning level is consistently modeled according to the previous aggregation scheme. Higher level decisions impose constraints on lower level actions, which in turn provide necessary feedback. In this top-down constrained approach, the assumption is that higher-level decisions have a more significant impact on the objectives.

The highest level of the hierarchical planning system allocates production capacity among product types by means of a linear programming based aggregate planning model. The horizon is normally longer than a year (15 months), in order to take into account demand fluctuations of products. Production and inventory holding costs are the only costs considered; set-up costs are ignored at this level. This model is intended to determine an optimal trade-off between inventory holding and overtime (production) costs, while leaving set-up costs out because of secondary impact. At this and all subsequent levels, the concept of a rolling horizon is employed, in order to perform repetitive open-loop optimization. The second step in the planning process is the disaggregation of production quantities for types (computed by the higher level) to determine the production quantities for families. This is performed by a heuristic, based on Economic Order Quantity (EOQ), and safety and overstock computation techniques. Capacity allocated to a product type is split among the families belonging to that type, for which the inventory is below the safety threshold; production volumes are chosen as close as possible to the EOQ. Finally, the third decision level consists of an item disaggregation heuristic, based on Equalizing of Run-Out Times (EROT). This technique maximizes the time between set-ups, by requiring the items of a family to run-out at the same time. Karmarkar [36] proves optimality of this technique.

Consistency between decisions at different levels is attempted to be ensured by the top-down constraints. However, sometimes the disaggregation is infeasible, i.e. top level constraints may yield an empty feasible set at a lower level. Gabby [20] considers an aggregate plan for which a feasible detailed plan (i.e. one without backlog) exists. However, this requires detailed demands to be known over the consistency horizon, thus, loosing some benefit drawn from the need of less detailed demand required in hierarchies. He derives necessary and sufficient conditions for a consistent disaggregation; see Gabby [20] for treatment of multi-item, single echelon, capacitated production problems.

Bitran and Hax (1977) [8] propose a computational improvement to Hax and Meal's model by reformulating family and item disaggregations as knapsack problems, for which they provided efficient solution algorithms [9]. Numerical results indicated that for low set-up costs, the solutions are near optimal. Bitran, Haas and Hax (1981) [6] prove that the EROT method is an optimal disaggregation scheme that minimizes the cost of initial inventory. This work improves the knapsack-based method in that: (i) at the family disaggregation level, the expected number of set-ups are minimized for a shorter horizon, allowing for greater responsiveness to seasonal variations, (ii) a "one step look ahead" procedure is required for disaggregation, and (iii) the families' production volumes are modulated in order to keep them close to their EOQs. The enhanced work was shown to produce superior results through a set of simulations.

Erschler, Fontan and Merce [17] derive necessary and sufficient conditions for consistency, based on mass balance equations. They propose the look ahead procedure to be extended to all periods, for which detailed demand is known. The necessary conditions are then employed to enhance the knapsack method, which was shown to perform better.

Graves [26], adopts a different approach to the problem and introduces feedback between decision layers. Based on the product aggregation scheme of Hax and Meal, the problem is formulated as a monolithic mixed integer program, which is decomposed by Lagrangean relaxation. This decomposition yields two subproblems: (i) aggregate planning (linear programming model), and (ii) disaggregation problem (lot sizing problems for each product type). The linking mechanism for these two subproblems is an inventory consistency relationship which is priced out by a set of Lagrange multipliers. The best values of the multipliers are found by an iterative procedure which may be interpreted as a feedback mechanism in the Hax-Meal framework.

Other extensions of the Hax-Meal model are directed towards incorporating it to multi-stage fabrication and assembly systems. Bitran, Haas and Hax (1982) [7] propose an extension of their previous work to a two-stage fabrication/assembly system. On an industrial test-bed, they successfully compare results obtained by the extended hierarchical planning system to those of an MRP system. However, in this model product families are subject to a rather restrictive definition. Apart from the advantages, some shortcomings of the Hax-Meal framework are summarized as follows [40]:

- The product aggregation scheme proposed is only relevant to a particular class of production systems, and models at each level are based on a typical cost structure.
- Detailed data requirement are reduced only if backordering is permitted.
- No randomness is taken into account and forecast errors have to be absorbed by safety stock.
- No spatial decomposition of the system is proposed.

MEIER's Hierarchical Control of a Production System

Meier [43] considers planning and control of a general multi-stage assembly/ disassembly manufacturing system to minimize a performance criterion composed of inventory holding costs, backlogging costs, and production costs. A two-layer "master" and "slave" hierarchy is developed. The master-level determines the amount of operations (per operation type) to be performed on machines, over given planning periods, via a linear programming model. For the treatment of large-scale problems, aggregation-disaggregation procedures are proposed. The slave controls the activities of the machine in real time, in order to satisfy the plan provided by the master. At this level, a set of priority rules are employed to determine the activities of the machines, as a function of the volumes of operations, and the inventory state. Discrepancies are possible if the slave cannot execute the master's decision precisely. This is mitigated by rolling horizons and repeated optimization. In this way, a close loop planning and control system is emulated, whereby the schedule on the rolling horizon yields to auto-adaptive planning corrections.

Hillion, Meier and Proth [33] present a top level model for a new approach to hierarchical production planning. At this level, the entities of relevance are production subsystems and part families. Aggregation procedures to determine these entities are detailed, and the problem is formulated as a linear program.

Optimality and Aggregation-Disaggregation Issues in Hierarchies

The suboptimality of the hierarchical approach has been addressed by Dempster, Fisher et al [14]. They demonstrate that multi-level decision problems can be modeled as multi-stage stochastic programs. Then, the analytical evaluation of any hierarchical system can be assessed with respect to the optimality of this stochastic program. An analytical evaluation, and a proof of asymptotic optimality of a simplified version of the hierarchical planning system of Armstrong and Hax [4] are presented. The job-shop consists of a set of parallel identical machines; the higher level decision is to determine the optimal number of machines, m. The lower level is concerned with the problem of scheduling n jobs on these m machines, in order to minimize the completion time. There are costs associated with purchasing machines, and costs proportional to the completion time. The performance of this hierarchical system is compared with a stochastic model, in which both costs are treated in a single model, and processing times are random with a known mean value. When the number of jobs tends to infinity, the performances of the two systems become identical. The application of such analytical results is dependent on the criteria and the model chosen. Also, owing to lack of accurate estimation of the quality of solutions provided by hierarchical systems, this evaluation method has found little application to other systems.

Aggregation-Disaggregation is another major issue concerned with the design of hierarchical systems. Infeasibility can be encountered during the disaggregation process. Krajewski and Ritzman [39] provide a survey of the problems and research in this field for manufacturing and service organizations. These disaggregation problems are encountered between aggregate planning at the top level and more detailed decisions of inventory control and scheduling at the bottom level. They recognize the lack of an interface mechanism, which diminishes the utility of the solution procedures of aggregate planning, inventory control and scheduling.

Aggregation schemes are of three types: (i) over time, or temporal, (ii) over products, and (iii) over machines, or spatial. The Hax and Meal framework only considers product and temporal aggregation. Except Meier [43], who includes aggregation of machines, there is no work addressing aggregation in all three dimensions.

Zoller [59], considers product disaggregation with a two-level economic model, in which aggregate production is determined by minimizing a cost function. The product mix and sales price are determined at the lower level in order to maximize profit, assuming that the demand volume depends on sales price and that the function relating the two variables is known. An iterative algorithm to reach optimal solutions of the low-level problem is presented. Gelders and Kleindorfer [21] [22], present a formal model of a one-machine job-shop scheduling problem with variable capacity, considering trade-offs between overtime and detailed scheduling costs. The scheduling problem considers minimizing the sum of weighted tardiness and weighted flow-time costs for a given capacity plan. Various lower-bounding structures for the problem are analyzed, and a branch-and-bound scheme is outlined. Computational experience with the algorithm in presented in [22].

Erschler, Fontan and Merce [17] consider the consistency of the disaggregation process in hierarchical planning, and necessary and sufficient conditions for consistency of decisions in a two-level structure are presented. A sub-set of these conditions improves upon the disaggregation procedure of Bitran and Hax [8], [9].

Axsater [5] addresses a double aggregation over products and machines, which results in an aggregate planning problem in terms of product groups and machine sub-systems at the aggregate level. This aggregate plan may not be possible to disaggregate. He presents perfect aggregation conditions, i.e. necessary and sufficient conditions on the aggregation matrices, where it will be possible to disaggregate any aggregate plan. If perfect aggregation is not possible, an approximate solution can be reached through a mathematical formulation of an approximation problem.

2.2.2 Hierarchical Systems for Flexible Manufacturing

Flexible Manufacturing Systems (FMS) offer challenging planning and scheduling problems due to their versatility, flexibility, quick changeover time, and association of automated material handling systems along with the need for efficient resource utilization (brought about by high capital investment). Morton *et al* [47] and Stecke [54] identify some issues that render FMS scheduling different from traditional manufacturing systems. Thus, the need for different principles in FMS design, planning, scheduling and control. Stecke [54], identifies FMS related problems, and suggests mathematical models useful to their analysis.

Conceptual Models in Production Management

O'Grady [49] provides an introduction to automated manufacturing systems with their characteristic features. Production Planning and Control structures of traditional and automated manufacturing systems are identified. Three major hierarchical frameworks: AMRF's (Automated Manufacturing Research Facility of the NIST), CAM-i (Computer Aided Manufacturing International Inc.), and one developed by their team have been described. None of these architectures have been implemented in their entirety.

The National Institute of Standards and Technology (NIST) within its Automated Manufacturing Research Facility (AMRF), has developed a five level hierarchical structure: Facility, Shop, Cell, Workstation, and Equipment levels of control. This decomposition is functional in nature [34], and is fixed (i.e. the structure is rigid and not system dependent). Their efforts [13], [35], focus on the design of a real-time production scheduler at the shop and cell levels, but not yet to the design of an integrated planning and scheduling hierarchy for the management of the entire factory.

Artificial Intelligence Based Approaches

Villa *et al* [55] propose a hierarchical framework to model FMS control. They define an FMS as a structure composed of a physical system, an information system and a decision-and-control system. The task performed by the latter can be divided into periodic planning and event-driven control. The authors suggest a decomposition, based on physical insights rather than mathematical techniques, to develop a tree-like structure with decision makers at each node. They present a hierarchical control structure by integrating mathematical tools from Control Theory with relational tools derived from Artificial Intelligence. This matching is suggested to provide an effective implementation of Expert Control System. This framework leading to an "Integrated Control Structure" is based on the same spatial decomposition of the physical system and frequency-band partition of events, whereas the decision-making units use AI tools to solve these problems.

Shaw and Whinston [52] address the application of generic artificial intelligence techniques to the planning (process planning) and scheduling of flexible manufacturing systems. The complex scheduling problems in FMS are made tractable by nonlinear planning. Planning for conjunctive goals is referred to as nonlinear planning because the resulting plans are partially ordered. Nonlinear planning systems seek to break a problem into sub-problems, using the divide-and-conquer strategy, while taking interactions of sub-plans into account. They employ a two-level hierarchy to decompose an n-part-m-machine scheduling problem into n sub-problems, with each sub-problem defined as the routing of one part. The primary interactions between the sub-problems are their sharing of m machines. The objectives are to minimize makespan and avoid conflicts.

Doumeingts (1986) [15] identifies the significance and potential application of AI techniques in the field of CIM. The GARI system for process planning (University of Grenoble) and the ISIS (Intelligent Scheduling and Information System) system in scheduling (Carnegie Mellon University - Robotics Institute) are detailed below, and referenced as major contributions in this field.

PATRIARCH is a hierarchical planning and scheduling system developed at Carnegie-Mellon University. In [47], Morton and Smunt highlight issues relating to FMS scheduling, which should integrate: (1) hierarchical structure, (2) decision support capability, (3) advanced knowledge representation, and (4) accurate practical large-scale heuristics. The four levels of the PATRIARCH system include: strategic planning, capacity planning, scheduling, and dispatching.

Further, the Intelligent Systems Laboratory of the Carnegie-Mellon Robotics Institute has developed large scale scheduling systems with sophisticated knowledge representations (Fox [18] [19]). These systems, namely ISIS - a job shop scheduling system, and CALLISTO - a project scheduling system, attempt to integrate all levels of hierarchical production planning. The original work of ISIS utilized a reservation system and backward scheduling with beam search, which were only moderately successful for loading of machines. CALLISTO uses a forward dispatch approach, improved in Morton, *et al* [47].

The FMS version incorporates many of the concepts of CALLISTO and ISIS, along with the consideration of multiple machines and dispatching priorities from detailed planned schedules.

The ISIS was followed by the development of a successor system called OPIS (Opportunity Intelligent Scheduler [53]). OPIS employs constraint propagation techniques to update schedule descriptions and detect inconsistencies introduced. These systems were tested using simulated data from actual manufacturing environments.

Scheduling an unreliable FMS

Hildebrant [30], Hildebrant and Suri [31], present a methodology and a multi-level algorithm for scheduling and real-time control of Flexible Manufacturing Systems (FMS) with unreliable machines connected by automatic transportation means. The control consists of satisfying a given demand of part types in order to minimize the makespan (production cycle time). The hierarchical decomposition of the problem leads to a three level hierarchy. At the high level, a multi-class queuing model is used to compute the completion time of tasks as a function of the utilization of machines and fixtures, which accounts for the possibility of failures. The solution results in the allocation of resources to parts and routes. The intermediate level, Even-Flow Algorithm, consists of a dynamic programming algorithm which sequences parts into the system so as to minimize deviations of these discrete inputs from the assumed homogeneous stream (of the high level). The lower level, consists of simple dispatching rules in order to determine the sequence of parts on all machines, given the loading sequence from the previous level.

Kimemia [37] and Kimemia and Gershwin [38], present a Multi-layer hierarchical control algorithm for the optimal control of a stochastic FMS. The manufacturing system consists of machines capable of processing a variety of parts, and are subject to failure. Changeover times are assumed negligible. The problem consists in meeting production requirements while the machines fail and are repaired at random times. The failure/repair process is assumed to be memoryless, hence the machine state is modeled as a Markov chain. Part production requirements are stated in terms of a steady rate. This, along with the assumption that the mean time between changes in machine state is longer than the operation processing times, enables a continuous model for part flows. Under these assumptions, a three-level controller is devised. Additionally, there exists a highest level of the control scheme for off-line calculation of the control policies or decision tables (to be used at subsequent levels), which completes a long-term feedback loop when the system parameters change. This resembles adaptive control.

At the highest level resides the Flow Control Level that determines the short-term part production rates based on a hedging point strategy. The demand, level of downstream buffers, reliability of work stations, and sharing of resource capacity are taken into account in the process. Statistical estimates of failure/repair, anticipating downtimes, are incorporated in this model. The intermediate level is the Routing Control Level that calculates the route splits, i.e. the proportion that entering parts must follow among the alternative paths possible for processing. The objective is to meet the production rates dictated by the high level, while minimizing congestion and delay within the system; the system is modeled as a network of queues. The lowest level of the control, the Scheduling Controller, is composed of dispatching rules to load parts into the system. The objective is to maintain the flow rates specified by the route controller.

Maimon and Gershwin [41] modify the flow control problem to allow for the consideration of alternative routings at this level. The consideration of loading and routing together eliminate the need for the intermediate level. This basic methodology, was extended by Akella, Choong and Gershwin [2], and tested on a printed circuit card assembly line at IBM to report superior performance. Gershwin (1986) [24] proposed an additional layer to the hierarchy in order to determine the set-up frequencies. The generalization of the entire work to address general events relating to production system, with a time-scale decomposition (or frequency separation) is presented by Gershwin (1988) [25]. A hierarchical structure based on the characteristics of the production system is suggested. The levels of the hierarchy correspond to classes of events that occur with distinct frequencies.

Xie [57] extends Gershwin's work to address non-identical parallel machines. Quadratic approximation of the cost functions are used, and a new technique to compute the parameters is proposed. Extensions to incorporate machine failures of a wide range are presented in Xie [58]. The failures are clustered near some discrete points on the failure spectrum, in order to define the hierarchical model. Each level in the hierarchy corresponds to a discrete point on the failure spectrum. Thus, a hierarchical controller of a multiple time scale type is proposed. Further, the concept of system configuration is introduced in [58]. In a particular configuration (or set-up of the entire set of machines), the system can produce a sub-set of part types to be produced. For the management of such systems, a three level hierarchy is proposed: (1) Sequencing of Configurations, (2) Flow Control, and (3) Real-time Sequencing of parts.

This class of work primarily addresses the scheduling and control aspects of manufacturing systems. The downstream demand is assumed known and continuous. Hence, the production planning aspects are not addressed. It has a sound theoretical foundation, but seems difficult to be extended for larger and more complex manufacturing systems that can be found in practice.

2.3 Motivation for Generalized Hierarchical Design

The progress in the field has been promising. The work done thus far has the following general deficiencies: (i) aggregation schemes are relevant to a particular class of production systems, (ii) consistency issues remain unresolved for the most part, (iii) randomness is generally not addressed, (iv) the architecture, relating to the number of levels and models at each level are too rigid, or (v) the models may be developed for very small problems of academic interest. The literature has failed to address the issues relating to a systematic hierarchical design as relevant to the specific production management problem at hand. However, there exists a fair scope of extension to the ideas proposed and development of new ideas. This chapter presents an approach to the Design of Hierarchical Production Management Systems.

3. Principles in Hierarchical Design

In general a complex global decision problem, D^o, represented by its model M^o, can be approached by a n-level hierarchy. The representation of this n-level hierarchical system is presented in figure 3.1. The hierarchical approach to decision making decomposes the global problem, D^o, in a series of sub-problems D^i. Each D^i is solved at level i with a model M^i. The solution to the global problem is obtained when the individual problems, i.e., n through 1 of the hierarchy are solved sequentially.

In order to systematically outline the procedure for constructing a hierarchy, some preliminary remarks are necessary. Each level is associated with a model and a decision making problem. A model consists of a set of entities. An entity is a symbol which represents a physical or conceptual object. Each entity is associated with a set of attributes. Each attribute (of an entity) can be assigned a value (constant or variable) from a value set associated with it. Entities may be related to each other. An entity is usually an identifiable "object," e.g. machines, parts; it may also be a less perceivable "object," like part families or aggregated machines. A relationship is an association among entities. For instance, an operation in a production process relates two entities: a machine, and a part. Attributes are used for characterizing an entity, e.g., the contents of a buffer is an attribute of the entity "buffer." Attributes can be qualitative or quantitative variables or constants. The value of the attribute "content" can be assigned a value from the value set $V = \{1,2,3,4,5\}$, if the buffer has a maximum capacity of 5 discrete parts. Thus, a typical entity-relation type model similar to Chen [10], and Hilger [32], can be adopted.

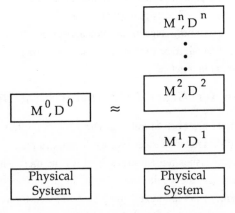

Figure 3.1: n-level Hierarchical System

Decision making consists of determining a set of optimal controls from a set of feasible controls, such that, the specified tasks or constraints are verified over a horizon, and some criteria are optimized. Thus, it requires to specify: (i) a set of constraints specified by the upper level (strategic constraints in the case of the highest level), (ii) a set of feasible controls (decisions) that can be applied, (iii) a set of criteria or objectives that are to be considered (a quantitative criterion or multi-criteria), and (iv) a decision or optimization horizon. More formally, we mathematically define the decision making problem D^i as a four-tuple: (T^i, U^i, C^i, H^i), where:

- T^i is a set of constraints specified by the upper level, i+1 (strategic constraints in the case of the highest level, n).

- U^i is a set of feasible controls (decisions) that can be applied.

- C^i is a set of criteria or objectives that are to be considered in D^i.

- $H^i \in R^+$, is the horizon of D^i. It is the period of time over which D^i has to be solved.

Decision making can be viewed as an optimization problem that consists of selecting a control X^i, $X^i \in U^i$, such that, the tasks or constraints T^i are verified over the horizon H^i, and the criteria C^i are optimized. We then say that X^i is the optimal control, or in other words, X^i is an optimal solution of D^i.

It is important to indicate that the decision making problem D^i and the model M^i are highly related. For example: (i) the criteria relevant to a level should be a function of the attributes of the entities in M^i, (ii) the values of some attributes of some entities should be modifiable within the horizon H^i by the application of a control Y^i, $Y^i \in U^i$, (iii) the value of a criterion should be modifiable within H^i also by the application of a control Y^i, $Y^i \in U^i$, etc.

In this section, we present the principles that are employed to construct a hierarchy of such sub-problems as a means to global decision making. The decomposition principles help in decomposing the overall problem into sub-problems, while aggregation principles help obtain more aggregate models from detailed ones. Each sub-problem bears a structure of model and structure of decision making problem. We base the design of the hierarchy on these principles, which will be presented later in section 5.

3.1 Aggregation Issues

Aggregation-Disaggregation is a major issue concerned with the design of hierarchical systems. Infeasibility can be encountered during the disaggregation process of aggregate production plans. Krajewski and Ritzman [39] is a survey of the problems and research in this field from a manufacturing, as well as a service organization perspective.

Aggregate production planning is essentially performed to decide resource/work force levels, the fashion in which the company reacts to demand forecasts, and planning required to allow for changes in the product mix. Translating demand forecasts for a wide range of products into resource requirements is a difficult task, which is further complicated by the uncertainty of demand forecasts. Owing to these complexities, production planning is performed hierarchically. Aggregate production planning can be applied to any level of the hierarchy, where it is designed to address product families (groups of products) with different degrees of detail or level of aggregation. The higher levels in the hierarchy are more aggregate, in fact very often in practice the highest level considers a single aggregate measure, like dollar value at cost, surface, volume, weight, to represent the entire product line. As we progressively go down the hierarchy, the level of aggregation of the product groups decreases, and more detailed planning decisions are performed. Very often, spatial aggregation, i.e. aggregation of production facilities is also performed in addition to the aggregation of products.

What the appropriate aggregating schemes should be is not always obvious in practice. It depends on the context of the particular planning problem, the range of products, and the level of aggregation required. The scheme is very often chosen based on a typical cost structure (see Hax and Meal [29]), which may not be applicable to the entire gamut of production topologies. From the point of view of resource level requirements, it is natural to consider products having similar processing requirements, so that they can be well represented by a common entity. Such approaches have been applied in practice as well as adopted in the literature (see Meier [43]). Most of these approaches tend to ignore the other attributes of products like holding cost and backlogging cost which are essential in the optimal allocation of resources among competing products. Very often in practice, optimality is not addressed explicitly; the production planning process is performed based on experience and/or with the help of support systems like MRP II. However, with industry having to face stiffer competition, the need for better production planning and allocation of resources is becoming increasingly important.

Nagi [48] presents the underlying aggregation theory. A two-level hierarchy is developed for holding and backlogging costs employing this aggregation scheme, and optimality is demonstrated in a particular case. In the general case, the product and machine aggregation is summarized as follows. In addition, temporal aggregation is also performed (see next section).

Products are aggregated into families by a modified version of the K-mean algorithm in cluster analysis (Hartigan 1975 [28]). Product entities are represented in IR^{m+c} (m+c dimensional real space) by a point. The axes represent the attributes of the product entities; m axes are used to represent the processing time required by the product entity on the m machine entities, and c axes are used to represent the attributes relevant to the criteria (at that level). For instance if the criteria are earliness and tardiness, c equals 2, and the axes represent per unit holding cost and backlogging cost, respectively. Each point is weighted by the long term production volume of the corresponding product. Then, the K-mean algorithm is employed to determine the clusters or product families. The modification to the K-mean algorithm is that clusters are permitted iff the Euclidean distance from the center of a cluster to its corresponding points does not exceed a user specified parameter, σ. This parameter is helpful in controlling the compactness of clusters. The attributes of the product family can be computed by the center of gravity of the cluster.

The advantages of aggregating parts into families in this manner allow for some uncertainty in the demand to be absorbed. The aggregate product family can be substituted by any product belonging to this family; this reduces variances, while retaining similar processing durations and holding costs in the schedule. Furthermore, it reduces the level of detail required for future production periods, i.e. forecasts in terms of aggregates are sufficient, and detailed product forecasts are not required.

Aggregation of machines into aggregate machines is also performed by the modified K-means algorithm. In this case, machines are represented in IR^N by a point. N axes are used to represent the processing time required by the N families on a machines. Then, the K-mean algorithm is employed to determine the clusters or aggregate machines, provided the Euclidean distance from the center of a cluster to its corresponding points does not exceed a user specified parameter. The processing time of a family on a cell is the maximal processing time of that family on the component machines of the corresponding cell. Worker entities can be aggregated according to the machine aggregation.

3.2   Temporal Aggregation

To clarify the term, an example of temporal aggregation in a two level hierarchical production planning structure is shown in Figure 3.2. The high level plans monthly production for all product entities over the entire planning horizon (for example - fifteen months), and the low level disaggregates the monthly results to a weekly plan for the product entities for the first month.

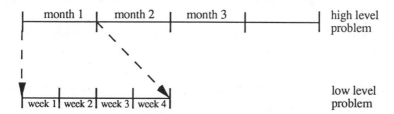

Fig. 3.2: Temporal aggregation in a two level hierarchical structure

There exists very little literature on the temporal aspect of aggregation in hierarchical production planning systems. Abraham *et al.* (1985) [1], have suggested the need of temporal aggregation in production planning and control of manufacturing systems, and Nagi [48] has indicated the need for a temporal aggregation scheme, where similar input variables (product demand) are aggregated on a time scale. So far, to our knowledge, a general temporal aggregation scheme for production planning systems has not been addressed in the literature.

In a recent paper, Harhalakis *et al* (1993) [27] address the problem of temporal aggregation in production planning. A single facility with multiple part types is considered. The planning horizon consists of a sequence of elementary time periods, and the demand for all part types is assumed to be known over these periods. The production planning problem consists of minimizing the holding and backlogging cost for all part types. Due to usual errors in demand forecasting, and due to the large size of the linear programming problem commonly encountered in such problems, there is a need for aggregating the production variables over the time horizon (typically, from weekly to monthly) to result in a hierarchical structure. They consider a two-level hierarchy composing a sub-problem at each level, and propose an iterative technique which solves these sub-problems in sequence. *A posteriori* bounds are developed, which are useful in evaluating the performance of the iterative algorithm. Quick lower and upper bounds of the original problem are also developed. Numerical results for numerous test cases are also presented.

In view of the lack of literature on the temporal aspect of aggregation in production planning systems, the application of the temporal aggregation in some other problem domains is reviewed below.

Engles, Larson *et al.* (1976) [16] consider the hydro and thermal generation scheduling problem using an iterative dynamic programming procedure. The output of the monolithic problem is the hourly resource allocation for a scheduling horizon of at least one year. A two-layer hierarchical structure is considered; the high layer which schedules all resources weekly over one year, and the lower layer which disaggregates the weekly results to schedule the resources hourly for one week. These low and high layer horizons are selected because the demand is characterized by weekly cycles and the maintenance of resources and the decisions of acquiring a new resource are made yearly. The technique of successive approximations is used, whereby each resource is rescheduled at the low layer, and then the high layer problem is solved. Since the objective function is convex, the iterative procedure converges to the optimum. However, this procedure is problem specific, and cannot be applied or extended easily to general hierarchical production planning problems.

Monts (1991) [45] proposes a temporal aggregation scheme to evaluate the marginal energy cost ($/mega watt-hour). The horizon for analyzing the given hourly marginal cost was set to one year. Due to the temporal structure of the marginal costs, aggregation along a time scale was performed according to the type of season (Winter, Spring, Summer or Fall) and the day type (weekday or weekend). This aggregation of the hourly marginal cost is not static in nature, and depends on the variance of the hourly marginal cost over a day. Both, two-layer and three-layer aggregation schemes are proposed, based on the hourly marginal cost variation. Monts's scheme is also problem specific, and does not always lead to the optimal solution.

Wei and Mehta (1980) [56] addressed the information loss due to temporal aggregation in a dynamic system where the influence of an input variable on an output variable is distributed over several time periods. These models are known as distributed lag models. A proper time scale for the temporal aggregation is proposed, based on the time horizon of the aggregated information required. A Monte Carlo study on some commonly used forecasting least square estimators was presented, and the information loss in terms of the prediction of the input variables was examined.

Cunningham *et al.* (1992) [12] examine the effect of temporal aggregation on money and interest rates. Aggregation was performed based on time averaging of the input variables, which was observed to cause a significant magnitude change in the output variables. Rossana *et al.* (1992) [50] consider the modeling of real wages of the manufacturing industry, and present the effect of temporal aggregation on the output variables.

3.3   Decomposition of Controls

In this section, we present the principles that are employed in decomposing the overall problem into sub-problems, i.e. in the construction of the hierarchy. Time-scale decomposition is a technique developed for the analysis of dynamic systems in which different components of the state vector have very different dynamics. In this decomposition, the modes of the system are partitioned into classes, in such a way that each class is either fast or slow, with respect to the other classes (see Chow and Kokotovic [11], and Sandell et al [51]). The literature in control theory essentially treats multi-level hierarchies. In the multi-layer structures related to hierarchical management systems, the controller is decomposed into algorithms operating at different time intervals. All the layers of the controller directly affect the process but the higher ones control its slower aspects only: they intervene less frequently, with longer optimization horizons, and are based on more aggregate models. The variables manipulated at the higher layers are more aggregate. Unfortunately, this technique has not been developed substantially in the multi-layer literature.

Gershwin (1988) [25], employs the frequency separation principle as the central idea of hierarchical decomposition of production scheduling. As in other multiple time-scale systems, at one end of the scale, there are quantities that are treated as static, and other variables are divided into groups (classes) according to their speed of dynamics. Owing to this grouping, it is claimed that computation of the behavior of these systems is simplified. The essential idea is: when dealing with any dynamic quantity, treat quantities that vary slower as static; and model quantities that vary faster in a way that ignores the details of their variations (represented by averages).

In our methodology, we employ similar concepts of time-scale or frequency decomposition, blending it with the aggregation aspects at the higher levels (level ≡ layer) of the hierarchy. This is intended to also address the planning related hierarchy, that requires a broader spectrum of activities (of different time-scales) to be considered. Higher levels of the hierarchy are more aggregate (smaller dimension), allowing for longer horizons and elementary periods associated with them. This enables the higher levels to address slower aspects of the system over longer horizons, which would not have been possible without this model reduction.

Thus, in our multi-layer structure for hierarchical production management systems, the controller is decomposed into algorithms (levels) operating at different time intervals. Higher levels control the slower aspects of the system (i.e. address activities of longer duration), they intervene less frequently (because they are concerned with less frequent random events), with longer optimization horizons, and are based on more aggregate models. Progressively, the lower levels address faster aspects of the system over shorter optimization horizons (and associated elementary periods), while becoming more detailed.

While the design of the hierarchical controller is based on controllable activities (controls) of different time-scales, it is also intended to address uncontrollable activities. These uncontrollable activities, also are termed as random events, are of different time-scales too. Each level of the controller, treats activities (controllable of uncontrollable) with longer durations (slower) as static, and treats activities that vary faster in a way that ignores their variational details by representing them as averages. Finally, the remaining activities (not too fast or too slow) of comparable durations are addressed at a particular level. The controllable activities are planned at this level (controls), while the uncontrollable ones are absorbed. As will be shown later, this decomposition of activities or controls also directly impacts on the calculation of the attributes of the entities at the different levels of the hierarchy.

4. Inputs to the Design Process

Production planning problems differ in nature and complexity. They depend on the manufacturing system under consideration as well as the characteristics of demand. Issues relating to the manufacturing system concern: i) dimensionality, ii) type/characteristics of products and production methods, and iii) characteristics and disruptive stochastic events associated with the resources. Manufacturing systems with a large number of machines and a variety of parts tend to present more complex planning problems. Product characteristics like number of levels in the Bills-of-Materials (BOM), number of component parts types, routings, and overall lead times, have significant impact on the complexity of the planning problem. Type of production, e.g. process, mass, batch, or jobbing production has also relevance in this regard. Furthermore, the randomness of work-center failures and labor absenteeism are some other important factors.

Issues associated with the demand characteristics are: i) dynamics of the demand, e.g. trends, seasonality, ii) stochastic nature of demand, and iii) accuracy, nature and horizon of forecast. Changes in product mix, cyclicity, seasonality and trend of the demand are the demand dynamics that effectuate complexity in the planning process. On the contrary, stable and regular demand does not require sophisticated planning tools. Randomness of demand, uncertainty, frequent order cancellations, changes and expediting also impact on the complexity of the problem. Finally, the length of the horizon over which forecasts are provided, and their accuracy play an important role in planning. Forecasts that are provided in the form of aggregates require conversion/translation to detailed production. Thus, planning decisions are usually performed hierarchically.

In view of the above comments, it seems that no single planning architecture can be employed for all planning problems. The architecture of the planning system, the number of levels in the hierarchy, the criteria of relevance, and the control to be applied are problem dependent.

The design of the planning hierarchy requires a variety of information specifying the manufacturing system as well as the managerial goals and decisions. In this section, we present a fairly comprehensive list of characteristics that attempts to encompass a large gamut of manufacturing and management systems. By this, we do not claim it to be exhaustive. Nonetheless, depending on the nature of the manufacturing systems and assumptions relating to it, most inputs relevant to the planning problem can be chosen among these.

4.1   Manufacturing System Details

The manufacturing system details can essentially be classified into the following major heads:

1) Work-center details: These consist of the number and types of work-centers or machines available in the system. If work-centers are prone to failure, then the various failure modes and associated distributions of time between failures and time to repair are of relevance (also see figure 4.1). The other important characteristics can include: (i) maintenance requirements (scheduled and unscheduled), (ii) available capacity (duration of uptime), and (iii) operating and idle costs per unit time. Addition of new machines to the system can also be permitted, but it is assumed to be a decision that is beyond present consideration (it is a systems design/strategic issue).

Given several failure modes of the machine, based on their Mean Time To Repair (MTTR), they can be classified along a time-scale as in figure 4.1. They can also be clustered, or classified into groups of failures, where each group is represented by a characteristic MTTR.

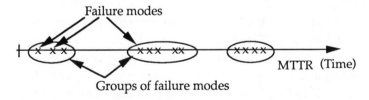

Figure 4.1: Time-scale classification of various failure modes

2) Worker details: These details are of relevance in the case of labor intensive production. These consist of the number and types (i.e. specific trades) of workers that can be made available in the system. In this case, we can assume that worker levels can be changed, i.e. workers can be hired and fired. Each worker type has a certain hiring and firing cost and lead time associated with it. Furthermore, each worker type has associated regular and overtime costs per unit of time. If the workers default, then the various default modes and associated distributions (and averages) of time between absenteeism and duration of absence are of relevance. Finally, the set of machines that can be operated by a particular worker type is of concern.

As in the case of machines, based on the mean duration of absence, the worker default modes can be classified along a time-scale. They can also be clustered, or classified into groups of defaults, wherein each group is represented by a characteristic absence duration.

3) Product details: These consist of the set of product types (finished products) that are manufactured in the system. For each product, the Bills-Of-Materials (BOM), identifying the set of semi-finished parts and raw materials should be known. These two sets constitute the manufactured parts in the system, which are hereby simply referred to as parts. For each part, a set of primary and alternative routings or production processes are assumed known. For a unit of each product, the cost for holding it in inventory per unit time (holding cost), and the penalty cost for violating the due-date by a unit time (backlogging cost), are important attributes.

4) Routing details: These consist of the sequence of operations. The set of work-centers the operation can be performed on as well as the set-up and run times associated are important. Further, if the operation is labor intensive one, the labor time required should also be specified.

5) Historical or Forecast data: Historical information in the case of existing manufacturing systems, or forecasted projections for new manufacturing systems, about the long term production requirements (or demand) of products are essential in designing the hierarchy.

Such are the details relating to the manufacturing system that can be of interest to the design of the hierarchy. Since the planning hierarchy remains in place for a relatively long duration, as compared to the production lead-times of products, the design is based on principal choices among alternatives. That is, for designing the hierarchy we employ principal routings, perform preferred assignments of workers to machines, etc. This is in order to reduce complexity, while basing it on long term assumptions relating to the use of the hierarchy for planning. For existing systems, where proportion of assignments to alternative choices is statistically known (e.g. the long term proportion of parts produced using a particular routing among alternatives), we can also employ this information in a weighted manner. Therefore, the details presented so far may take a more simplistic form. Furthermore, from detailed part routings, *representative routings for products* are constructed as follows. The total production time (work-content) of a product at each work-center is accrued for the finished product as well all its dependent components (as indicated in its BOM) and represented in a cumulative manner. Note that in such a representative routing, the sequence of work-centers is immaterial; it only represents the total work-content required of each work-center.

4.2 Managerial Inputs

Knowledge of the managerial decisions and objectives are necessary in the design of the production management hierarchy. The managerial inputs details can essentially be classified into the following major heads:

1) Controls, U^0: Controls are decisions that can be performed (by a human or the management tool) during the production planning process. Decision to sub-contract products, hiring or firing, decisions of overtime levels are related control. Performing set-ups, or loading parts are controls performed more frequently than those mentioned earlier.

Each control is associated with a period or duration of time which is elapsed before the results of the control can be observed (or expected). This duration depends on the type of control, and the speed of the system to accomplish the action. This duration is referred to as the response time of the system to this control. For instance, production of a product has a response time of the order of its cumulative lead time. Hiring of new workers is a control which generally has a longer response. Thus, each control can be assigned an order of magnitude for its response time in the particular manufacturing system.

2) Criteria, C^0: Criteria are the management objectives that have to be optimized during production planning. Minimization of production costs, distribution costs, inventory costs, backlogging costs, hiring and firing costs, regular and overtime labor costs, and set-up costs may be relevant. It is important to indicate that the criteria to be considered in the planning problem bear a strong relationship with the controls involved. For instance, if hiring and firing of workers is a control, then minimization of the corresponding costs should be criteria to be considered in production planning. However, if the worker level is assumed to be fixed in the problem, the hiring and firing costs are of no significance.

5. Systematic Design of the Hierarchy

In this section, we present the design procedure for the construction of a planning hierarchy. This design procedure is attempted to be presented in an algorithmic form, but it requires significant human interaction and decision making. Thus, it cannot be viewed as an entirely automatic process. This construction is performed in a bottom-up manner, because the physical system is the only detailed information available. The basic design approach is first presented conceptually. Later, we will present a more precise and comprehensive design algorithm.

We begin at the bottom level (level 1), and consider the entities, which are physical entities, e.g. products, machines, and workers. Therefore, the model is the physical one (and known), and we need to formulate the decision making problem. As mentioned earlier, the end result of this planning hierarchy is to determine the types and number of products to produce in each elementary period of the level 1 horizon. Thus, the natural direction to be adopted is to select the appropriate elementary period, and the horizon length. The elementary period is determined based on several factors that include product lead times, intervals between shipment and inventory updates, to mention a few. For instance, if the products have lead-times of the order of minutes or hours, and the inventory is updated daily, shipments are performed and orders are accepted daily, then a day is an appropriate elementary period. Since the elementary periods at this level are usually of a fairly short duration, not all controls can be effectuated during this. Thus, only a set of controls that have a response time at least an order of magnitude less than the length of the elementary period are considered. This elementary period is also intended to absorb random events having a response time of the order of the controls, i.e. an order of magnitude less than the length of the elementary period. Random events of much shorter response duration are considered as averages, while those of much longer response duration are considered in their actual state (and are absorbed at some other higher level of the hierarchy). This point will be further illustrated in the operation of the hierarchy (section 6). Note that capacities of resources at each level is computed by subtracting from the available time, a summation of the expected durations of non-productive activities (random events like failures) tying up the resources and having mean durations that are much less than the duration of the current elementary period.

The set of controls to be considered defines the complexity of the decision making problem (number of variables and constraints), after which the length of the horizon can be ascertained; this is ascertained either by the complexity of the problem or the solution time permissible. Thus, owing to the detailed nature and large dimension of level 1, its horizon is limited by the complexity of the problem. Finally, the criteria to be considered are determined based on the controls at this level. This completes the definition of the decision making problem at the bottom level. Note that at this time the next higher level is not known, so the high level constraints to this problem are not know exactly although their general form is understood.

The need to address planning at a longer horizon than the current one, and to consider all possible controls requires the development of higher levels. The essential idea is to reduce the dimension and the complexity of the current model by temporal and entity aggregation. We employ the approach proposed in Nagi [48] to accomplish these aggregations, in a manner that is consistent with the criteria at the bottom level. That is, attributes of product entities that bear relevance to the criterion under consideration, as well as some other related attributes are considered in the aggregation. For example, if the criteria considered are inventory and backlogging costs, the attributes that bear relevance to these criteria are the holding cost and backlogging cost per unit product and unit time. In addition, attributes relating to the processing time on each machine are also considered. The resources are also aggregated, and their capacities are represented to reflect the absorption of the random events at level 1 as well. After aggregation, the attributes of the newly derived entities are computed. The precise details of these computations for an example are presented in section 7. Here, we allude to computation of some attributes of aggregate product entities as follows: (i) the holding cost and backlogging costs for are computed as a weighted average of the respective holding cost and backlogging costs of the detailed constituent product entities; the long term production quantities are employed as weights (see Nagi [48] for details), (ii) the processing times are computed similarly as weighted averages of the processing times of the constituent product entities (see Nagi [48] for details), and (iii) the long term production quantities are the summation of the long term production quantities of the component products. In this manner, we develop the model at the next upper level (level 2).

Having determined the model of the level under consideration, we need to define the decision making problem. Once again, we select the appropriate elementary period, and the horizon. The elementary period is based on the following guide-lines: it is usually a multiple of the lower level elementary period, and is no greater than the horizon at the lower level. Since the elementary periods at this level are of longer duration than those of the lower level, some additional controls can be effectuated during this. The earlier (lower level) controls are also considered, but now for the aggregate entities. Once again, the entire set of selected controls should have a response time at least an order of magnitude less than the length of the current elementary period. Furthermore, since the elementary period at this level is longer than that of the lower level, some additional random events (with a response time an order of magnitude less than the length of the current elementary period, but greater than the response times of previously considered random events) can be absorbed. Random events of much shorter response duration are considered as averages, while those of much larger response duration are considered in their actual state (and are absorbed at some other higher level of the hierarchy).

The set of controls to be considered then defines the complexity of the decision making problem (number of variables and constraints), after which the length of the horizon can be ascertained; this is ascertained by the complexity of the problem and the solution time permissible. Thus, owing to the reduced dimension at this level, the horizon is longer than that of the lower level. Finally, the criteria to be considered are determined based on the controls at this level; notice that, the form of the criteria now refer to the aggregate entities. This completes the definition of the decision making problem (DMP) at the current level.

The subsequent higher levels are designed in a similar fashion, i.e. aggregating and defining the model followed by defining the decision making problem. The process is repeated until the following conditions are obtained: (i) the desired horizon is obtained, (ii) the desired degree of aggregation is accomplished, (iii) all controls have been addressed,

158

and (iv) all criteria have been addressed. Note that based on this design procedure, the controls as well as the criteria automatically assume a hierarchy among themselves. The flowchart of the design process is presented in figure 5.1. In this flowchart, the block corresponding to "Finalize," refers to the finalization of the exact form of the high level constraints for the decision making problem D^{i-1}, which can be formulated only after the level i entities are determined. An example describing this procedure is presented in section 7.

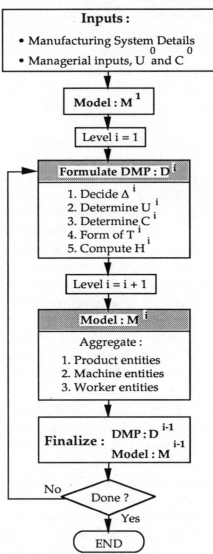

Figure 5.1: Flowchart for the Design of the Planning Hierarchy

6. Operation of the Hierarchy

After the construction or design of the hierarchy is complete, it can be employed for solving the planning problem and decision making. Unlike the design process, which is bottom-up, the operation of the hierarchy is top-down. The first sub-section consists of the top-down computation of the solutions, while the second sub-section is devoted to the reaction of the hierarchy to random events.

6.1 Top-Down Solution Procedure to the Planning Problem

In this section, we present the operation of the hierarchy for top-down decision making in planning. We assume that the decision making problems and the associated algorithms are known at this stage. The solution procedure of the planning problem begins at the top-most level of the hierarchy, say level n. This level is the highest, hence the upper level constraints, T^n are strategic constraints in this case; they could also be constraints emanating from demand forecasting. The demand at this level is expressed in terms of the aggregate product entity at level n. For some early period(s), it can be taken from the bottom up aggregation of the detailed existing customer demand, while later periods it is forecasted demand. Given this demand, this level solves its DMP over the horizon H^n in order to optimize the criteria C^n.

The solution (usually optimal) of this problem X^n, truncated over the lower level horizon H^{n-1}, is then transmitted to the next lower level (n-1). This is in the form of its upper level constraints T^{n-1} for level n-1. This level also accepts additional information from demand forecasting regarding the proportions of level n-1 product entities composed in the level n product entities, for each elementary period Δ^{n-1}; in the absence of such information, the default is the long term/historical ratios. The primary function of level n-1 is, given the high level constraints T^{n-1} to determine the set of controls belonging to U^{n-1}, such that the criteria C^{n-1} are optimized over H^{n-1}.

The top-down solution procedure continues for lower levels of the hierarchy in a similar manner. Each level i solves its DMP over the horizon H^i in order to optimize the criteria C^i. The solution of which, X^i, truncated over the lower level horizon H^{i-1}, is then transmitted to the lower level (i-1). This takes the form of the upper level constraints T^{i-1} for level i-1. This level also accepts additional information from demand forecasting (detailed customer demand in the case of level 1) regarding the proportions of level i-1 product entities composed in the level i product entities, for each elementary period Δ^{n-1}; the default is the long term/historical ratios. The objectives of level i-1 are determining the appropriate controls corresponding to this level (belonging to U^{i-1}), such that the criteria C^{i-1} are optimized over H^{i-1}.

The final result of this top-down computation is the production levels of product types in the elementary periods Δ^1 over the horizon H^1, under resource capacity constraints.

Rolling horizon mechanism:

In a monolithic architecture, the rolling horizon mechanism is employed in order to progressively take future information into account while only a portion of the solution is implemented. This is a method to emulate an infinite horizon in practice. The hierarchical architecture also employs this methodology, but the mechanism is different due to the presence of several levels. Each level in the hierarchy works on a rolling horizon basis. That is, each level recomputes its solution after every elementary period corresponding to it. In effect, only the portion corresponding to the first elementary period is executed at every computation. Consider that all the levels compute their solutions in a top-down fashion as explained earlier. Now, only the first elementary period of the lowest level (i.e., level 1), i.e. Δ^1 is implemented on the shop-floor (or, subsequent planning/scheduling hierarchy). Thereafter, level 1 recomputes a new solution on $[\Delta^1, \Delta^1 + H^1]$. Of course there is some vagueness for the high level (level 2) constraints for the last elementary period of level 1, because the level 2 solution is only specified after every Δ^2, where $\Delta^2 > \Delta^1$. But, this problem can be averted by assuming linearity of production during Δ^2. As soon as level 1 has implemented enough elementary periods (of duration Δ^1) such that we are at the beginning of the second Δ^2 period, the level 2 solution is recomputed over $[\Delta^2, \Delta^2 + H^2]$. The process goes on for all subsequent levels of the hierarchy. Notice that, level i problem is solved Δ^{i+1}/Δ^i times more frequently that the level i+1 problem; the lowest level problem is solved Δ^n/Δ^1 times more frequently than the highest level problem.

6.2   Reaction of the Hierarchy to Random Events

In the previous section, we presented the top-down solution procedure and the rolling mechanism. This corresponds to a perfect situation, where each lower level respects the higher level decision and no discrepancies arise during the execution of plans on the shop-floor. In practice, however, there are a number of disturbances and random events that make it difficult to strictly respect higher level decisions. This calls for a bottom-up feedback procedure, in that each level transmits the difference between the planned and accomplished states to the next higher level. This is done in order that the higher level takes this discrepancy into account during the subsequent calculation. Such bottom-up feedbacks are transmitted by a level that detects some discrepancy to its next higher level until this discrepancy is absorbed by some level. Usually such a feedback procedure originates at the bottom level of the hierarchy, due to random events on the shop floor (or at the subsequent planning/scheduling hierarchy), and is transmitted upwards until it can be absorbed by some level. Unusually high demands, or cancellation of orders also constitute random events that trigger the need for recomputation of the higher level(s).

However, there are certain other random events in the production planning environment that have a response time comparable to the duration of elementary periods at some level of the hierarchy. For example, a worker strike, or a severe machine breakdown can have durations comparable to durations of a high level elementary period of the hierarchy. In this case, when a random event arrives in the system, the response time of the system to this event is evaluated by the some mechanism external to our system (e.g., in the case of a failure, the maintenance department estimates the time to repair). If the exact duration of this activity cannot be assessed, it is based on the expectation of the duration. This is then compared to the duration of the elementary periods of the various hierarchical levels. This activity is assigned to the level i such that Δ^i is an order of magnitude greater than the duration of this activity. Now, for all the levels including and below i (i.e. for all levels j such that $j \le i$), the state of the system is assumed to be constant (e.g. the machine is assumed unavailable in the case of a machine failure), and solutions are recomputed top-down. On the completion of this random activity (i.e. the machine is repaired), the state of the system is revised (i.e. the machine is assumed available until the next failure of this type). The solutions are recomputed top-down from level i onwards for the revised system state. In this way, random events are addressed at an appropriate level of the hierarchy.

Thus, consistent with this methodology, and with reference to time-scale decomposition, we make the following remark. Capacities of resource entities at each level (say level i) of the hierarchy are calculated in a manner that considers all activities occupying this resource, with a duration much less than the elementary period at this level (Δ^i) on the average. That is, the resource capacity at that level is reduced by an amount that corresponds to the average duration that these activities are going to occupy the resource. If an activity of duration larger than Δ^i were to occupy this resource, the resource is assumed to have zero capacity (or assumed unavailable) in levels i through 1, until the completion of the activity. The underlying theory for this remark draws its motivation from time-scale decomposition, and the extension is that we manage to blend it with the aggregation or the multi-layer hierarchy. The overall operation of the hierarchy is summarized in figure 6.1.

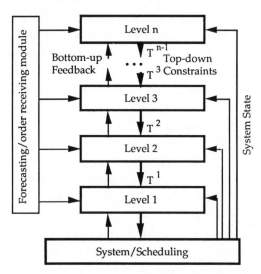

Figure 6.1: Overall Operation of the Planning Hierarchy

7. An Example

In this section, we present a simplified example to demonstrate the design procedure of the planning hierarchy as presented in section 6. The reason we term it as simplified is because the actual details specifying the manufacturing system (as mentioned in section 4) are not detailed; instead, we assign a relative order to the relevant quantities of interest. Furthermore, the system is assumed to be deterministic. Thus, the operation of the hierarchy in the case of random events cannot be demonstrated by this example.

7.1 Inputs to the Design Process

We present the relevant inputs to the design of the hierarchy for the example considered. Note that, we assign a relative order of time to the relevant controls/activities of interest, while ignoring the actual details of the system.

Manufacturing system details:

Here, we only present the details relating to the manufacturing system deemed essential for the design stage of this simplified example:

1) Work-center details: We consider m machines available in the system. They are reliable (or mean time between failures and mean time to repair are too short to appear at any planning level, e.g. their duration is of the order of a few minutes), and do not require to be maintained. Operating and idle costs per unit time are not considered.

2) Worker details: We assume a set of w worker types required in the system. In this case, we can assume that worker levels can be changed, i.e. workers can be hired and fired. Each worker type has a certain hiring and firing costs and lead time associated with it. Let us assign that the lead times for hiring and firing are of the order of 2-3 weeks, while overtime can be planned on less than a week (but greater than a day). Each worker type has associated regular and overtime costs per unit of time. The workers do not default. Let a working shift consist of 8 hours per day, and a maximum of 2 shifts per day are possible. However, a different set of workers must be used in each shift of a single day, i.e. the same set of workers cannot work for both shifts of the same day. The maximal overtime level is 25% of the regular hours per shift, i.e. 2 hours per shift. Overtime costs 25% higher than regular costs for all worker types.

3) Product details: We consider n product types (finished products) that are manufactured in the system. We ignore details regarding the Bills-Of-Materials (BOM) and components, and assume that a representative product routing is available for each part. A holding and backlogging cost is associated to each product type; the penalty is specified over a duration of one working day.

4) Routing details: Each product has a representative routing which reflects the cumulative work-content (processing time) required by that product type at each machine. Set-up times are ignored (or assumed too short to consider at the planning levels).

Furthermore, the time required from each worker type is also assumed known. We attribute the total lead time for products to be of the order of 10-15 minutes.

5) Historical or Forecast data: Historical information about the long term production requirements (or demand) of products is assumed available.

Managerial Inputs:

The details related to the managerial inputs are as follows:

1) Controls, U^0:

- Production of products (response time O(10-15 minutes))
- Hiring and Firing of workers (response time O(2-3 weeks))
- Decision of worker overtime levels (response time O(2-3 days))

Where, O($\cdot$) represents the order $\cdot$ units of time.

2) Criteria, C^0:

- Minimization of Production costs:
  - Hiring and firing costs
  - Overtime costs
- Holding and backlogging costs for products

These criteria are the management objectives that have to be optimized during the production planning. Note that the criteria considered in the planning problem bear a strong relationship with the controls considered. For example, if hiring and firing of workers is a possible control, the corresponding costs are relevant in the objective function; otherwise, the costs are irrelevant.

7.2 Design of the Planning Hierarchy

We present the design procedure for the construction of a planning hierarchy below.

$\underline{D^1}$: Decision Making Problem at Level 1

1. Elementary period, $\Delta^1 = 1$ day

   - 1 day > lead time for products, O(10-15 minutes).
   - 1 day = time interval after which orders are delivered to stock, shipments are made and inventory is updated.
   - 1 day >> mean time for an important class of machine failures, if any.

2. Controls, U^1

   Production levels for products in elementary periods of duration Δ^1.
   Note, all other controls have response time larger than a day.

3. Criteria, C^1

   Holding and backlogging cost.

4. Constraints, T^1

The intrinsic constraints are essentially resource (machine and worker) capacities. The exact higher level constraints T^1 are not known because the high level remains undefined at this time. The general form of these constraints are aggregate production constraints related to the production levels of aggregate products (families) that must be respected by the current level during the disaggregation process. Furthermore, aggregate resource capacity levels may also represent high level constraints.

The complexity of the decision making problem can be determined in the following manner. For each elementary period, there are n variables corresponding to the production levels of products, and n variables corresponding to the inventory costs of these products. We roughly compute the number of constraints for each elementary period as follows. There are n(2) aggregate production constraints (equality constraints); where, n(i) is the number of product related entities at level i (i=2 in this case). There are m machine capacity constraints. There may be some additional constraints (see next chapter). Finally, there are the non-negativity constraints. Based on this information the complexity of the DMP over one elementary period is roughly assessed.

The next step consists of determining the length of the horizon H^1, or the number of elementary periods Δ^1 that can be included in the horizon H^1. This should preferably be performed after the next higher level model M^2 is determined, and the exact formulation of D^1 is known.

5.   Horizon, $H^1 = 2$ weeks (assumed)

- Let us assume that 2 weeks nearly equal the horizon during which customer orders are firmed up and affect production/stock.
- Let us assume that this is based on the complexity of the DMP (which is proportional to the number of elementary periods = 10 in the horizon). Let us finally assume that this is based on the memory limitations of the computer and solution time requirements (which is < 15 minutes).

M^2: Model at Level 2

Once again, at a conceptual level, we present the procedure to develop the model at level 2. Employing the aggregation rules proposed in section 5.6.1 with appropriate values of parameters, let us assume that we obtain n(2) number of product related entities, m(2) machine related entities, and w(2) worker related entities. Let the clustering results find approval when evaluated in the light of physical or practical considerations.

Now, the high level constraints for the level 1, T^1 can be formulated. Thus, the decision making problem at the level 1, D^1 can be completed at this stage.

D^2: Decision Making Problem at Level 2

1   Elementary period, $\Delta^2 = 1$ week

- Δ^2 is a multiple of Δ^1 (5 days is a multiple of 1 day).
- $\Delta^2 \leq H^1$ (2 weeks).

- Δ^2 < response time of decision of worker overtime levels (response time O(2-3 days)).

2 Controls, U^2

   1 Production levels for product entities at level 2 on machine entities at level 2 for elementary periods of duration Δ^2.

   2 Decision of worker overtime levels (response time O(2-3 days)).

3 Criteria, C^2

   1 Holding and backlogging cost for product entities at level 2.

   2 Minimization of overtime costs.

4 Constraints, T^2

The intrinsic constraints are essentially resource (machine entities at level 2 and worker entities at level 2) capacities. As before, the exact higher level constraints T^2, are not known because the high level remains undefined at this time. The general form of these constraints are aggregate production constraints, related to the production levels of product entities that must be respected by the current level during disaggregation. Furthermore, aggregate resource capacity levels may also represent high level constraints.

The complexity of the decision making problem can be determined in the following manner. For each elementary period, there are n(2) variables corresponding to the production levels of product entities, n(2) variables corresponding to the inventory costs of these product entities, and w(2) variables corresponding to the worker overtime levels. We roughly compute the number of constraints for each elementary period as follows. There n(3) aggregate production constraints (equality constraints). There are w(2) worker capacity constraints, and w(2) worker overtime level constraints. Finally, there are the non-negativity constraints. Based on this information the complexity of the DMP over one elementary period is roughly assessed.

The next step consists of determining the length of the horizon H^2. Once again, it is preferable to ascertain this when the next higher level model M^3 is determined, and the exact formulation of D^2 is known. However, for this simplified example, we assume that the general form of these should suffice.

5 Horizon, H^2 = 3 months

- H^2 (3 months) > H^1 (2 weeks).
- Let us assume that this is based on the complexity of the DMP (which is proportional to the number of elementary periods = 12 in the horizon). Let us assume that this is based on the memory limitations of the computer and solution time requirements (which is < 15 minutes).

M^3: Model at level 3

Once again conceptually we present the procedure to develop the model at level 3. Employing the aggregation rules proposed in section 5.6.1 with appropriate values of parameters, let us assume we obtain n(3) number of product related entities, m(3) machine related entities, and w(3) worker related entities. Let the clustering results find approval when evaluated in the light of physical or practical considerations.

Now, the high level constraints for the level 2, T^2 can be formulated. Thus, the decision making problem at the level 2, D^2 can be completed at this stage.

$\underline{D^3}$: Decision Making Problem at Level 3

1   Elementary period, Δ^3 = 4 weeks (roughly 1 month)

- Δ^3 is a multiple of Δ^1 (a period of 4 weeks is a multiple of 1 week).
- $\Delta^3 \leq H^2$ (3 months).
- Δ^3 < response time of hiring and firing workers (response time O(2-3 weeks)).

2   Controls, U^3

　　1   Production levels for product entities at level 3 on machine entities at level 3 for elementary periods of duration Δ^3 .

　　2   Decision of worker overtime levels (response time O(2-3 days)).

　　3   Decision of hiring or firing workers (response time O(2-3 weeks)).

3   Criteria, C^3

　　1   Holding and backlogging cost for product entities at level 2.

　　2   Minimization of overtime costs.

　　3   Minimization of hiring and firing costs.

4   Constraints, T^3

The intrinsic constraints are essentially resource (machine entities at level 3 and worker entities at level 3) capacities. As before, the exact higher level constraints T^3, are not known because the high level (if any) remains undefined at this time. However, we can anticipate termination at this stage because all the criteria and controls have been addressed. So, T^3 are likely to take the form of strategic constraints.

The complexity of the decision making problem can be determined in the following manner. For each elementary period, there are n(3) variables corresponding to the production levels of product entities, w(3) variables corresponding to the worker entity levels, and w(3) variables corresponding to the worker overtime levels. We roughly compute the number of constraints for each elementary period as follows. There are w(3) worker capacity constraints, w(3) worker level constraints, and w(3) worker overtime level constraints. Finally, there are the non-negativity constraints. Based on this information the complexity of the DMP over one elementary period is roughly assessed. The next step consists of determining the length of the horizon H^3.

5   Horizon, $H^3 \approx$ 12 months (48 weeks)

- H^3 (12 months) > H^2 (3 months, or 12 weeks).

- Let us assume that this is based on the complexity of the DMP (which is proportional to the number of elementary periods = 12 in the horizon). Let us assume that this is based on the memory limitations of the computer and solution time requirements (which is < 15 minutes).

Check for Termination conditions

At this stage:

- all controls belonging to U^0 have been addressed.

- all criteria belonging to C^0 have been addressed.

- H^3 > duration that can absorb the most significant disturbances, both internal as well as external (e.g. demand seasonality over a year).

- At level 3, the desired degree of aggregation is accomplished.

Thus, the hierarchy consists of three levels, and the structure can be represented in fig. 7.1.

Level	Δ	H	Controls, U	Criteria, C	Constraints, T
1	1 day	2 weeks	Production of products Agg. Worker hours	Holding & Backlogging	Aggregate Production
2	1 week	3 months	Prod. of agg. products Overtime levels	Holding & Backlogging Overtime costs	Aggregate Production Number of workers
3	4 weeks	12 months	Prod. of agg. products Overtime levels Hiring & Firing	Holding & Backlogging Overtime costs Hiring & Firing costs	Global worker levels (Number of shifts, & maximal overtime)

Figure 7.1: Summary of Decision Making Problems in the Hierarchy

8. Conclusions

This chapter attempts to define efficient architectures and operation modes for production management. The motivation for these is derived from the criticality of proper management for an overall desired performance of a production system. The chapter details the lack of existing work and literature addressing some important facets of production planning problems in a generic fashion. The need for a problem specific hierarchy is advocated over a fixed hierarchy trying to serve as a panacea for all production planning problems. In response to this reality, a multi-layer hierarchical approach to production planning problems is adopted, wherein the architecture (in particular, the number of levels) is strongly based on the specific physical production system and the complexity of the decision making problem at hand.

The scope of production management problems addressed in this chapter, and the purpose of the corresponding hierarchy, is to perform typical tactical level planning decisions: all the way from aggregate production planning over a long horizon, down to the detailed production planning over a relatively short horizon. The focus of the chapter is on the design and operation aspects of such Hierarchical Production Management Systems.

Regarding the design of the system, which starts with a given physical system, the relevant inputs required for the design process are detailed. Product and machines are aggregated based on developed aggregation schemes for a planning problem with holding and backlogging penalties. These schemes provide consistent solutions during the disaggregation process in multi-layer hierarchies. Temporal aggregation is also introduced in the theory. These three aggregation dimensions for model reduction have been developed and blended with the time-scale decomposition of activities, in order to provide a solid theoretical foundation of the architecture. Thus, a systematic stepwise design approach for the construction of the hierarchy has been presented.

The design process results in the appropriate number of layers, their models, entities, definitions, planning horizons, and domains. Each level is modeled by a set of pertinent entities to that level, the links between these entities, a set of attributes associated with each entity, and the value set associated with each attribute. This is followed by the formal definition of the Decision Making Problem (DMP) at each level, along with a set of possible controls (decisions), a set of constraints, and one or more optimality criteria over a given horizon. The lengths of horizons are derived in such a way that each level is able to absorb most of the random events associated with it, under the existing hardware (memory, depending on the DMP dimension) and software (solution time permissible for the DMP) limitations.

The operation of the hierarchy in an unreliable environment is also explained. In particular, the rolling horizon mechanism, and the reaction of the hierarchy to random events has been detailed. Solution algorithms to resolve some related optimization problems (especially, disaggregation problems) have been presented. The mechanism for top-down constraint propagation is an important aspect for the operation of the hierarchy. Problem solutions are attempted in a sequential manner; the solution of any problem in sequence determines some parameters in the subsequent problem.

This chapter does not cover the entire gamut of planning problems and their related assumptions. Employing the concepts proposed, the hierarchy can be extended to both higher or strategic level decisions, as well as lower or operational level decisions. Strategic decisions have lead-times that are generally higher than tactical decisions, and are performed at a higher degree of abstraction (aggregation). Thus, they lend themselves adaptable to the methodology presented in this chapter. Encompassing operational decisions may require extensions to the methodologies presented here, or even developing some other ones. Finally, on the execution side, it would be interesting to implement a real time closed loop control process in the future.

It is hoped that this methodology can be applied to other types of large-scale complex decision making problems (such as communication system). Considering more examples, and validating their architectures can be further justification to the methodology developed here. It is believed that such improved tools in a manufacturing environment for company-wide decision making and management, which are well integrated with other functions, can provide an answer to the much required competitive edge to the U.S. manufacturing sector.

9. References

[1]   Abraham, C., Dietrich, B., Graves, S., Maxwell, W. and Yano, C. (1985): "A Research Agenda for Models to Plan and Schedule Manufacturing Systems," presented at an NSF workshop as "Scheduling the Factory of the Future: A Research Planning Session," Decision Sciences Department, University of Pennsylvania, March 1985, and Boston TIMS/ORSA meeting.

[2]   Akella, R., Choong, Y. and Gershwin, S.B. (1984) : "Performance of Hierarchical Production Scheduling Policy," IEEE Transactions on Components, Hybrids, and Manufacturing Technology, Vol. CHMT-7, No. 3, 1984.

[3]   Anthony, R.N. (1965) : "Planning and Control Systems : A Framework for Analysis," Harvard University, Graduate School of Business Administration, MA.

[4]   Armstrong, R.J. and Hax , A.C. (1977) : "A Hierarchical Approach for a Naval Tender Job-Shop Design," Applied Mathematical Programming, Chap. 10, S.P. Bradley, A.C. Hax and T.L. Magnanti. Addison-Wesley, 1977.

[5]   Axsater, S. (1981) : "Aggregation of Product Data for Hierarchical Production Planning," Operation Research, Vol. 29, No. 4, 1981.

[6]   Bitran, G.R., Haas, E.A. and Hax, A.C. (1981) : "Hierarchical Production Planning : A Single Stage System," Operations Research, Vol. 29, No. 4, 1981.

[7]   Bitran, G.R., Haas, E.A. and Hax, A.C. (1982) : "Hierarchical Production Planning : A Two Stage System," Operations Research, Vol. 30, No. 2, 1982.

[8]   Bitran, G.R. and Hax, A.C. (1977) : "On the Design of Hierarchical Planning Systems," Decision Sciences, Vol. 8, 1977.

[9]   Bitran, G.R. and Hax, A.C. (1981) : "Disaggregation and Resource Allocation using Convex Knapsack Problems with Bounded Variables," Management Science, Vol. 27, No. 4, 1981.

[10]   Chen, P, P-S. (1976) : The Entity-Relationship Model - Towards a Unified View of Data," ACM Transactions on Database Systems, Vol. 1, No. 1, pp. 9-36, 1976.

[11]   Chow, J.H. and Kokotovic, P.V. (1985) : "Time Scale Modeling of Sparse Dynamic Networks," IEEE Transactions on Automatic Control, Vol. AC-30, No. 8, 1985.

[12]   Cunningham, Thomas J. Hardouvelis, Gikas A. (1992): "Money and Interest Rates: The Effects of Temporal Aggregation and Data Revisions," *Journal of Economics and Business*, 44/1, 19.

[13]   Davis, W.J. and Jones, A.T. : "A real-time production scheduler for a stochastic manufacturing environment," Int. J. Computer Integrated Manufacturing, Vol. 1, No. 2, pp. 101-112, 1988.

[14] Dempster, M.A.H., Fisher, M.L., Lageweg, B., Jansen, L., Lenstra, J.K. and A.H.G. Rinnoy Kan (1981) : "Analytical evaluation of Hierarchical Planning Systems," Operations Research, Vol. 29, No. 4, 1981.

[15] Doumeingts, Guy (1986) : "Artificial Intelligence Concept Techniques for Computer Integrated Manufacturing," Flexible Manufacturing Systems: Methods and Studies, edited by Kusiak, A., Elsevier Science Publishers B.V., 1986.

[16] Engles, L., Larson, R.E., Peschon, J., Stanton, K.N., (1976): "Dynamic programming applied to hydro and thermal generation scheduling," IEEE tutorial course on application of optimization methods in power system eng., IEEE, NY.

[17] Erscheler, J., Fontan, G. and Merce, C. (1986) : "Consistency of the Disaggregation Process in Hierarchical Planning," Operations Research, Vol. 34, No. 3, 1986.

[18] Fox, M. and Smith, S.F. (1984): "ISIS: A Knowledge-Based System for Factory Scheduling," Expert Systems, Vol. 1, No. 1, pp. 25-49, 1984.

[19] Fox, M., Greenberh, M., Sathi, A., Mattic, J. and Rychnener, M. (1983): "CALLISTO: An Intelligent Project Management System," working paper, Intelligent Systems Lab, The Robotics Institute, Carnegie-Mellon University, November, 1983.

[20] Gabby, H. (1975) : "A Hierarchical Approach to Production Planning," TR-120, Operations Research Center, M.I.T, Cambridge, MA, 1975.

[21] Gelders, L. and Kleindorfer, P.R. (1974) : "Coordinating Aggregate and Detailed Scheduling in the One-Machine Job Shop: Part I. Theory," Operations Research, Vol. 22, No. 1, pp. 46-60, 1974.

[22] Gelders, L. and Kleindorfer, P.R. (1975) : "Coordinating Aggregate and Detailed Scheduling in the One-Machine Job Shop: Part I. Theory," Operations Research, Vol. 23, No. 2, pp. 312-324, 1975.

[23] Gelders, L.F. and van Wassenhove, L.N. (1982): "Hierarchical Integration in Production Planning : Theory and Practice," Journal of Operations Management, Vol. 3, No. 1, 1982.

[24] Gershwin, S.B. (1986) : "Stochastic Scheduling and Set-ups in Flexible Manufacturing Systems," Proceedings of the 2nd ORSA/TIMS Conference on Flexible Manufacturing Systems : OR Models and Applications, K. Stecke and R. Suri eds., 1986.

[25] Gershwin, S.B. (1988) : "Hierarchical Flow Control : A Framework for Scheduling and Planning Discrete Event Manufacturing Systems," IEEE Proceedings : Special Issue on Discrete Event Systems.

[26] Graves, S.C. (1982) : "Using Lagrangean Techniques to solve Hierarchical Production Planning Problems," Management Science, Vol. 28, No. 3, 1982.

[27] Harhalakis, G., Mehra, A., Nagi, R. and Proth, J.M. (1993) : "Temporal Aggregation in Production Planning," T.R. 93-54, Institute for Systems Research, University of Maryland, 1993.

[28] Hartigan, J.A. (1975) : "Clustering Algorithms," John Wiley & Sons, 1975.

[29] Hax, A.C. and Meal, H.C. (1975) : "Hierarchical Integration of Production Planning and Scheduling," in Studies in the Management Sciences, M.A. Geisler, ed., Vol. 1, Logistics, North Holland - American Elsevier, 1975.

[30] Hildebrant, R.R. (1980) : "Scheduling and Control of Flexible Manufacturing Systems whose Machines are Prone to Failure," Ph.D. Thesis, M.I.T. Dept. of Astronautics and Aeronautics, August 1980.

[31] Hilderbrant, R.R. and Suri, R. (1980) : "Methodology and Multi-level Algorithm for Scheduling and Real-time Control of Flexible Manufacturing Systems," Proceedings of 3rd. International Symposium on Large Engineering Systems, Memorial University of Newfoundland, July, 1980.

[32] Hilger, J. (1988) : "Langage de Production: Description Coherente du Court Terme (premier partie)," INRIA, France, Research Report No. 912, 1988.

[33] Hillion, H., Meier, K. and Proth, J.M. (1987) : "Production Subsystems and Part-families: The Tope Level Model in Hierarchical Production Planning Systems," Operational Research'87, G.K. Rand, ed., Elsevier Science Publishers B.V. (North-Holland), © IFORS, 1988.

[34] Jackson, R.H.F. and Jones, A.W.T. : "Hierarchical Control and Real-time Optimization in Automated Manufacturing Systems," National Bureau of Standards Technical Report NBSIR 86-3503, 1986.

[35] Jones, A.T. and Mclean, C. : "A proposed Hierarchical Control Model for Automated Manufacturing Systems," Journal of Manufacturing Systems, Vol. 5, No. 1, pp. 15-25, 1986.

[36] Karmarkar, U.S. (1981) : "Equalization of Run-Out Times," Operations Research, Vol. 29, No. 4, 1981.

[37] Kimemia, J. (1982) : "Hierarchical Control of Production in Flexible Manufacturing Systems," Ph.D. Thesis, M.I.T. Dept of Electrical Engineering and Computer Science, 1982.

[38] Kimemia, J. and Gershwin, S.B. (1983) : "An Algorithm for the Computer Control of a Flexible Manufacturing System," IIE Transactions, Vol. 15, No. 4, pp. 353-362, 1983.

[39] Krajewski, L.J. and Ritzman, L.P. (1977) : "Disaggregation in manufacturing and service organizations: survey of problems and research," Decision Sciences, Vol. 8, No. 1, 1977.

[40] Libosvar, C. (1988) : "Hierarchical Production Management : The Flow Control Layer," Ph.D. Thesis, University of Metz, France, 1988.

[41] Maimon, O.Z. and Gershwin, S.B. (1988) : "Dynamic Scheduling and Routing for Flexible Manufacturing Systems that have Unreliable Machines," Operations Research, Vol. 36, No. 2, 1988.

[42] Maxwell, W., Muckstadt, J.A., Thomas, J. and VanderEecken, J. (1983): "A Modeling Framework for Planning and Control of Production in Discrete Parts Manufacturing Systems and Assembly Systems," Interfaces, Vol. 13, 1983.

[43] Meier, K. (1989) : "Commande Hierarchisee d'un Systeme de Production," Ph.D. Thesis, University of Metz, France, 1989.

172

[44] Mesarovic, M.D., Macko, D. and Takahara, Y. (1970) : "Theory of Hierarchical, Multilevel, Systems," Academic Press, Inc. New York, 1970.

[45] Monts, K. (1991): "An Empirical Procedure for the Temporal Aggregation of Electric Utility Marginal Energy Costs," *IEEE Transactions on Power Systems,* 6 (2).

[46] Morton, T.E., Mark, F. and Sathi, A. (1984): "PATRIARCH; A Multilevel System for Cost Accounting, Planning, Scheduling," working paper GSIA, Carnegie-Mellon University, May, 1984.

[47] Morton, T.E. and Smunt, T.L. (1986) : "A Planning and Scheduling System for Flexible Manufacturing," Flexible Manufacturing Systems: Methods and Studies, edited by Kusiak, A., Elsevier Science Publishers B.V., 1986.

[48] Nagi, R., (1991) : "Design and Operation of Hierarchical Production Management Systems," Ph.D. Dissertation, Ph.D. 91-13, Institute for Systems Research, University of Maryland, 1991.

[49] O'Grady, P.L. (1986) : "Controlling Automated Manufacturing Systems," Kogan Page Ltd., 1986.

[50] Rossana, R.J., Seater, J.J. (1992): "Aggregation, Unit Roots and the Time Series Structure of Manufacturing Real Wages," *International Economic Review,* 33 (1).

[51] Sandell, N.R., Varayia, P., Athans, M.A. and Safonov, M. (1978) : "A Survey of Decentralized Control Methods for large Scale Systems," IEEE Transactions on Automatic Control, Vol. AC-23, No. 2, 1978.

[52] Shaw, M.J.P. and Whinston, A.B. (1986) : "Applications of Artificial Intelligence to Planning and Scheduling in Flexible Manufacturing," Flexible Manufacturing Systems: Methods and Studies, ed. Kusiak, A., Elsevier Science Publishers B.V.

[53] Smith, S.F., Ow, P.S. and Matthys, D.C. (1989) : "OPIS : An Opportunistic Factory Scheduling System," Proceedings Int. Symposium for Computer Scientists, Bejing, China, August, 1989.

[54] Stecke, K. E. (1986) : "Useful Models to Address FMS Operating Problems," Flexible Manufacturing Systems: Methods and Studies, edited by Kusiak, A., Elsevier Science Publishers B.V., 1986.

[55] Villa, A., Mosca, R. and Murari, G. (1986) : "Expert Control Theory: A Key for Solving Production and Control Problems in Flexible Manufacturing," Proceedings of the 1986 IEEE Conference on Robotics and Automation, San Francisco, CA.

[56] Wei, W.W.S., Mehta, J. (1980): "Temporal aggregation and information loss in the distributed lag model," *Analyzing Time Series,* Proceeding of the international conference, Guerney, Channel Islands, Publisher: North-Holland, Amsterdam.

[57] Xie, X.-L. (1989) : "Real Time Scheduling and Routing for Flexible Manufacturing Systems with Unreliable Machines," Recherche opérationelle/Operations Research, Vol. 23, No. 4, pp. 355-374, 1989.

[58] Xie, X.-L. (1989) : "Controle Hierarchique d'un Systeme de Production soumis a Perturbations," Ph.D. Thesis, University of Nancy I, France, 1989.

[59] Zoller, K. (1971) : "Optimal Disaggregation of Aggregate Production Plans," Management Science, Vol. 17, No. 8, 1971.

7   Hierarchical Control Approach to Managerial Problems for Manufacturing Systems

George Kapsiotis and Spyros Tzafestas

1.   Introduction

Production Planning for multi-product CIM systems is a simple problem to identify but very difficult to solve optimally. This problem is generally concerned with the specification of the number of units to be produced from each item for a forthcoming planning horizon, so that a total demand is fulfilled while operational costs are minimized (or equivalent profit is maximized). One of the major difficulties that complicates the problem is that the production subsystem is only one part of the whole process of supply, manufacturing, promotion and distribution of products to the end-customers. Interactions of production with other elements of the system can be easily identified. For example, the production-inventory relationship is well known. Setting appropriate inventory levels is a major issue in aggregate planning. However, aggregate production decisions also generate constraints for purchase, transportation and distribution. The whole process is also related to other functions, among which marketing plays a vital role. Clearly promotional activities influence the inventory control system and conversely a new inventory control system may affect the demand pattern that faces a manufacturing company.

Since the classical work of HMMS (Holt et al.,[43]) there has been published a voluminous amount of related literature which discusses optimization models for separate production, inventory and marketing decisions. These models implicitly assume that decision making in these areas is separable (decouplable) and they treat the decisions from other sectors either as constraints upon, or as input data to their own decision model. However it is clear that there is a strong need to make the interactions between decision areas explicit and study the effects and costs of simultaneous versus separate sequential decision making by the firm.

In modern manufacturing enterprises, the supply process of the required raw materials and (or) semi-product inventory ensures a continuous operation of the production function. For such multi-product fast growing business the fundamental problem of inventory control, to achieve balance between the advantages of holding large inventories and those of holding small ones, becomes much complicated and increasingly difficult. Single-level conventional methods are not so effective when tackling the important issue of inventory integration, scheduling, production and marketing process into a coherent coordinated planning system. This need is being recognized by some researchers and some attempts of solving the problem have already been made[3,5,42,75,107]. One major approach is to use aggregate production planning. The most important point for such a design is the second stage of this procedure, namely the disaggregation process, which has to determine plans for each individual product. Special care must be given to ensure the concistency and feasibility of the resulting plans [6,34-35,75]. Within a different framework one can use a multi-product model where all items are considered jointly, or in a less aggregated detail [12,65,107]. To this end we feel that normative decision models which use multi-level techniques and depart from the classical aggregate approach can offer significant advantages when dealing with large and complex problems.

The work presented here is structured as follows : In section 2 a survey of the related production planning literature and hierarchical system's theory is given. In the first subsection production the planning literature is reviewed, and in the second, basic hierarchical concepts are outlined for both linear and non-linear systems. In section 3 the problem of optimizing the operation of an autonomous manufacturing/supply chain is considered. The chain consists of an upper level manufacturer and a hierarchy of indermediate suppliers/subsuppliers that provide the required materials. In section 4 some markering issues are considered. The Marketing subsystem plays a vital role in modern integrated manufacturing systems and its activities must be taken explicitly into account for an efficient globally optimum strategy. In section 5 a composite inventory-marketing problem is considered. The problem is treated in its most general form and appropriate hierarchical algorithms are developed. In section 6 an integrated multi-product dynamic model is given which considers explicitly the participation of all organizational units in the process of the long-term strategic planning. Finally, in section 7, some concluding remarks are given followed by a substantial bibliography.

All results presented in sections 2 through 6 are supported by simulated examples.

It is noted that management of the transportation/distribution system for the manufactured products is not considered in the present work. Though transportation and logistics distribution decisions contribute equally to the final operational cost, a quite different mathematical technique - namely the Generalized Network technique - is required for a complete analysis. For a complete discussion of this subject refer to [96,97]. Logistic system design is also treated in [9,39,87].

2. An Overview of the Related Literature

2.1 Production Planning Literature.

The problem in its general form can be stated as: *Find a composite plan of the manufacturing firm (production vector plus marketing decisions) so that an objective goal is attained, while a set of operational and/or physical constraints are satisfied.*

The production vector may consist of a single variable, usually a production rate variable, or include a work force size, production capacity level, and other production control variables. However, the marketing-production models so far developed affect only the demand sequence and do not alter the production capacity of the firm. Although in pure production planning models the sales are usually subjected to some degree of forecasting, combined functional models explicitly assume that the firm can manipulate this sequence, which is no longer assumed to be given, but is generated according to marketing decisions.

The demand of an industial good depends mainly upon price, advertising, quality, service-level and distribution network [51,56]. The role of each of these parameters varies according to the specific type and stage of the life-cycle of the product under consideration. However, in the related literature attention was primarily focused on advertisement and price. Though of equal importance, it is more difficult to build normative models that quantify the effects of the other factors upon the production planning decisions. Our interest here focuses on advertisement decisions that become more and more important in contemporary competitive market environments [56].

Some of the most important works that treat the price of the product as the marketing subsystem control variable, are those of Pekelman [72,73], Kunreuther [52-53], Feichtinger and Hartl [30], Thompson et. al. [104], Tuite [106] and Vanthienen [108].

Some early works on dynamic models include the work of Thomas [102-103] who considers the problem of simultaneously making price and production decisions. In [102] he considers a monopolist having a fixed plus linear production cost and facing deterministic demand. Bergstom and Smith [12] treat an extension of the HMMS model [43] to the multi-product case, but allowed the firm to choose the demand in each period, instead of forecasting it. Kotler [51] on the other side, was among the first workers who initiated modelling of marketing-inventory systems. He studied the interaction between marketing policies and economic ordering quantities. On a quite different framework Mariani et al. [65] provide a modified maximum principle which allows them to solve a deterministic multi-product inventory problem and find the number of joint replenishements, the order quantities, and the times at which orders should be placed so that total cost is minimized. Leitch [55] considers the HMMS production model plus a linear lagging model to represent advertisement effects. However, the purpose of such a promotional policy is used to avoid peak-load production costs by shifting seasonal demand for a product, so that a kind of "smoothing" in demand patterns is achieved. The total value of demand is not being changed. More recent works include that of Parlar [71] who uses the Nerlove-Arrow model for the marketing subsystem and provides a numerical example for the dynamic constrained problem developed. Choi S.B. [17] uses a continuous-time model dealing with perishable inventories and a distributed time lag generalization of the Vidale-Wolfe equation. He presents a suitable algorithm and treats a special case application for constant decay rate function.

There are some other relevant works in a somewhat different setting which consider models for production and marketing where the solution process may take place in a decentralized mode, i.e. each function chooses its own variables.

Damon and Schramm [23] developed a three component model which allows marketing, production and finance to participate in the planning. After the global solution is obtained, the relative profitability of such a simultaneous formulation, compared to a sequential one, is demonstrated through a numerical example. Such an approach is also adopted and applied in section 5 of the present paper. Welam [110] tried to model the marketing side using response functions that have been presented in the marketing literature. In order to limit the bias in comparing these two solutions (since promotional data was unknown for the HMMS paint factory system under study), deviations from the actual values of price and promotion were involved. Freeland [33] analyzes the single-period marketing production planning problem in a functionally decentralized firm. He shows that in a decentralized mode the information needed for the marketing subsystem to optimally plan its policies, is the correct marginal cost of production. He suggests an iterative scheme for determining this marginal cost under reasonable assumptions. Abad [1-2] provides the same finding for the dynamic single-product case, using a continuous time formulation. He presents a decentralized procedure based on the goal-coordination principle of control theory.

Considering the mathematical principles applied for the solution of such complex problems, the work of Drew [26] is very important, as he has been among the first to apply multi-level techniques to managerial problems, thus providing some insight into their potential merits. Also the work of Younis and Mahmoud [107] is also of great importance in our approach. They explicitly consider the coupling between the inventory-production departments and give a four-level hierarchical algorithm to obtain optimal policies for each one of them.

Finally, other related works that discuss dynamic production-planning models and concepts can be found in [7,10-11,14,20, 27-29, 36,38,46,64,74,77,93,105].

2.2 Hierarchical Techniques Review
Linear Systems.
The mathematical models of many physical and engineering systems are frequently of high dimensionality and involve interacting dynamic components. The information processing demand and the requirements for experimenting with these models for control purposes, are

usually excessive. It is therefore natural to seek for techniques that reduce the computational effort. Hierarchical approaches and methodologies of large scale systems provide such techniques through the manipulation of system structure in an appropriate way.

Hierarchical or multi-level control schemes were first introduced in the work of Mesarovic et. al. [66] which is an important landmark in the development of an integrated theory of hierarchical systems. The essential idea in most hierarchical techniques is to solve, independently, decomposed subsystem problems at the first level for a set of variables called "coordination variables", which are provided by a second level hierarchy. At the second level these coordination variables are iteratively improved using information from the first level. From a computational point of view, the result is that as the size of the coordination vector is increased, the second-level problem becomes simpler, thus leading to less computation at the second level. These ideas have been proved very successful when handling linear systems with quadratic costs. Two major techniques have been applied for such systems, namely: **Goal Coordination** and **Interaction Prediction** methods, their major difference being that the coordination vector of the latter constitutes both the interaction vector and the Lagrange multipliers associated with the interconnection constraints. For discrete systems that involve time-delays the multi-level technique developed by Tamura [89-90], which is based on the *Lagrange Duality Theory* is very effective. Recently, another algorithm has been proposed by Cai [16] which makes use of the interaction prediction principle. Hierarchical control techniques can also be used when tackling non-linear optimization problems, where, even for small size problems, centralized single-level methods, present insurmountable difficulties. For this reason there has been much interest in the development of decomposition-coordination techniques for solving such problems [59-63,82-85]. It has been argued that such techniques have computational advantages both from the point of view of the accuracy that can be achieved and in terms of storage requirements and computational time criteria [84-85,94].

A characteristic feature of many natural and man-made systems is their hierarchical structure. In such cases the advantages offered by the hierarchical approach are obvious. Applications of multi-level techniques to a diverse variety of disciplines are reported in the related literature [2,26,31,68,84-85,89-90].

Non-linear Systems

The solution of non linear optimization problems via single level techniques leads ultimately to the solution of a non linear two-point boundary value problem (TPBV). Such TPBV problems invariably require successive approximation techniques for their solution [36]. As we have already mentioned, the application of a hierarchical algorithm implies decomposition of the system into several interacting subsystems, which provide the basis for the formulation of independent optimal control subproblems, linked together by a higher level coordinating unit.

In most algorithms the key variable of the first level is the costate vector which, each time the hierarchical iteration is switched to the first level, must be obtained either directly (non linear case) or indirectly (linear case). The first attempt to incorporate the costate vector in the coordination variables was made in the "generalized gradient method" of Mahmoud et. al. [58]. This method avoids complicated first-level problems. Soon after that, Hassan and Singh [41] made a modification of this scheme and called it **costate prediction**. Costate prediction was originally applied to discrete systems, because, for accuracy reasons, its application to continuous time systems requires a fine discretization that leads to an undesirable computational burden. Generally speaking, the optimization approach to nonlinear control systems using hierarchical algorithms is a reinjection-type decomposition-coordination that uses a combination of a pseudomodel and the prediction principle. The key idea is to write the functional criterion under minimization in a separable form (if it is already in separable form this step is omitted) and to write the nonlinear dynamic equations in a form of a linear part which is block-wise separable and another term which contains the non linearities, the delays and the interaction terms. The role of the higher hierarchical level here is to fix the nonseparable part in the criterion function (if it exists) and the nonlinear part in the dynamic equation. This results in a set of low-order dynamic optimization problems to be solved at the lowest level in the hierarchy. The higher level will successively approximate the specified variables to their optimal values.

Coordination approaches have been rather successfully applied to linear and non linear discrete time systems. However, not so many practical algorithms are available for time delay systems. One exception is the time delay algorithm of Tamura [89] which, in principle, can also be used for non linear systems, although its main success has been treating linear constrained time delay systems. The time-delay systems are some of the most difficult systems to which optimal control can be applied. Application of the maximum principle to such systems (either continuous or discrete) leads to two-point boundary value problems with delayed and advanced arguments in the state and the costate vectors, respectively. This undesirable characteristic makes optimization of such systems impossible and the designer must resort to suboptimal techniques such as parameter imbedding, sensitivity or series expansion, and system transformation [45].

In [45] a costate prediction method is developed for non linear discrete time systems with time-delay and a non separable cost function. Decomposing appropriately the original system, a two level algorithm has been developed which at the lower level avoids the solution of a two-point boundary value problem, since the costate vector is included in the coordination vector as a cordination variable. However the resulting algorithm treats only a fixed time- delay both in the control and state variables, though the authors do mention that multi-delay systems can be handled too.

Another major difficulty in non linear systems, besides the existing distributed time-delays, is the presence of inequality constraints. For unconstrained non linear problems there are efficient hierarchical algorithms that are mainly based on the costate prediction approach [59,84-85]. The constrained case has been considered in [59-60] where the treatment of constraints is implicit, thereby yielding constrained lower order problems after a suitable decomposition. In [62] Mahmoud et. al. treat the constrained problem. The form of the constraints (that is met in most of the applications) is of the type $u_m \leq u \leq u_M$, where u_m and u_M are respectively the minimum and the maximum allowable levels of the control signal. The control constraints are treated explicitly in [62], and three different techniques, namely slack variable method, intervention parameter, and penalty function methods are discussed. In a similar work which refers to an application of multi-level techniques to a supply-inventory production problem, Younis and Mahmoud [107] develop a costate prediction algorithm that also treats constraints on the state variables (of the same form as the control inequalities) via a penalty function approach. However, in the last two references only unretarded systems are treated and no time delay is introduced.

Recently the authors developed a multi-level algorithm for interconnected non-linear subsystems, which may involve multi-time delays and constrained control and state variables [99]. The algorithm unifies all previously reported techniques combining their features appropriately and treating the problem in its most general form. Also an application from the Production planning literature, involving automation acquisition, is treated using the previous algorithm.

3. Coordinated Control of Supply-chains

3.1 The problem - Introduction
Consider a manufacturer producing M final products and requiring $m_1,...,m_n$ raw materials. The demand for these materials is assumed to be known (or can be forecast) and possibly time varying. This means that the production subsystem of the enterprise initiates orders and provides plans for the sequence m1,...,mn. A combined treatment of these two problems can be found in [107]. The inventory department of the manufacturer initiates orders to its lower hierarchical level which consists of the suppliers of the required materials. Similarly, suppliers send orders one level down to their subsuppliers, which provide the (tertiary) raw material necessary to produce their own (secondary) raw material. This procedure can be repeated as many times as needed.

For the sake of simplicity and with no loss of generality, one can distinguish three levels in this process. Level I consists of the final manufacturer that orders the final raw materials to a number of (primary) suppliers. Level II consists of the suppliers (S) that provide the final raw material to the manufacturer, by ordering appropriate quantities to the downstream level of the chain. Finally, level III consists of the subsuppliers (SS) that provide the upper level (suppliers) with the ordered goods.

Thus the structure of the chain has the form of Fig.1.

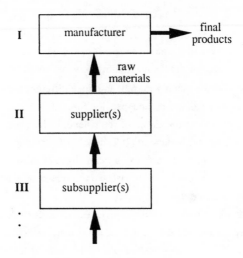

Fig. 1 : Structure of the Manufacturing/supply chain

It is noted again that depending on the specific chain, l-levels may be present. This structure of the supply chain is well recognized [21] but can be tailored according to the needs of any specific problem. Clearly, such a complex structure cannot, in general be handled with conventional single level methods. Some type of coordination may be needed, depending upon the specified objectives either local or global, and the relationship of the partners engaged in the chain.

3.2 Scenario I : Optimization at the manufacturer level

Consider again a manufacturer that produces M final products using m1,...,mn raw materials. Suppose that each of the raw materials is provided by a distinct supplier. The same is also assumed to be true for the supplying process of each supplier, i.e. a distinct subsupplier provides the required material to each supplier. Under this assumption we have the structure shown in Fig. 2.

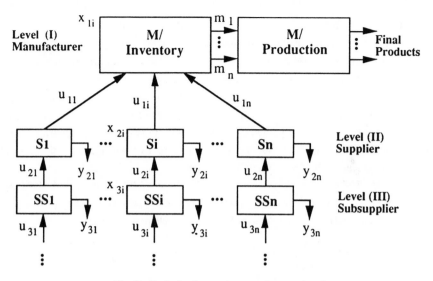

Fig. 2 : Optimization at the manufacturer level

Here x_{ji} is the inventory of the i-material (i=1,...,n) at the j-level (j=1,...,l), u_{ji} is the order plan that the j-level requires for the i-material, and y_{ji} is the local demand of the supply chain. In our case this is considered to be zero, since we are mainly dealing with an autonomous supply chain where all the procured materials are fed to the upper level. Note that here we assume that each raw material can be procured by a single (sub)supplier just one level down the hierarchy. If for each secondary raw material, at the supply level, n_i different subsuppliers are required, then u_{2i} should be replaced by $\sum u_{2ij}$. But this does not cause any problem in our analysis.

Writing down the inventory-balance equations at each level of the supply chain (assuming l=3 levels) one obtains:

$$x_{i1}(k+1) = x_{1i}(k) + u_{1i}(k-\theta_{1i}) - m_{1i}(k)$$
$$x_{2i}(k+1) = x_{2i}(k) + u_{2i}(k-\theta_{2i}) - u_{2i}(k) \qquad (3.1)$$
$$x_{3i}(k+1) = x_{3i}(k) + u_{3i}(k-\theta_{3i}) - u_{3i}(k)$$
$$\vdots$$

$$i = 1,...,n \qquad\qquad k = 1,...,k_f$$

where $m_{1i}=m_i$ is the sequence of demand pattern for material i, provided in principle by the production department at the manufacturer level. The delay θ_{li} includes (where necessary) both the delivery process delay/lag and the delay introduced by the production process itself, when it takes place at each supply (subsupply) production planning subsystem. In other words, in our model the production process (if any) at each level of the chain is determined by the delay it induces where this procedure is involved. Note that we are not seeking a detailed production schedule for each subsystem, which would call for a coupled model, but for effective inventory ordering policies that ensure effective operation of the specified system.

The number of levels of the chain and the value of i, which defines the horizontal length of the chain can vary according to the specific problem at hand. Also, more complicated relationships can be envisaged, e.g. the case where the inventory of supplier i depends on the inventory of supplier i+1, thus allowing some interaction among them, or the case of interdependent procurement of raw materials and so on.

The manufacturer should keep for each of the supplied materials a safety stock x_i^* against unexpected delivery delays or increase of the rate of consumed products. Assuming quadratic inventory and ordering costs the manufacturer problem can be stated as:

$$\min \sum_{i=1}^{n} \sum_{k=1}^{k_f} h_i(x_i(k) - x_i^*(k))^2 + r_i u_i^2(k) = \min \sum_{i=1}^{n} \sum_{k=1}^{k_f} J_i(k) \qquad (3.2)$$

subject to :

$$x_i(k+1) = x_i(k) + u_i(k - \theta_i) - m_i(k)$$

$$0 \le u_i(k) \le U, \quad 0 \le x_i(k) \le X, \quad i = 1,...,n \qquad (3.3)$$

where, for simplicity, the index 1 has been dropped (global constraints for the total inventory and ordering can be imposed too [26-48]). The delays are introduced by the down-stream subsystems (lower levels) in supplying the required quantities. Each delay may be distributed in some time interval, but here it is considered to be fixed. Under these conditions it is implicitly assumed that the manufacturer's inventory subsystem is affected only by the delay θ_{1i} introduced by its lower level hierarchy, and it makes no difference whether the other parts of the chain are optimizing or not their operation. Once the manufacturer has determined the optimal order pattern $u_i(k)$, this sequence can be employed at level (II) (inventory equations) of the chain. These quantities are the actual deliveries for the suppliers. Therefore one observes that each level of the chain is confronted with the same problem (assuming the same type of costs) and the same algorithmic scheme can be used to solve the corresponding problem. In this way sequential solution propagates downstream until final overall supply profile is obtained .

Clearly, this approach, as posed, assumes that the manufacturer, being the leader in the supply chain, simply announces his policy (providing order patterns $u_i(k)$), and then the suppliers minimize their own operational cost treating $u_i(k)$ as an uncontrollable input to their system. In other words, no cooperation between the various levels of the chain is permitted. The upper level hierarchy decisions are regarded as requirements that have to be satisfied by the lower-level ones. In the same way the outcome of these decisions becomes the objective framework for the next level downstream the hierarchy.

Clearly, this approach permits each upper level to minimize its own cost effectively. But this is done at the expense of each immediate downstream level which may be confronted with more costly order pattern, than if this order pattern were realized by local decisions.

The solution of this problem would be using the multi-level approach of Tamura [89]. Define the dual of the problem (3.2), (3.3), i.e.:

$$\max_{p_i} M_i(p_i)$$

$$M_i(p_i) = \min_{u_i, x_i} \sum_i \sum_{k=1} h_i (x_i(k) - x_i^*)^2 + r_i u_i^2(k) +$$

$$+ p_i(k) [x_i(k+1) - x_i(k) - u_i(k - \theta_i) + m_i(k)] = \qquad (3.4)$$

$$\min_{u_i, x_i} \sum_i \sum_k \{ J_i(k) + p_i(k-1) x_i(k) - p_i(k) x_i(k) - p_i(k+\theta_i) u_i(k) +$$

$$p_i(k) = 0, \quad k \leq 0 \wedge k \geq k_f + 1$$

If $p_i(k)$ is given and fixed, one gets the following independent minimization problems :

k = 1

$$\min_{u_i(1)} \{ J_i(1) + p_i(0) x_i(1) - p_i(1) x_i(1) - p_i(1+1_i) u_i(1) \}$$

$x_i(1) = x_{i1}$ given , $p_i(0) = 0$

k = 2,...,k

$$\min_{u_i(k), x_i(k)} \{ J_i(k) + p_i(k-1) x_i(k) - p_i(k) x_i(k) - p_i(k+\theta_t) u_i(k) \}$$

k = k_f + 1

$$\min_{x(k_f)} \{ J_i(k_f + 1) + p_i(k_f) x_i(k_f + 1) \}$$

$(J_i(k_f+1) = $ terminal cost, if any)

which can be solved analytically. At the upper level, one has to update $p_i(k)$ according to :

$$p_i(k+1) = p_i(k) + k \nabla_{p_i(k)} M_i(p_i)$$

k>0 step length, where

$$\nabla_{p_i(k)} M_i(p_i) = x_i(k+1) - x_i(k) - u_i(k - \theta_i) + m_i(k) \qquad (3.5)$$

and all values of $x_i(k)$ and $u_i(k)$ are known from the previous level. Since costs are convex and state-space equations are linear, convergence is assured.

3.3 Scenario II : Multi-level coordinated optimization

Now consider the alternative formulation where the objective function to be minimized is the total cost of all levels involved. This means that all members of the chain act as a team trying to meet the required constraints, i.e. the suppliers/subsuppliers are not regarded simply as followers but participate actively in decision plans. A type of coordination that distributes local tasks and ensures overall optimum performance is necessary. It can be assumed that this supreme coordinator consists of some top management at each level of the chain that appropriately interacts with each partner, receiving and transmitting a suitable amount of information.

For simplicity, let us rewrite the inventory balance equations for one product supply process:

$$x_1(k+1) = x_1(k) + u_1(k-\theta_1) - m_1(k) \tag{3.6}$$

$$x_i(k+1) = x_i(k) + u_i(k-\theta_i) - u_{i-1}(k)$$

$$i = 2,...,l \quad , \quad k = 1,...,k_f$$

where the supply chain has l levels. The global performance criterion is :

$$\sum_{i=1}^{l} \sum_{k=1}^{k_f} h_i (x_i(k) - x_i^*)^2 + r_i u_i^2(k) = \sum_{i=1}^{l} \sum_{k=1}^{k_f} J_i(k) \tag{3.7}$$

The state of the ith subsystem is affected by its local control action (delayed θ_i time intervals) and also the control of the previous (i-1)th subsystem.

Define $z_i(k)=u_{i-1}(k)$. Then the overall Lagrangian is :

$$L = \sum_i \sum_k J_i(k) + b_i(k) [z_i(k) - u_{i-1}(k)] +$$

$$+ p_i(k) [x_i(k+1) - x_i(k) - u_i(k-\theta_i) + z_i(k)] \tag{3.8}$$

and so the optimality conditions give:

$$\frac{\partial L}{\partial b_i} = 0 \Rightarrow z_i - u_{i-1} = 0 \tag{3.9}$$

$$\frac{\partial L}{\partial z_i} = 0 \Rightarrow b_i + p_i = 0 \Rightarrow b_i = -p_i$$

Now, assuming that $b_i(k)$ is fixed and known for i=2,...l one obtains the following subproblems:

$$\min \sum_k J_i(k) - b_{i+1} u_i(k) \tag{3.10}$$

$$x_i(k+1) = x_i(k) + u_i(k-1_i) - z_i(k)$$

$$i = 2,...,l \quad , \quad b_{l+1}(k) = 0 \quad , \quad z_1(k) = m_1(k)$$

where $z_i(k)$ is known from the solution of the previous subsystem.

The above procedure is schematically represented by the coordination scheme of Fig. 3.

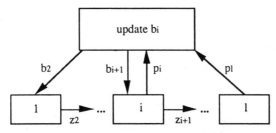

Fig. 3 : Coordination scheme for the supply chain

For each of the subproblems i=1,2...,l the previous temporal decomposition can be used, which gives k_f+1 single variable minimizations. At the upper level we have to update b_i by the simple substitution-coordination scheme :

$$b_i^{k+1} = b_i^k + k \{ - p_i^k - b_i^k \} \tag{3.11}$$

where i=1,2...,l and k is th step length

Here, p_i is the adjoint multiplier obtained from the solution of the subproblems. From (3.9) we know that $b_i = -p_i$ and p_i is the adjoint variable of x_i, i.e. p_i is the marginal value/cost of an additional unit of inventory (according to the interpretation of the adjoint variables). Inspecting the modified local criterion (3.10) one can see that in this case each subsystem tries to minimize its local operational cost plus the cost incured for each additional unit of inventory that the (i+1)th subsystem has to keep, such that the orders of subsystem i are satisfied. This conclusion seems quite reasonable since a global cost has to be minimized and the interactions among subsystems have to be taken into account.

3.4 Scenario III : Decentralized Optimization

From the previous formulation it is clear that the disturbance that affects the whole chain, is the demand which originates from the manufacturer's production department and imposes some constraints on the order policy of each level, so as to satisfy all the generated demand. If each decision maker could freely choose his demand level, then a desired value z_i^* would be specified according to the specific conditions that dominate the market environment. For each of the supply chain levels the uncontrollable disturbance input comes from the previous subsystem ordering. It would be desirable that each level should try to eliminate this effect by enforcing this previous subsystem's order pattern to be as close as possible to the above mentioned precalculated value z_i^*.

One way to accomplish this task, is to introduce deviation terms $(z_i(k)-z_i^*)^2$ (where $z_i=u_{i-1}$) into the local objective function of each subsystem, which now becomes:

$$J_i(k) = \sum_{k=1}^{k_f} h_i(x_i(k) - x_i^*)^2 + r_i u_i^2 + s_i(z_i(k) - z_i^*)^2 \tag{3.12}$$

With this modified criterion, the overall Lagrangian becomes:

$$L = \sum_{i=1}^{1} \sum_{k=1}^{k_f} J_i(k) + p_i(k+1) [x_i(k+1) - x_i(k) - u_i(k-1_i) + z_i(k)] + \qquad (3.13)$$

$$+ b_{i+1}(k) [z_{i+1}(k) - u_i(k)]$$

Then the necessary conditions for optimality yield:

$$x_i(k+1) = x_i(k) + u_i(k) - z_i(k) \qquad (3.14)$$

$$z_{i+1}(k) = u_i(k)$$

$$2 h_i (x_i(k) - x_i^*) + p_i(k) - p_i(k+1) = 0$$

$$2 r_i (u_i(k) - u_i^*) - p_i(k+1_i) + b_i(k) = 0$$

$$2 s_i (z_i(k) - z_i^*) - p_i(k+1) + b_i(k) = 0$$

Assuming now, that b_i is given and fixed, one has to solve 1 subproblems with respect to u_i, x_i, z_i with the aid of decomposition methods as before. At the upper level one has again to enforce interconnections to be satisfied, by say a simple substitution. In this way an optimum global solution can be obtained for which every part engaged in the process accepts some upper level directions so that demand interconnections constraints are satisfied and all costs are manipulated effectively.

However, instead of seeking a global optimum solution, here we are seeking a suboptimal one, which can be obtained in a decentralized way. Each subsystem solves its local problem and then feeds forward this solution. The key point to achieve this, is that instead of inserting quadratic terms in the local criterio, depending upon deviations of the interactions zi (steming of the previous subsystem) from a prespecified value, we introduce quadratic terms in the locally dependent interconnection variables z_{i+1}.

Thus the local criterion becomes :

$$J_i(k) = \sum_{k=1}^{k_f} h_i (x_i(k) - x_i^*)^2 + r_i u_i^2 + w_i(k) (z_{i+1}(k) - Z_{i+1}(k))^2$$

where w_i, and Z_{i+1} can be found by the same procedure as in [83]. For the case where $s_i=0$, after some manipulation one obtains:

$$w_i(k) = \{ q_{i+1}^{-1}(k+1) + A_{i+1}(k) q_{i+1}^{-1}(k) A_{i+1}^T(k) + B_{i+1}(k) r_{i+1}^{-1}(k) B_{i+1}^T(k) \}^{-1}$$

$$Z_{i+1}(k) = x_{i+1}^*(k+1) - A_{i+1}(k) x_{i+1}^*(k)$$

where A_{i+1}, and B_{i+1} are the (i+1)th subsystem matrices. This suboptimal decentralized solution scheme allows possible conflicts between the various parts of the process, and all simulated examples have shown that it takes only a small percent of the computation time of the optimal solution. Also the loss of optimality is proven to be very small (depending upon the conflict of the objectives for each part in the chain). Thus one can use this approach for solving the problem in a repetitive manner, as more accurate demand forecast becomes available in each time period.

3.5 Simulation Examples

All previous algorithms corresponding to the three different scenarios have been implemented computationally and extensive simulation results have been carried out to verify our conclusions. In all cases the computation time for scenarios/cases (II) and (III) was only a percentage of the time required for the global optimum solution. The degree of suboptimality depended on how severe was the conflict for each of the subsystems involved in the chain. In figures (4a) to (7b) a complete simulated numerical example is given, for a chain consisting of three levels. The induced delay for each subsystem is one time period. The safety stock x_i^* is assumed to be zero (note that this fact restricts the conflict among the decision makers involved in the process). The external demand supplied by the production department of the manufacturer level has a sinusoidal form as shown in figure 4a. The cost parameters are : $h_1=1$, $h_2=2$, $h_3=3$,$r_1=1$, $r_2=10$, $r_3=15$. In figures (4a) and (4b) the form of the optimal solution is presented. In figures (5a),(5b) the trajectories of subsystem (I), which correspond to the scenarios (II) and (III) respectively are given. Similarly in figures {(6a),(6b)} and {(7a),(7b)} the trajectories for subsystems (II) and (III) are given for the two scenarios. The ratio of the computational time for this example was found to be 18 to 1 for the optimal and suboptimal solution. The value of the global cost for the overall system was 1.803×10^5, while scenarios (II) and (III) gave 1.820×10^5 and 1.835×10^5 respectively. Thus the loss of optimality is about 0.95% for the decentralized solution, scenario (III), and 1.77% for scenario (II). Since the computation time is almost identical in these two cases, the one corresponding to the decentralized suboptimal solution is certainly preferable.

4. Marketing Issues

Effective strategic planning of "Integrated Manufacturing Systems" requires the explicit consideration of existing interdependencies among their functional/operational subsystems. Pure production planning models, on the one hand, simplify this problem by assuming that the final product(s) demand is a variable externally given or forecasted. On the other hand, functional models consider the product(s) demand as an internal variable of the system, effectively controlled by the marketing subsystem variables. This approach permits the management of the system to develop an overall optimal design plan. Although the complexity of the problem is enormously increased, the cost savings that can be achieved (over the simpler sequential approach) are worth the model design. We remark here that the final outcome of the marketing subsystem activities does not result from an homogenous input that forms the final demand level of the product(s). This input includes a set of variables that affects the sales of the firm in a known or an unknown way, and is referred to as the "marketing-mix" variables.

optimal solution

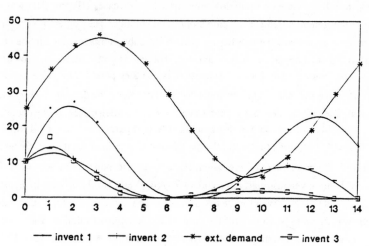

optimal solution

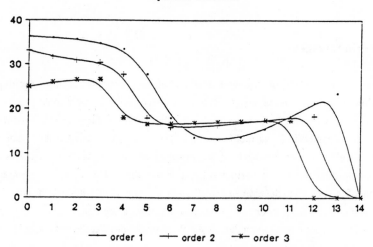

Fig. 4a,4b : simulation example

subsystem I

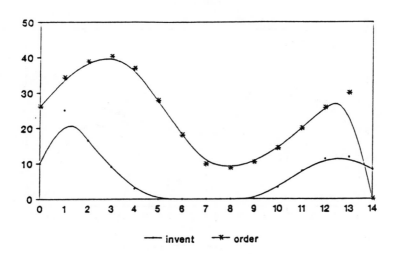

subsystem I

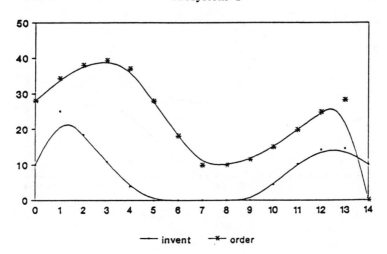

Fig. 5a,5b : simulation example

subsystem II

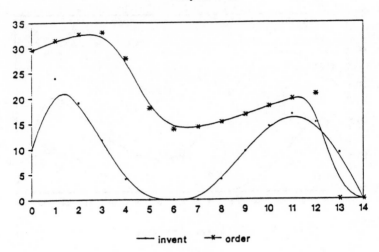

subsystem II

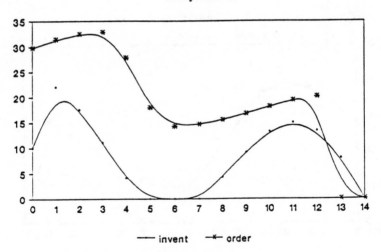

Fig. 6a,6b : simulation example

subsystem III

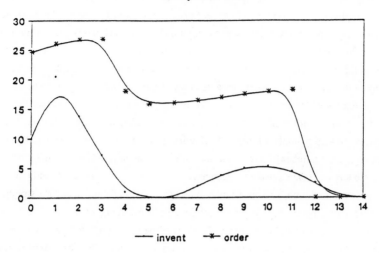

subsystem III

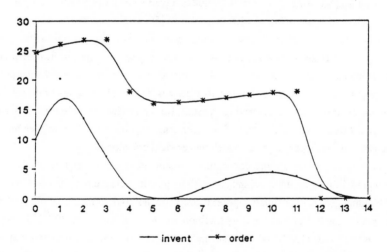

Fig. 7a,7b : simulation example

Many efforts have been made by the marketing scientists over the last years with respect to the development of apropriate models describing the dependence of the product's demand over this set of variables. However, since the marketing effort is far from being an homogenous input the market's response to variations in the level of any marketing input is in general conditioned by the level of the other activities. Variation of two or more marketing activities at the same time may have greater or less effects than the sum of the separate effects. Furthermore there are several other characteristics of the marketing environment, besides the marketing-mix variable interaction that make it difficult to predict and control the effect of marketing actions (competitive effects, delayed response, multiple products, functional interactions, multiple goals of the manufactiring firm). However, despite the existing complexity, a large number of successful measurement models, decision-making models and theoretic models were developed and efficiently implemented by researchers and practitioners. This obvious contradiction is removed by a plain fact stated in [56] : "managers simply cannot afford not to take a model- building approach in their fast-moving, competitive environment". Under some assumptions and further specifications of the marketing process and the marketing environment, related models can describe the reality quite well. The development of further model-based aids suitable for the marketing strategy development will provide more profitable and healthier environmets for those organizations that can use them correctly.

Certainly, one of the most important promotional tools of modern marketing management is *advertising*. The potential value of advertisement can be attested by the yearly increased advertising spending. As Lilien et.al. say [56] : "No one knows what advertising really does in the marketplace. However, what advertising is supposed to do is fairly clear : advertising is undertaken to increase or to acheive company sales and/or profits over what they would be otherwise". More specifically, the purpose of advertising is to enhance potential buyer's responses to the organization and its offerings by providing suitable information, by channeling desires, and by supplying reasons for preferring the offer of a particular organization. Over the last years quite a few advertising-response models have been built; they fall into two general classes : a priori models and econometric models.

A priori models provide a more conceptually sound set of characteristics. Models of this category draw heavily on intuition and although its practitioners are not oblivious to data, the model building goal is to postulate a general structutre, not to describe a specific application. Representative models of this building tradition are those of Vidale-Wolfe [109], Nerlove-Arrow, Little , Simon (1982) and Mahajan and Muller (1986) [56]. **Econometric models** are better related to available data. Econometric models usually start from a specific data base (e.g. time series of sales or share and advertising). These models include those of Bass (1969), Bass and Clarke (1972), Rao and Miller (1975), and Eastlack and Rao(1986). A recent review of the subject can be found in Hanssens et. al. (1990). Besides the previous two categories, some other related works can be found in the literature on mixed (hybrid) models that start with rather complicated a priori models and endeavor by statistical methods to fit and evaluate them. Finally, we note here that Little (1979) [57], and Saunders (1987) [76], reviewing these studies, summarize a set of propositions about the shape of the relationship of the "effort" i.e. the level of the marketing instrument (advertisement) and the "effect" i.e. sales/awareness or any other appropriate measure of response. The above discussion has set the basic framework of the strategic role of the marketing subsystem in the activities of Integrated Manufacturing Systems and the problems associated with it. A short review of the related literature where composite production-marketing models are discussed, was given in section 2. For the development presented in the following sections, a hybrid model that describes the market's response function was chosen. This model is an extension of the Koyck-type rational distributed lag model used to model the marketing process by Leitch [55]. The model is consistent with empirical evidence indicating that the effect of advertisement is cumulative and that its residual influence will decline with time unless the effort is continued. Furthermore, it has been found successful in terms of goodness of fit and predictive ability, also because of its linearity it can be easily manipulated in an involved computation scheme. Its mathematical form is as follows :

$$s_{t+1} = (1-\lambda) s_t + d + a_0 A_t + a_1 A_{t-1} + ... + a_k A_{t-k}$$

where

s_t = level of sales in period t

A_t = level of advertising expenditures in period t

d = sales level at zero advertising

a_i = advertising effectiveness in the period in which the expenditure is made, λ = decay rate of sales.

Estimation of the model order and its parameters is assumed to be done in a consistent manner repeatedly, in each planning period (if required), resulting in a time-varying relationship that can better capture dynamically changing environments.

5. The Composite Inventory-Marketing Problem : An Hierarchical Control Approach

5.1 Problem Formulation

We consider now a composite inventory/marketing problem which is defined as follows. The firm purchases units of a product and then uses its promotional policy to achieve a desired average sales level. The objective is to minimize the total firm's cost.

Inventory Submodel

The Inventory/Advertisement (I/A) model consists of two submodels. The first submodel, denoted by I, stands for the inventory (storage) department of the company, which purchases and sells one or more products. At times $k = 1, 2,...,$ k decisions have to be made by the management board about the replenishment of the inventory (stock) quantity $x_i(k)$ of the ith product. To this end, orders $u_i(k)$ are initiated at a cost $q_i(u_i(k), k)$. The cost of holding $x_i(k)$ units as inventory of the ith product is $h_i(k)$. In the time interval $(k, k+1)$ the inventory is replenished by sales $s_i(k)$. Of course, $s_i(k)$ is not known in advance, but it is a function of the company's promotion policy. Thus the fundamental inventory subproblem is to achieve a balance between large stocks and small ones. Large stocks require large quantities to be ordered which result in a high purchasing cost and to a large amount of sales, if a proper promotional policy is adapted for achieving the planned market share. Small stocks imply small ordering costs, but the sale level is constrained by the current available stocks. It is assumed that the deliveries to the company are made in a distributed way, thus imposing corresponding delays to be considered. The products are considered to be independent i.e. the orders of one product does not affect the orders of any other product. Backlogging is not allowed, and the delays are assumed fixed and known. Finally, the case where product losses at a constant rate r_i are possible due to deterioration, damage, and/or other factors, is considered.

On the basis of the above, the "I" submodel has the form

$$x_i(k+1) = x_i(k) + \sum_{m=0}^{\theta_1} g_{im} u_i(k-m) - s_i(k) - \tau_i x_i(k) \tag{5.1}$$

for $i = 1,2,...,N$, where N is the number of different products, g_{im} is a coefficient that shows which part of the order for the ith product made at time k-m, is delivered at time k, θ_1 is the maximum delivery period (which is known and the same for all products, if not the same then there are appropriate zero coefficients in the model for the ith product), and all the other symbols are as described above. It should be remarked that obtaining good estimates of all parameters involved in the "I" model (5.1), requires a lot of work and time. Our purpose here is to study the quantitative impact of these parameters and variables on an integrated environment, and to determine the "best policy" to meet these requirements. These parameters are assumed to be known. Of course, in practice one may have to consider many more factors and to solve the problem in a repetitive manner.

Marketing Submodel

Here it is assumed that the objective of the company is to achieve the planned market share, with a minimum operational and advertisement cost. No attempt will be made to maximize the company's profit which would require the knowledge of the products prices over the whole planning horizon. It is further assumed that the firm can influence and shift the media preference with proper products advertising. Competition is not taken into account, but the model could be extended to do so. It is also assumed that the effect of advertising is cumulative and that its residual influence declines with time, unless the effort is continued. Zero advertising does not necessarily result in zero sales, for we assume an exogenous demand (d_i). Under these assumptions the advertising model (A) becomes:

$$s_i(k+1) = s_i(k) - \lambda_i s_i(k) + \sum_{m=0}^{\theta_2} h_{im} A_i(k-m) + d_i(k), \quad 0 \le \lambda_i < 1 \tag{5.2}$$

where $s_i(k)$ denotes the sales of product i at time k, λ_i is the delay factor that sales diminish due to forgetting, $A_i(k)$ is the advertisement effort at time k made for product i, and h_{im} is the residual effect of advertisement effort at time k (h_{im} may be a decreasing sequence, e.g. h_{im} = $(h_i)^m$ as in Leitch [55]). This submodel is quite flexible and possesses the majority of Little's key features. As for the "I" submodel, knowledge of the parameters and variables preassumes much effort and large data bases with historical data.

The problem

The objective of the whole operation is now to minimize the following cost function

$$J = \min \sum_{i=1}^{N} \sum_{k=1}^{k_f} \{ h_i(x_i(k), k) + q_i(u_i(k), k) + c_i A_i^2(k) + d_i(s_i - s_{id})^2 \} = \tag{5.3}$$

$$= \min \sum_{i=1}^{N} J_i(k)$$

where s_{id} is the desired sales level of the ith product. It is assumed that products are not subtituted one for another, nor complementary.

Individual constraints of this model are:

$$0 \le x_i(k) \le X_i \quad , \quad 0 \le u_i(k) \le U_i$$
$$0 \le A_i(k) \le A_i \quad , \quad i = 1,...,N$$

Global constraints are imposed because of limited warehouse capacity of the enterprise and constrained advertisement budget , namely:

$$\sum_{i=1}^{N} x_i(k) \le M \quad , \quad \sum_{i=1}^{N} A_i(k) \le A$$

where M is the maximum overall storage capacity, A is the maximum overall advertisement expenditure budget. The constraint $s \ge 0$ needs not to be posed as it is obviously reduntant because of the form of the model. Though the individual cost items $h_i(x_i(k), k)$ and $q_i(u_i(k), k)$ may be general enough, later on they will be restricted to the widely accepted square approximations, as the advertisement ones.

5.2. Multilevel Solution of the Composite I/A Problem

5.2.1 Algorithm I

In this section the coupled inventory-marketing problem will be solved by using multi-level control algorithms. The dual optimization method [54,89] is first employed. The basis of the dual optimization method is to formulate the Lagrangian of the problem and then to solve the dual problem. Specifically the dual problem is :

$$\max_{\lambda} \Phi(\lambda) \quad , \quad \text{where } \Phi(\lambda) = \min \ L(x_i, s_i, u_i, A_i, \lambda), \text{ and}$$

$$L = \sum_{i=1}^{N} \sum_{k=1}^{k_f} J_i(k) + \sum_{k=1}^{k_f} \lambda_1(k) \sum_{i=1}^{N} \left[x_i(k) - \frac{M}{N} \right] + \sum_{k=1}^{k_f} \lambda_2(k) \sum_{i=1}^{N} \left[A_i(k) - \frac{A}{N} \right] \quad (5.4)$$

After some manipulation the function L becomes :

$$L = \sum_{i=1}^{N} \left\{ \sum_{k=1}^{k_f} J_i(k) + \sum_{k=1}^{k_f} \lambda_1(k) \left(x_i(k) - \frac{M}{N} \right) + \sum_{k=1}^{k_f} \lambda_2(k) \left(A_i(k) - \frac{A}{N} \right) \right\}$$

Therefore the N independent minimization problems are :

$$\min \sum_{k=1}^{k_f} \{ G_i(x_i(k), k) + h_i(u(k), k) + R_i(A_i(k), k) + d_i(s_i(k) - s_{id})^2 \quad (5.5)$$

$$+ \lambda_1(k)[x_i(k) - M] + \lambda_2(k)[A_i(k) - A] \}$$

(where the constant terms have been omitted) subject to :

$$x_i(k+1) = (1 - r_i) \, x_i(k) + \sum_{m=0}^{\theta_1} g_{im} \, u_i(k-m) - s_i(k) \quad (5.6a)$$

$$s_i(k+1) = (1 - \lambda) \, s_i(k) + \sum_{m=0}^{\theta_2} h_{im} \, A_i(k-m) + d_i(k) \, , \ 0 \le \lambda < 1 \quad (5.6b)$$

$$x_i(1) = x_{i1} \quad , \quad s_i(1) = s_{i1}$$
$$0 \le x_i(k) \le X_i \, , \quad 0 \le A_i(k) \le A_i \, , \quad 0 \le u_i(k) \le U_i$$

Thus at the lower level we have to solve the N independent problems given by (5.5),(5.6a,b). The solutions for the lower level are then used to obtain an improved $\lambda(k)$ trajectory in order to maximize the dual functional $\Phi(\lambda)$. $\Phi(\lambda)$ can be maximized using one of the standard gradient methods. For example at the l-iteration we have

$$\lambda^{l+1}(k) = \lambda^l(k) + a \nabla \Phi_{\lambda(k)}(k) \quad (5.7)$$

where "a" is the step length, and

$$
\nabla \Phi_{\lambda(k)}(k) = \begin{pmatrix} \displaystyle\sum_{i=1}^{N} x_i(k) - M \\ \displaystyle\sum_{i=1}^{N} A_i(k) - A \end{pmatrix}
$$

is the gradient of the dual function. The key-point of this method is that instead of solving the N problems posed by (5.5),(5.6a,b) one can go a stage further by decomposing the subproblems with respect to the time indices. This approach [89] allows the solution of problems with multiple delays in the state and control vectors without extending the dimension of these vectors.

Though our analysis is valid without having to specify the convex form of the corresponding costs, we will continue to assume that these costs have the widely adopted form of quadratic functions. Thus we set

$$
J_i(k) = h_i x_i^2(k) + q_i u_i^2(k) + c_i A_i^2(k) + D_i (s_i(k) - s_{id})^2
$$

with appropriate unit cost parameters h_i, q_i, and c_i. Equations (5.6a,b) can be rewritten as:

$$
x_i(k+1) = \sum_{m=0}^{\Theta} A_{im} x_i(k-m) + \sum_{m=0}^{\Theta} g_{im} u_i(k-m) - s_i(k) \tag{5.8}
$$

where
$$
A_{im} = 1 - r_i, \quad m = 0
$$
$$
A_{im} = 0, \qquad\quad m \geq 1
$$

and

$$
s_i(k+1) = d_i + \sum_{m=0}^{\Theta} B_{im} s_i(k-m) + \sum_{m=0}^{\Theta} h_{im} A_i(k-m) \tag{5.9}
$$

where
$$
B_{im} = 1 - \lambda_i, \quad m = 0
$$
$$
B_{im} = 0, \qquad\quad m \geq 1
$$

Note that one can have different delays ϑ_1 and ϑ_2 in (5.8) and (5.9) with appropriate zero coefficients.

The Langrangian of the i-subproblem of (5.2) is max $L_i'(p_i)$, where

$$L_i'(\mathbf{p}_i) = \min_{x_i, u_i, s_i, A_i} \sum_{k=1}^{k_f} \{h_i\, x_i^2(k) + q_i\, u_i^2(k) + c_i\, A_i^2(k) + D_i\,(s_i(k) - s_{id})^2 +$$

$$+ p_{i1}(k)\left[x_i(k+1) - (1-r_i)\, x_i(k) - \sum_{m=0}^{\Theta} g_{im}\, u_i(k-m) + s_i(k) \right] +$$

$$+ p_{i2}(k)\left[s_i(k+1) - d_i - (1-\lambda_i)\, s_i(k) - \sum_{m=0}^{\Theta} h_{im}\, A_i(k-m) \right] +$$

$$+ \lambda_1(k)\,[x_i(k) - M] + \lambda_2(k)[A_i(k) - A]\} =$$

$$= p_{i1}(k_f)\, x_i(k_f+1) + p_{i2}(k_f)\, s_i(k_f+1) + \sum_{k=1}^{k_f} \{J_i(k) + p_{i1}(k-1)\, x_i(k) - (1-r_i)\, p_{i1}(k)\, x_i(k) -$$

$$- \sum_{m=0}^{\Theta} p_{i1}(k+m)\, g_{im}\, u_i(k) + p_{i1}(k)\, s_i(k) + p_{i2}(k-1)\, s_i(k) - (1-\lambda_i)\, p_{i2}(k)\, s_i(k) -$$

$$- \sum_{m=0}^{\Theta} p_{i2}(k+m)\, h_{im}\, A_i(k-m) - p_{i2}(k)\, d_i + \lambda_1(k)\,[x_i(k)-M] + \lambda_2(k)\,[A_i(k)-A]\} \quad (5.10)$$

where $p(0) = 0$

$p(k) = 0, \qquad k \geq k_f+1.$

Defining the Hamiltonian as:

$$H_i(x_i(k),\, u_i(k),\, \mathbf{p}_i,\, s_i(k),\, A_i(k)) = \qquad\qquad (5.11)$$

$$J_i(k) - (1-r_i)\, p_{i1}(k)\, x_i(k) - \sum_{m=0}^{\Theta} p_{i1}(k+m)\, g_{im}\, u_i(k) + p_{i1}(k)\, s_i(k) -$$

$$- (1-\lambda_i)\, p_{i2}(k)\, s_i(k) - \sum_{m=0}^{\Theta} h_{im}\, p_{i2}(k+m)\, A_i(k) - p_{i2}(k)\, d_i$$

equation (10) becomes:

$$L_i = p_{i1}(k_f)\, x_i(k_f+1) + p_{i2}(k_f)\, s_i(k_f+1) + \sum_{k=1}^{k_f} \{ p_{i1}(k-1)\, x_i(k) +$$

$$+ p_{i2}(k-1)\, s_i(k) + H_i(k) + \lambda_1(k)\,[x_i(k) - M]$$

$$+ \lambda_2(k)\,[A_i(k) - A] \}$$

For fixed p_{i1}, p_{i2} the computation of $L_i'(\mathbf{p}_i)$ can be carried out by minimizing the function independently for each time index $k = 1,2,...,k_f+1$, i.e.

k = 1:

$s_i(1)$, $x_i(1)$ initial conditions (known)

$$\min_{u_i, A_i} \ \lambda_1(1)(x_i(1) - M) + \lambda_2(1)(A_i(1) - A) + h_i x_i^2(1) + q_i u_i^2(1) + c_i A_i^2(1) +$$

$$+ D_i(s_i(1) - s_{id})^2 - (1 - r_i) p_{i1}(1) x_i(1) - \sum_{m=0}^{\Theta} p_{i1}(1+m) g_{im} u_i(1) -$$

$$- (1 - \lambda_i) p_{i2}(1) s_i(1) - \sum_{m=0}^{\Theta} p_{i2}(1+m) h_{im} A_i(1) - p_{i2}(1) d_i =$$

$$= \min_{u_i(1), A_i(1)} \{c_0 + q_i u_i^2(1) + c_i A_i^2(1) - c_{11} u_i(1) - c_{12} A_i(1)\} \qquad (5.12)$$

where c_0, c_{11} and c_{12} are constants easily identified from the previous step.

k = 2,...,k_f:

$$\min_{u_i, A_i, x_i, s_i} \{\lambda_1(k)(x_i(k) - M) + \lambda_2(k)(A_i(k) - A) + H_i(k) +$$

$$p_{i1}(k-1) x_i(k) + p_{i2}(k-1) s_i(k)\} \qquad (5.13)$$

k = k_f + 1:

$$\min_{x_i, s_i} \{p_{i1}(k_f) x_i(k_f + 1) + p_{i2}(k_f) s_i(k_f + 1)\} \qquad (5.14)$$

If we have a quadratic terminal cost $F_i(k_f + 1)$ it is added in equation (5.14). Note that in this case one can have an almost - analytical solution which is found by solving problems (5.12), (5.13) and (5.14). The complete algorithm can be represented as shown in Fig.1

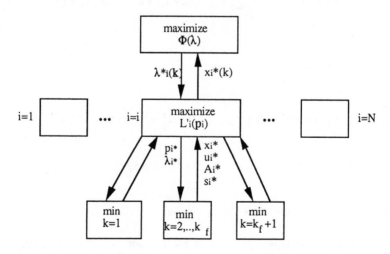

Fig. 8 : Hierarchical solution of the I/A problem by Algorithm 1.

5.2.3 Algorithm II

The above approach is essentially based on Tamura's algorithm applied to distributed delay systems. In the final stage, the marketing (M) and inventory (I) department send to the upper level their decisions (x_i*, u_i*, s_i*, A_i*). This assumes good communication channels within the enterprise and close cooperation so that the decisions in the two areas are planned in a centralized way. Actually this is the main point that has been criticized by several authors. As Freeland [33] points out, instead of the necessary cooperation of the M/I-P departments, one often observes some conflict between them. This is primarily due to two reasons: (a) poor communication and coordination, and (b) poor designed evaluation and incentive measures by the management organization.

Thus, following the approach of Abad [2], a decentralized procedure which permits each department to solve its problem independently will be given here, and then, with the coordination of a supreme unit, e.g. general management, will provide the overall solution. Inserting the global constraints into the objective function, one gets after some manipulation as in the previous section, the following N independent subproblems:

$$\sum_{k=1}^{k_f} \{ J_i(k) + \lambda_1(k) [x_i(k) - M] + \lambda_2(k) [A_i(k) - A] \}$$

Then, letting $z_i(k) = s_i(k)$ one can rewrite the state equations:

$$x_i(k+1) = (1 - r_i) x_i(k) + \sum_{m=0}^{\theta_1} g_{im} u_i(k-m) - z_i(k)$$

$$s_i(k + 1) = (1 - \lambda_i) s_i(k) + \sum_{m=0}^{\theta_2} h_{im} A_i(k - m) + d_i(k) , \quad 0 \leq \lambda_i < 1$$

$$z_i(k) - s_i(k) = 0$$

In this case the Lagrangian L_i becomes:

$$L_i = \sum_{k=1}^{k_f} J_i(k) + \lambda_1(k) [x_i(k) - M] + \lambda_2(k) [A_i(k) - A] + n_{i1}\{x_i(k + 1) - (1-r_i) x_i(k) -$$

$$- \sum_{m=0}^{\theta} g_{im} u_i(k - m) + z_i(k)\} + n_{i2} \{ s_i(k + 1) - (1 - \lambda_i) s_i(k) -$$

$$- \sum_{m=0}^{\theta} h_{im} A_i(k - m) - d_i(k)\} + b_i(k) \{z_i(k) - s_i(k)\}$$

Now assuming that $[b_i(k), z_i(k)]$ is given and fixed one obtains the following subproblems:

Inventory subproblem:

$$\min \sum_{k=1}^{k_f} h_i\, x_i^2(k) + q_i\, u_i^2(k) + \lambda_1(k)[(x_i(k) - M] + b_i(k)\, z_i(k)$$

$$\text{s.t.} \quad x_i(k+1) = (1 - r_i)\, x_i(k) + \sum_{m=0}^{\theta} g_{im}\, u_i(k-m) - z_i(k)$$

Marketing subproblem:

$$\min \sum_{k=1}^{k_f} \{ c_i\, A_i^2 + D_i\, (s_i(k) - s_{id})^2 - b_i(k)\, s_i(k) \}$$

$$\text{s.t.} \quad s_i(k+1) = (1 - \lambda_i)\, s_i(k) + \sum_{m=0}^{\theta} \lambda_{im}\, A_i(k-m) + d_i(k)$$

Thus, by considering as coordination variable the vector $[b_i, z_i]$ one can decompose the overall problem into two independent problems ((M) and (I)) which can be solved distinctly by the two departments. To ensure optimality at the upper level, one has to satisfy the constraints:

$$\frac{\partial L}{\partial z_i} = 0 \Rightarrow b_i = -n_{i1}$$

$$\frac{\partial L}{\partial b_i} = 0 \Rightarrow z_i = s_i$$

Thus, at each iteration of the upper level, the coordination vector is updated as:

$$b_i^{l+1} = -n_{1i}^l \quad , \quad z_i^{l+1} = s_i^l \qquad \text{(l-iteration)}$$

The information flow of this algorithm has the form shown in Fig. 2

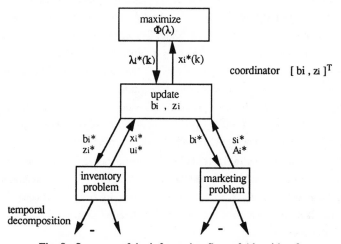

Fig. 9 : Structure of the information flow of Algorithm 2

Essentially, this is an interaction prediction approach which has been proven to have very good convergence characteristics. Actually, this is an extension of Abad's work [2] to our multi-product, distributed time-delay case. One could use the goal-coordination method with b_i as the coordination variable but this would lead to a singular solution with respect to z_i (z_i enters linearly in both the objective function and the state equations). Then one would have to modify the scheme either by inserting $(z_i - s_i)^2$ in the cost criterion as a penalty function, or by adopting some other method to overcome this difficulty. This is actually a more straightforward approach which needs a simple substitution at the second level of the algorithm, avoiding the gradient method which is more time consuming. If there is no binding constraint for u_i, one can provide the same findings as Abad for b_i which is regarded as the marginal cost of production for the production department (if u_i is considered as the production rate control variable). This is again consistent with Freeland's work [17] who has found that, for the static case, the only information that the two departments have to exchange is the correct value of the marginal production cost, and suggested an iterative scheme to compute it.

5.3 Simulation Examples

The algorithms described in the previous section were implemented and applied to several simulation examples in order to analyze the effect on the optimal solution caused by the model parameter changes/values. In figures (10a,b) the case of two products is considered. The model parameters are $h_1=0.1$, $D_1=1$, $q_1=0.5$, $c_1=0.4$, $s_{d1}=60$ for the first product and $h_2=0.1$, $D_2=2$, $q_2=0.5$, $c_2=0.5$, $s_{d2}=70$ for the second one. Inventory and sales decay parameters are assumed, $r=0.01$, $\lambda=0.02$, for both products. No time delay is introduced in the delivery or the advertisement process. The response of the system is given in figures (10a,b). The deviation from the desired sales is bigger for the first product since less weight has been attributed to this goal by the management board. The effect of distributed time delay in the supply and advertisement process is shown by figures (11a,b), where the same system is simulated with a distributed time delay of order two. Note the shifting of the inventory pattern and the reccesion of the order profile at the final time intervals.

In figures (12a,b) another two-product case is considered, with the following model parameters $h_1=0.01$, $D_1=1$, $q_1=0.1$, $c_1=0.1$, , $s_{d1}=70$, $h_2=0.01$, $D_2=2$, $q_2=0.1$, $c_2=0.1$, $s_{d2}=70$, $r=0.015$, $\lambda=0.025$. The upper bound of ordering equals 65 units and remains active for most of the simulated time. In figures (13a,b) the same model is simulated, where a global constraint is added because of limited warehouse capacity ($M=60$). Note again the pattern of inventory as the global constraint becomes active. A sensitivity analysis has been carried out for this model, but lack of space does not permit us a detailed presentation. The results were verified by both algorithms.

subsystem I

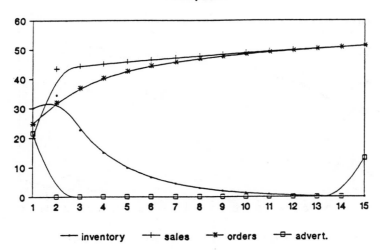

subsystem II

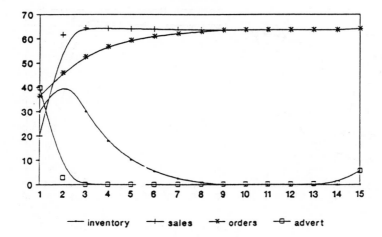

Fig. 10a,b : Simulation example

subsystem I

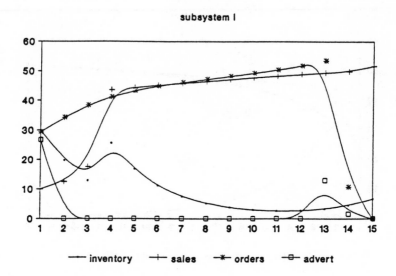

subsystem II

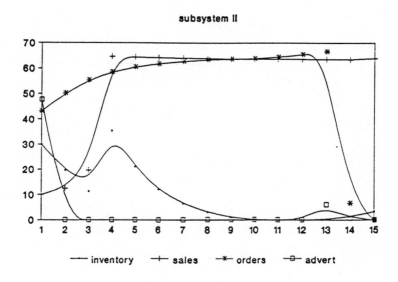

Fig. 11a,b : Simulation example

subsystem I

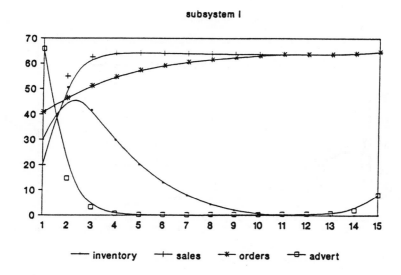

subsystem II

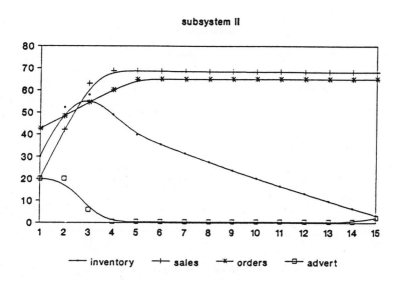

Fig. 12a,b : Simulation example

206

subsystem I

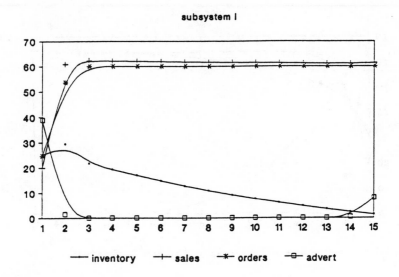

subsystem II

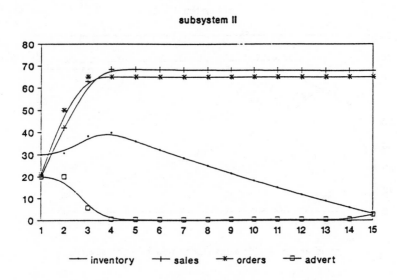

Fig. 13a,b : Simulation example

6. An Integrated Multi-Product Dynamic Production Planning Model for CIM Operation

6.1 The basic inventory-production model

A simplified scheme for the model is given in Fig.14

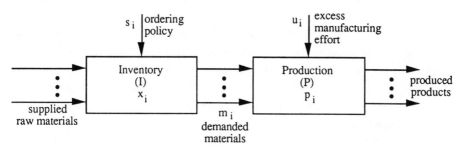

Fig.14 : Inventory - Production Model

It consists of two subsystems: Inventory Subsystem (I) and Production Subsystem (P). Clearly the two functions are interelated. Production influences inventory by initiating orders and inventory influences production by shortage of raw materials. Safety stocks held by the enterprise provide protection against:

(a) delayed or interrupted supply s_i of raw material

(b) unpredictable temporary acceleration in the consumption rate of the inventoried items due to a sudden increase of the end-product demand

(c) both (a) and (d).

Defining $x_i(k)$, $s_i(k)$ and $m_i(k)$ as the inventory level, supply level and demand level of raw material i at the kth period, respectively, we have:

$$x_i(k+1) = x_i(k) - r_i x_i(k) + \sum_{m=0}^{\theta_i} g_{im} u_i(k-m) - m_i(k) \qquad i=1,...,N \qquad (6.1)$$

where possible distributed time delays of the ordering process and inventory deterioration at a constant rate r_i are assumed. Physical constraints can be imposed to the model (6.1) so that

$$0 \le x_i(k) \le X_i \qquad \text{and} \qquad 0 \le u_i(k) \le U_i \qquad i=1,...,N \qquad (6.2)$$

Note that the lower bounds may be non-zero too.

According to [107], inventoried items are demanded and consumed by the (P) system. It is assumed that the level of production (amount of produced pieces) at each period is affected by the production process capacity (control effort over manufacturing), unpredicted market demand and loss and/or rejection of units due to quality control. Thus we have:

$$p_i(k+1) = h_i(m_1(k),...,m_N(k)) - q_i p_i(k) + u_i(k), \qquad 0 \le U_i(k) \le U_i \qquad (6.3)$$

where $p_i(k)$ is the production level and $u_i(k)$ is the excess manufacturing control effort. The form of the function $h_i(.)$ is assumed to be estimated from statistical data as provided in a numerical example in [107]. Younis and Mahmoud [107], introduced department costs to be minimized, and developed an algorithm which decouples the two subproblems by fixing the coupling terms $m_i(k)$. However, the resulting algorithm updates these sequences using information only from the Inventory department and then passing it to the Production subproblem.

Here we view $m_i(k)$ as a controlled variable that determines the flow rate from one department to another, and is analogous to the already planned production order volume. Thus it can be viewed as a decision variable that couples the two functions. Then we can assume that for a range of values of parameters and for a range of operational constraints, this function h_i has a linear form (with or without time delays depending upon the specific structure of the manufacturing process).

On the order hand, if the variable m_i is still regarded as in [107], then either a linearization procedure could be involved, or more effectively, a linear relationship can be constructed from the available statistical data assuming a Moving Average (MA) model and using some appropriate identification method. This would induce possible time delays in the model for the variable m_i, its order being equal to the order of the identification model. Since the above relationship may be time varying, the proposed managerial plans should be updated by solving the formulated problem after each planning period (or after a number of planning periods) on a rolling horizon basis. Once the problem to be solved was put in a linear form, such a rolling horizon procedure is a preferable managerial conception and causes no extra difficulties. In the next section a different approach is presented which needs only a small percentage of the current compuration scheme.

To sum up the above discussion, we argue that with any of the two considerations a linear function (with or without time delays) can be well justified. The same evidence, though within a different setting, invoking a linear relationship is also used in [8].

The corporate objective to be optimized is the total cost which includes inventory deviations, production deviations, ordering, manufacturing effort and excess production orders volume. Obviously, the last one is only a small percentage of the other costs, i.e. it has a much smaller value of the cost coefficient.

Assuming quadratic forms we have:

$$J = \sum_i \sum_k h_i (x_i(k) - x_i^*)^2 + w_i (p_i(k) - p_i^*)^2 + k_i s_i^2(k) + d_i u_i^2(k) + d_{m_i} m_i^2(k) \qquad (6.4)$$

where h_i, w_i, k_i, d_i, d_{mi} are the corresponding cost parameters. Note that from the convergence point of view the inclusion of m_i^2 in the cost function with a very small coefficient is equivalent to the inclusion of a quadratic term $(m_i - m_i^*)^2$ [4].

The combined inventory supply-production problem is to minimize (6.4), subject to (6.1), (6.2) and (6.3).

A global optimum solution is desired. One of the multi-level techniques developed in this field can be invoked. Because of the possible existence of multiple time delays and imposed constraints, Tamura's and Cai's approaches are the most appropriate ones [16,89-90]. The detailed algorithms are well described in the literature of this field [84-85,89-90]. Global warehouse or production capacity constraints can also be imposed to the previous model. A decomposition procedure can be then used as in the previous section.

The above model is very flexible in the sense that time varying market environments can be captured and handled effectively. The computation time needed for the solution is very reasonable (depending upon the size of the problem) so that revised plans can be reelaborated on a rolling horizon procedure. Finally, the model can be used as a decision making tool for developing alternative solutions, or evaluating some already existing plans.

In the following figures (Figs 15,16) a numerical example is given, where this methodology has been used. A typical case of a manufacture producing two products and using four raw materials is considered. The specific production subsystem is as follows:

Product A:
$$p_1(k+1) = 0.6m_1(k) + 0.5m_2(k) + 0.4m_3(k) + 0.3m_3(k) - 0.05p_1(k) + u_1(k) \qquad (6.5)$$

Product B:
$$p_2(k+1) = 0.3m_1(k) + 0.01m_3(k-1) + 0.5m_2(k) + 0.3m_3(k) + 0.5m_4(k) - 0.04p_2(k) + u_2(k)$$

Some product rejection is assumed because of scrappage and quality control. Safety stock is zero $(x_i^*=0)$, and the inventory constraint $1 \leq x_i \leq 15$, $i=1,...,4$ is imposed. The desired production schedule has the following form:

$$p_1^* = \begin{cases} 20 & 0 \leq k \leq 13 \\ 15 & 14 \leq k \leq 26 \\ 25 & 27 \leq k \leq 39 \\ 20 & 40 \leq k \leq 52 \end{cases} \qquad p_2^* = \begin{cases} 10 & 0 \leq k \leq 13 \\ 5 & 14 \leq k \leq 26 \\ 0.5 & 27 \leq k \leq 52 \end{cases}$$

Clearly, the first product exhibits a kind of seasonality while the second one is gradually retiring from the market. The optimal path for the related variables of the first two products are shown in Figs. 15 and 16.

6.2 Extension of the previous model to include marketing activities

For our purposes the role of marketing is defined rather narrowly to consist of estimating final product demand for any given level of advertising expenditure, for every time period encompassed by the planning horizon of the firm. Thus, in the proposed integrated model the contribution of marketing is the generation of a controlled demand function that depends upon the efficiency of the firm's advertising $A(k)$.

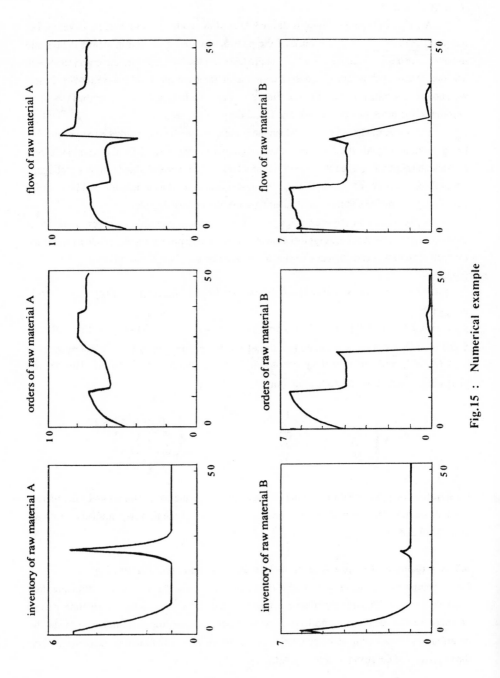

Fig.15 : Numerical example

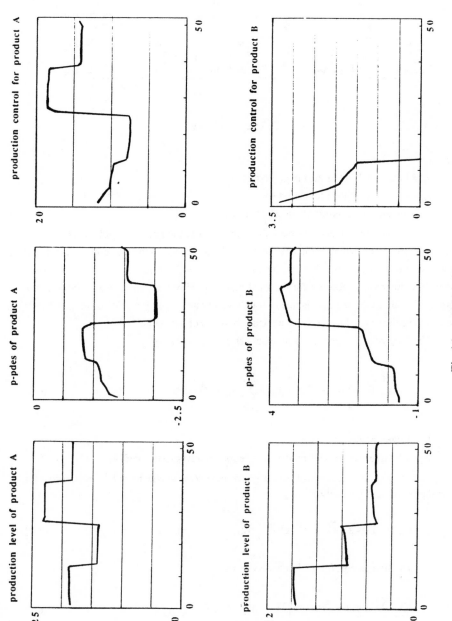

Fig.16 : Numerical example

A pictorial representation of the integrated model is given in Fig.17 :

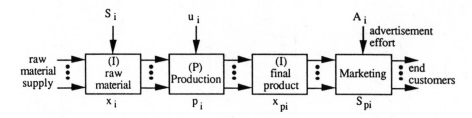

Fig. 17 : Integrated model

Here

x_{pi}= inventory level of the i final product

s_{pi}= sales level of the i final product

A_i= advertisement effort of the i final product

It is assumed that the demand the manufacturing firm is facing, is controlled through the advertising procedure which influences both buyers preference and final product adoption. Exogenous demand (if any) that can be forecasted is not considered, although it should obviously be taken into account for the final planning. However, since this function is not controlled explicitly by the management, it is beyond our scope. The price of the product is assumed to be determined by the structure and conditions of the market.

The form of sales-advertisement relationship is as follows:

$$s_{pi}(k+1) = (1-\lambda_i)\, s_{pi}(k) + \sum_{j=0}^{\theta_i} b_{ij} A_i(k-j), \qquad i = 1,...,m \qquad (6.6)$$

where

λ_i = forgetting/decay constant

b_{ij}= effectiveness for the i-product at j-period, of advertisement effort.

To complete the equations of the integrated model we add the inventory relationship for the final product $x_{pi}(k)$, namely :

$$x_{pi}(k+1) = x_{pi}(k) + p_i(k) - s_{pi}(k), \quad i = 1,...,m, \quad 0 \le x_{pi} \le X_{pi} \qquad (6.7)$$

The incurred additional cost for the last two systems is:

$$\sum_i \sum_k d_{spi}\,(s_{pi}(k) - s_{pi}^*)^2 + q_{pi}\, x_{pi}^2(k) + r_{Ai}\, A_i^2(k) \qquad (6.8)$$

where d_{spi}, q_{pi} and r_{Ai} are appropriate cost coefficients. This cost consists of partial costs due to advertisement, final product inventory, and deviations from the prespecified desired sales plan. The total operational cost that has now to be minimized, is the sum of the costs (6.4) and (6.8).

Obviously, if the integrated overall problem is to be solved by a conventional single-level method, one needs a high computational power even in the case of a single product. The individual or global constraints imposed, complicate further its solution. Updating the firm's plan on the basis of the solution of the previous model with new inputted parameters, is prohibitive with single level algorithms. The interaction-prediction coordination technique, Tamura's time-delay algorithm, and Cai's algorithm are the most appropriate techniques for handling the integrated problem. Algorithmic details are not given here, but the required algorithms are fully described in the literature of the field [16,84-85,89-90]. A version of Tamura's algorithm has been implemented by the authors and extensive numerical examples have been worked out to get an insight of the dynamic interdependencies and clearly identify the effect of each parameter in the resulting solution.

In order to investigate the relative merits of the simultaneous versus the sequential model, the same problem was solved with both approaches and registered the resulting costs. The approach which suggested getting the overall solution was discussed previously. The sequential model consists of two subsystems. The first subsystem comprises all other functions except for the supply process of raw materials. Thus this subsystem aims to manipulate the production and the marketing decisions so that the induced costs are minimalized. Then the inventory-supply subsystem has to satisfy the determined demand of raw material (planned orders volume), by choosing its replenishment policy appropriately. Note that the sequential model does not offer the maximum allowable partition of the enterprise's subfunctions. One can first allow the marketing process to submit marketing plans by autonomously optimizing its performance, then the production process to calculate the required input, and finally the inventory process to specify ordering policies. Such a sequential model introduces more conflicts for each local objective, and obviously is more costly for the corporate manufacturing. This is the reason we have intentionally chosen the previous partition to examine the cost savings that may be achieved. Numerical examples have verified in all cases the cost saving when using the global approach. The cost differences ranged from 3% to 8% for the examined problems depending on how well in opposition the objective of each subsystem was, and also on the relative cost parameters. Optimal sales and production trajectories for both approaches are given in Fig.18 in the single product case. In this case, the cost reduction is 4% for the global approach. As noted in [23], the specific numbers are dependent upon the postulated parameter values of the model, but no attempt was made to bias the parameters towards enhancing the performance of the simultaneous model. We have rather tried the opposite by not considering the fully separated sequential model, but considering instead a two-component

214

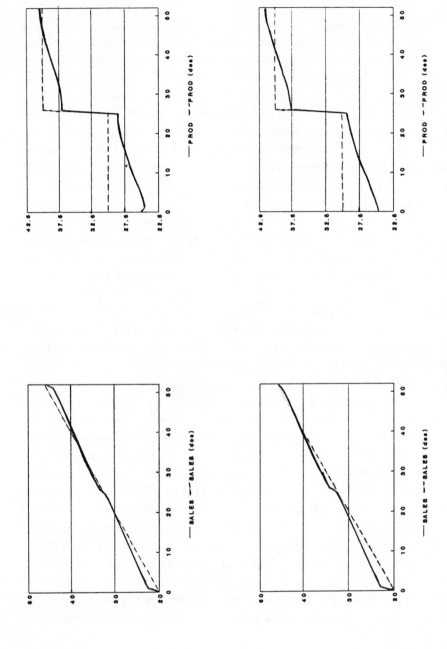

Fig. 18 : Sequential (above) versus simultaneous solution (below)

model. Certainly, every individual has to evaluate the conditions under which these benefits are worth the cost of model design, estimation/calibration and computer implementation.

7. Conclusions

The logistics chain deals with the total flow of goods from the acquisition of raw materials to the delivery of the finished products to the customer. Production is just one link in this chain. Traditionally, Operational Research models have focused on one of the functional areas. This has led to isolated computer packages, supporting managerial decisions in a specific area. As noted in [5-6], lack of integration and neglect of the fuctional policy interdependencies is a basic deficiency in production planning designs, prohibiting effective overall management

In the present work an analytical methodology for the long-term strategic planning problem of Integrated Manufacturing Systems was developed, through hierarchical control optimization techniques.

First, we have addressed the problem of effective operation of an autonomous manufacturing supply chain which consists of an upper level manufacturer and a 1-level hierarchy of local/intermediate suppliers/subsuppliers that provide the necessary materials. This structure is well adopted by some industries and provides the basis for our analysis.

Orders steming from the manufacturer level are regarded as a variable that is designated by the production department of the manufacturer, which in turn is originated from the induced demand of the distribution chain of the produced products. It was assumed that each level in the chain may add some delay in the materials procurement by some extra processing it undergoes, and/or by time intervened to supply this material. Each subsystem involved in the process has (incurs) some costs that have to be handled. These are mainly ordering and holding costs. Transportation costs may also be included, but then we have to resort to other optimization methods based on linear programming or network theory [96-97].

Under these assumptions three different scenarios have been discussed. In the first scenario, the manufacturer optimizes his own cost, irrespectivelly of the operational cost that his decisions may incur to the downstream hierarchy of the chain. This implies that the manufacturer can impose his policy to the (sub)suppliers who cannot bargain with him. In the second scenario cooperation exists among all the levels of the chain. The common objective is the minimization of the total operational cost. The coordination unit distributes some information and imposes some goals that have to be taken into consideration by each local decision maker. Finally in the third scenario each subsystem is free to act so that a modified calculated local criterion, which takes into account possible conflicts with downstream levels is optimized. Again, a kind of cooperation is needed. Each part tries to minimize its own cost plus weighted deviations of its local order profile from a calculated reference variable, that according to the management calculations, would render the (i+1)th subsystem with acceptable costs. The coordination task in this case seems to be more difficult (not from a mathematical point of view) and maybe some incentives would have to be discussed or imposed in order for level to cooperate. In this way a decentralization is achieved, permitting each subsystem to solve some modified local problems and to feedforward its solution to the next one.

Secondly, the composite multi-product inventory-marketing problem is discussed and analyzed. For such real large-scale management problems, the use of a single-level (centralized) solution may be prohibitive. Hierarchical coordination methods provide a safe guide to tackle such problems by decomposing them into a number of smaller subproblems and adjusting appropriately their individual solutions to provide an overall optimal plan. This type of multi-level hierarchical solution has been used to provide an algorithmic scheme for a marketing / inventory problem. The model thus developed incorporates most of the parameters that actually affect the marketing / inventory supply decisions. It constitutes an effective and flexible tool for quantitative decision making, that can identify the crucial points of the problem and suggest plans to overcome them.

The major difficulty in applying the algorithms of the paper lies in the assumption that the parameters of the model are known. However, there exist several efficient econometric estimation methods which provide exact or at least acceptable values for the parameters involved, provided that a data base with accurate measurements is available. Our intention here was to present a quantitative method that could serve as a guideline to management when planning/testing different scenarios.

From a technical point of view, a hierarchical procedure was adopted with proper system decomposition and coordination, so as to obtain an optimal solution of the overall I/A problem. Distributed time delays in both the supply and the advertisment processes, global constraints plus the capability of handling perishable goods were considered. Intensive promotion, subcontracting and rush delivery can be included by adjusting properly the model parameters. The model is amendable to adaptive parameter estimation techniques. Much of the existing literature treats the marketing / inventory decision making problems in a centralized way. However, for multi-period multi-product cases it may be impractical to solve the overall I/A problem by centralized techniques. The two algorithms proposed in this paper are appropriate for such cases and provide efficient procedures to achieve the management goals.

Finally an integrated dynamic multi-product production planning model was developed to incorporate basic functional interdependencies between production planning decisions and other functional policy areas, namely inventory and marketing decisions with their pertinent goal. It contributes to the development of optimal production policies in multi-product systems in a two-fold way:

(i) It builds upon the work of [107] discussing the coupling subfunction in a slight different framework and giving an optimal solution to the overall dynamic constrained problem. Then, the initial model is extended to include marketing decisions, thus offering a global model that provides a very flexible and effective tool for decision making by the firm's management board. The potential merits of simultaneous versus sequential planning are also discussed and critically evaluated.

(ii) It discusses the applicability of suitable existing multi-level optimization control techniques which are shown to be a valuable tool for obtaining solutions to large complex production planning problems [70]. For such cases, single-level centralized planning seems insufficient because of the excessive computational requirements even for very simple problems.

In recent years the number of models and applications on integrated production planning have been growing rapidly. Meaningful integrated models that link production with other major functions have been developed, which certainly require other than conventional single-level methods. However, no model portrays the whole reality. Each one balances realism and complexity against analytic tractability and computational ease, and no single modelling framework can reasonably be expected to be the best under all circumstances. The careful testing, evaluation, and implementation of such models is one of the biggest challenges for the successful and profitable operation of modern complex manufacturing industry and enterprise.

8. REFERENCES

[1]   Abad P.L., "An Optimal Control Approach to Marketing Production Planning", *Optimal Control Applications & Methods*, Vol. 3, pp. 1-14, 1982.

[2]   Abad P.L., "Approach to Decentralized Marketing-Production Planning", *Int. J. Systems Sci.*, Vol. 13, pp. 227-235, 1982.

[3]   Abad P.L., "Two Level Algorithm for Decentralized Control of a Serially Connected Dynamic System", *Int. J. Systems Sci.*, Vol. 16, pp. 616-624, 1985.

[4]   Aronson J. E. and G. L. Thompson, "A Survey on Forward Methods in Mathematical Programming", *Large Scale Systems*, Vol. 7, pp. 1-17, 1984.

[5]   Axsater S., "Coordinating Control of Production-Inventory Systems", *Int. J. Production Res.*, Vol. 14, pp. 424-436, 1976.

[6]   Axsater S. and J. Henrik, "Aggregation and Disaggregation in Hierarchical Production Planning", *Eur. J. Opl. Res.*, Vol. 12, pp. 338-350, 1984.

[7]   Axsater S., "Control Theory Concepts in Production and Inventory Control", *Int.J. Systems Sci.*, Vol. 16, pp. 161-169, 1985.

[8]   Baetge J. and T. Fischer, "Stochastic Control Methods for Simultaneous Synchronization of the Sort-Term Production Stock- and Price-policies when the Seasonal Demand is Unknown", in *Optimal Control and Economic Analysis*, G. Feichtinger Editor,1982.

[9]   Ballou R.H., *Business Logistics Mangement*, Prentice Hall, Inc., 1992.

[10]   Bensoussan A., Hurst E.G. and B. Naslund, *Management Applications of Modern Control Theory*, North Holland, 1974.

[11]   Bensoussan A, Crouchy M. and J.M. Proth, *Mathematical Theory of Production Planning*, North Holland, 1983.

[12]   Bergstrom G.I. and B.E. Smith, "Multi-item Production Planning : An Extension of the HMMS Rules", *Management Sci.*, Vol. 16, pp. 614-629, 1970.

[13]   Bhaskaran S. and S. Sethi, "Decision and Forecast Horizons in a Stochastic Environment : A Survey", *Optimal Control Applications & Methods*, Vol. 8, pp. 201-217, 1987.

[14]   Blattberg R.C. and A. Levin, "Modelling the Effectiveness and Profitability of Trade Promotions", *Marketing Sci.*, Vol. 6., pp. 124-146, 1977.

[15]   Buffa E.S. and J.G. Miller, *Production - Inventory Systems : Planning and Control*, Homewood - Irwin, 1979.

[16]   Cai X., "Multi-level Optimization of Large Scale Time-delay Systems and Optimal Hydrothermal Scheduling", *Control-Theory and Advanced Technology*, Vol. 1, pp. 217-238, 1985.

[17] Choi S.B. and H. Hwang, "Optimization of Production Planning Problem with Continuously Distributed Time-Lags", *Int. J.Systems Sci.*, Vol. 17, pp. 1499-1508, 1986.

[18] Choi S.B. and Hwang H., "Dynamic Optimization of Production Planning Problem with Periodic Operation", *Int. J. Systems Sci.*, Vol. 17, pp. 1163-1174, 1986.

[19] Choi S.B. and H. Hwang, "Optimal Control of Production aand Marketing Systems with Distributed Time Lags", *Optimal Control Applications & Methods*, Vol. 8, pp. 351-364, 1987.

[20] Christensen J. and W. Brogan, "Modelling and Optimal Control of a Production Process", *Int. J. Systems Sci.*, Vol. 1, pp. 247-255, 1971.

[21] *CIM for Multi Supplier Operations* (1989-1991), CMSO-Esprit Project, Final Deliverables, February 1992.

[22] Clarke D.G., "Econometric Measurement of the Duration of Advertising", *J.Marketing Res.*, Vol. 13, pp. 345-357, 1976.

[23] Damon W. and R. Schramm, "A Simultaneous Decision Model for Production, Marketing and Finance", *Management Sci.*, Vol. 19, pp.161-172, 1972.

[24] De S., Nof S.Y. and A.B.Whinston, "Decision Support in Computer-Integrated Manufacturing", *Decision Support Systems,* Vol. 1, pp. 37-56, 1985.

[25] Dockner E. and S. Jorgensen , "Optimal Advertising Policies for Diffusion Models of New Product Innovation in Monopolistic Situations", *Management Sci.*, Vol. 34, pp. 119-130, 1988.

[26] Drew S.A.W., " The Application of Hierarchical Control Methods to a Managerial Problem", *Int. J. Systems Sci.*, Vol. 6, pp. 371-395, 1975.

[27] Eilon S., "Five Approaches to Aggregate Production Planning", *AIIE Trans.*, Vol. 7, pp. 118-131, 1975.

[28] Eliashberg J. and R. Steinberg, "Marketing - Production Decisions in an Industrial Chanell of Distribution", *Management Sci.*, Vol. 37, pp. 981-1000, 1987.

[29] Eppen G., Gould F. and P. Pashigian, "Extension of the Planning Horizon Theorem in the Dynamic Lot Size", *Management Sci.*, Vol. 15, pp. 268-277, 1969.

[30] Feichtinger G. and Hartl R., "Optimal Pricing and Production in an Inventory Model", *Eur. J. Opl. Res.*, Vol. 19, pp. 45-56, 1985.

[31] Filip F.G., Donciulescu and L. Orasanu, "Multilevel Optimization in Production Control", *Symp. on Large Scale System Theory and Applications*, Toulouse, France 1980.

[32] Finlay P. , Nand C. and J. Martin, "The State of Decision Support Systems : A Review", *OMEGA*, Vol. 17, pp. 525-531, 1989.

[33] Freeland J.R., "Coordination Strategies for Production and Marketing in a Functionally Decentralized Firm", *AIIE Trans.*, Vol. 12, pp. 126-132, 1980.

[34]   Gaalman G.J., "Optimal Aggregation of Multi-item Production Smoothing Models", *Management Sci.*, Vol. 24, pp. 1733-1739, 1978.

[35]   Gelders F.L. and L.N. Wassenhove, "Production Planning : a Review", *Eur. J. Opl. Res.,* Vol. 7, pp. 101-110, 1981.

[36]   Graves S., "Using Lagrangean Techniques to Solve Hierarchical Production Planning Problems", *Management Sci.*, Vol. 28, pp. 260- , 1982.

[37]   Groover M.P., *Automation, Production Systems and Computer-Aided Manufacturing,* Prentice Hall, Englewood Cliffs, N.J., 1980.

[38]   Gunasekaran A., Babu A. S. and N. Ramaswamy, "Behaviour of a Production System for varying Marketing Decisions", *Int. J. Systems Sci.*, Vol. 20, pp. 241-251, 1987

[39]   Hameri A.P., "Logistics - State of the Art", *Working Paper*, Helsinki University of Technology, 1990.

[40]   Hanssens D.M., Parsons L.J. and R.L. Schultz, *Market Response Models : Econometric and Time Series,* Boston, Kluwer, 1990.

[41]   Hassan M.F. and M.G. Singh, "A Two-level Costate Prediction Algorithm for Nonlinear Systems", *Automatica,* Vol. 13, pp. 635-634, 1977.

[42]   Hax A.C. and H.C. Meal, "Hierarchical Integration of Production Planning and Scheduling", in : M.A. Geisler Ed., *Studies in Management Sciences, Vol. I, Logistics*, North Holland , 1975.

[43]   Holt C.C., Modigliani F., Muth G.F. and H.A. Simon, *Planning, Production, Inventories and Work-Force*, Englewood Cliffs, Prentice Hall, 1960.

[44]   Horsky D. and L.S.Simon, "Advertising and the Diffusion of New Products", *Marketing Sci.,* Vol. 2, pp. 1-17, 1983.

[45]   Jamshidi M. amd J.M. Brideau, "On the Hierarchical Control of Discrete-time Systems With Time Delay", *Large Scale Systems,* Vol. 7, pp. 33-46, 1984.

[46]   Jorgensen S., "Optimal Production, Purchasing and Pricing. A Differential Game Approach.", *Eur. J. Opl. Res.*, Vol. 24, pp. 64-76, 1986.

[47]   Kamien M.I and N.L Scwartz, *Dynamic Optimization :The Calculus of Variations and Optimal Control in Economics and Management Science*, North Holland, New York, 1981

[48]   Kapsiotis G. and S. Tzafestas, "Hierarchical Control Approach to a Composite Inventory-Marketing Problem", *Journal of Intelligent and Robotic Syastems: Theory and Applications* (in press)

[49]   Kapsiotis G. and S. Tzafestas, "Decision Making for Inventory/Production Planning Using Model Based Predictive Control", in *Parallel and Distributed Computing in Engineering Systems,* (S. Tzafestas, P. Borne, L. Grantinetti editors), pp.551-556, North Holland, 1992.

[50]   Kleindorfer P.R., Kriebels C.H., Thompson G.L. and G.G Kleindorfer , "Discrete Optimal Control of Production Plans", *Management Sci.*, Vol. 22, pp. 261-273, 1974.

[51]   Kotler P., *Marketing Decision-Making. A Model Building Approach*, Holt, Rinehart and Winston, Inc., New York, 1971.

[52]   Kunreuther H. and J.F. Richard, "Optimal Pricing and Inventory Decisions for Non-Seasonal Items", *Econometrica*, pp. 173-175, 1971.

[53]   Kunreuther H. and L. Schrage, "Joint Pricing and Inventory Decisions for Constant Priced Items", *Management Sci.*, pp. 732-738, 1973.

[54]   Lasdon L.S., *Optimization Theory for Large Systems*, Macmillan 1970.

[55]   Leitch R.A., "Marketing Strategy and the Optimal Production Schedule", *Management Sci.*, Vol. 21, pp. 301-312, 1974.

[56]   Lilien G., Kotler P. and K.S. Moorthy, *Marketing Models*, Prentice Hall International, 1992.

[57]   Little J.D.C. ,"Aggregate Advertising Models : The State of the Art", *Opns. Res.*, Vol 27, pp. 629-667, 1979.

[58]   Mahmoud M., Vogt W. and M. Mickle, "Multilevel Control and Optimization   Using Generalized Gradient techniques", *Int. J. Control* , Vol. 25, pp. 525-543, 1977.

[59]   Mahmoud M.S., "Optimal Control of Constrained Problems by the Costate Coordination Structure", *Automatica*, Vol. 14, pp. 31-40, 1978.

[60]   Mahmoud M.S., Vogt W.G. and M.H. Mickle, "Decomposition and Coordination Methods for Constrained Optimization", *J. Optimization Theory and Applications*, Vol. 28, pp. 549-584, 1979.

[61]   Mahmoud M.S., "Dynamic Mutilevel Optimization for a Class of Nonlinear Systems", *Int. J. Control*, Vol. 30, pp. 348-365, 1979.

[62]   Mahmoud M.S. and A.S. Fawsy, "On the Hierarchical Algorithms for Nonlinear Systems with Bounded Control", *Optimal Control Applications & Methods*, Vol. 5, pp. 275-288, 1984.

[63]   Mahmoud M.S, Hassan M.F. amd M.G. Darwish, *Large Scale Control Systems : Theories and Techniques,* Marcel Dekker, N. York, 1985.

[64]   Mak K.L., "Production, Scheduling and Inventory Control for Multi-Stage Manufacturing Systems", *Int. J. Systems Sci.*, Vol. 18, 1227-1245, 1987.

[65]   Mariani L. and B. Nicoletti, "Optimization of Deterministic Multi-product Inventory Model with Joint Replenishment", *Management Sci.*, Vol. 20, pp. 349-362, 1973.

[66]   Mesarovic M.D., Macko D. and Y. Takahara, *Theory of Hierarchical Multilevel Systems*, Academic Press, New York, 1970.

222

[67]   Monahan G.E., "Optimal Advertising with Stochastic Demand", *Management Sci.*, Vol. 29, pp. 106-117, 1983.

[68]   Monteiro P. and P. Correia, "Short-Term Production Scheduling : A new and more Efficient Algorithm and its Application to the Pulp and Paper Industry", *Information and Decision Technologies*, Vol. 18, pp. 241-263, 1992.

[69]   Montgomery D.B. and G.L. Urban, *Applications of Management Science in Marketing*, Prentice Hall, Inc., Englewood Cliffs, New Jersey, 1969.

[70]   Nachane D.M., "Optimization Methods in Multilevel Systems : A Methodological Survey", *Eur. J. Opl. Res.*, Vol. 21, pp. 25-38, 1984.

[71]   Parlar M., "A Problem in Jointly Optimal Production and Advertising Decisions", *Int. J. Systems Sci.*, Vol. 17, pp. 1373-1380, 1986.

[72]   Pekelman D., "Simultaneous Price-Production Decisions", *Opns. Res.*, Vol. 22, pp. 788-794, 1974.

[73]   Pekelman D., "Production Smothing with Fluctuating Price", *Management Sci.*, Vol. 21, pp. 576-590, 1975.

[74]   Porter B. and Taylor F. ,"Modal Control of Production Inventory Systems", *Int. J. Systems Sci.*, Vol. 3, pp. 325-331, 1972.

[75]   Saad G.H., "Hierarchical Production Planning Systems : Extensions and Modifications ", *J. Opl. Res. Soc.*, Vol. 41, pp. 609-624, 1990.

[76]   Saunders J. , "The Specification of Aggregate Market Models", *Eur. J. of Marketing*, Vol. 21, pp. 1-47, 1987.

[77]   Sethi S.P. and G.L. Thompson, *Optimal Control Theory : Applications to Management Science*, Martinus Nijhoff Publishing, Boston 1981.

[78]   Sethi S.P., "Dynamic Optimal Control Models in Advertising : A survey", *SIAM Rev.* 19, pp. 685-725, 1977.

[79]   Sethi S. P., "A Survey of Management Science Applications of the Deterministic Maximum principle", in : *Applied Optimal Control* ed. A. Bensoussan et al., pp. 33-68, 1978.

[80]   Silver E.A., "A Tutorial on Production Smoothing and Work Force Balancing", *Opns. Res.*, Vol. 15, pp. 985-1110, 1967.

[81]   Simon H. and K.H. Sebastian, "Diffusion and Advertising : The German Telephone Campaign", *Management Sci.* , Vol. 33, pp. 451-466, 1987.

[82]   Singh M.C. and M.F. Hassan, "Hierarchical Successive Aproximation Algorithms for Nonlinear Systems", Parts (I) and (II), *Large Scale Systems*, Vol. 2, pp. 65-79, 81-95, 1981.

[83]   Singh M.G., Drew A.W and J.F. Coales, "Comparisons of Practical Hierarchical Control Methods for Interconnected Dynamical Systems", *Automatica,* Vol. 11, pp. 331-350, 1975.

[84]   Singh M.G. and A.Titli, *Systems : Decomposition, Optimization and Control,* Pergamon Press, Oxford, 1978.

[85]   Singh M.G., *Dynamical Hierarchical Control,* North Holland, 1980.

[86]   Sterman J.D., "Modelling Management Behaviour : Misperceptions of Feedback in a Dynamic Decision Making Experiment", *Management Sci.,* Vol. 35, pp. 321-339, 1989.

[87]   Stock J. R. and D.M. Lambert, *Strategic Logistics Management,* 2nd ed., Homewood, Ill : R.D. Irwin, 1987.

[88]   Sweeney D. J., "Finding an Optimal Dynamic Advertising Policy", *Int. J. Systems Sci.,* Vol. 5, pp. 987-994, 1974.

[89]   Tamura H. ,"Decentralized Optimization for Distributed-lag Models of Discrete Systems", *Automatica,* Vol. 11, pp. 593-602, 1975.

[90]   Tamura H. and T. Yoshikawa, *Large-Scale Systems Control and Decision-Making,* Marcel-Dekker, N.Y., 1990.

[91]   Tapiero Ch.S., *Managerial Planning : An Optimum and Stochastic Control Approach,* Gordon Breach Science Publishers, New York, 1977.

[92]   Tzafestas S., "Systems, Management, Operational Research and Control - A Unified Look", *EURO III, Eur. Symposium on Opl. Res.,* Amsterdam, April, 1979.

[93]   Tzafestas S. ed., *Optimization and Control of Dynamic Operational Research Models,* North Holland, 1982.

[94]   Tzafestas S. and M.F. Hassan, "Complex Large Scale Systems Methodologies in Conjuction with Modern Computer Technology", *Control-Theory and Advanced Technology,* Vol. 2, pp. 105-130, 1986.

[95]   Tzafestas S., "Knowledge-Based Decision Support Systems for Planning in a Distribution Chain", *ESPRIT CMSO Report,* NTUA 1989.

[96]   Tzafestas S., Kapsiotis G. and S. Reveliotis, "A Dual Algorithm for Post-Optimization of the Generalized Network Optimal Flow Problem", *Foundation of Computing & Decision Sci.,* Vol. 16, pp. 39-54, 1991.

[97]   Tzafestas S., Kapsiotis G. and S. Reveliotis, "The Generalized Network Approach to Optimized Decision Making and Planning", In : *Computational Systems Analysis : Topics and Trends* (A. Sydow Editor), Elsevier/North Holland 1992.

[98]   Tzafestas S. and G. Kapsiotis, "Coordinated Control of Manufacturing/Supply Chains Using Multi-Level Techniques", *CIM Systems,* (under review).

[99]   Tzafestas S. and G. Kapsiotis, "A Unified Hierarchical Algorithm for Constrained Nonlinear Multi-Delay Interconnected Systems : Application to Production Planning", *Control : Theory and Advanced Technology*, (under review).

[100]   Tzafestas S. and G. Kapsiotis, "Integrated Versus Sequential Production, Inventory and Marketing Decisions in Multi-Product CIM Operations", *Information and Decision Technologies*, (under review).

[101]   S. Tzafestas and G. Kapsiotis, "An Integrated Multi-product Dynamic Production Model", *APMS'93, Intl. Conf. on Advances in Production Management Systems*, Athens, Greece, Sept. 1993.

[102]   Thomas J., "Price-Production Decisions with Deterministic Demand", *Management Sci.*, Vol. 16, pp. 747-750, 1970.

[103]   Thomas J., "Linear Programming Models for Production-Advertising Decisions", *Management Sci.*, Vol. 17, pp. B474-484, 1971.

[104]   Thompson G. L., Sethi S. P. and J. T. Teng, "Strong Planning and Forecast Horizons for a Model with Simultaneous Price and Production Decisions", *Eur. J. Opl. Res.*, Vol. 16, pp. 378-388, 1984.

[105]   Thompson G. L. and Sethi S.P., "Turnpike Horizons for Production Planning", *Management Sci.*, Vol. , pp. 229-241, 1980.

[106]   Tuite M.F., "Merging Marketing Strategy Selection and Production Scheduling : A Higher Order Optimum", *J. Industrial Engineering*, pp. 76-84, 1968.

[107]   Younis M.A. and M.S. Mahmoud, "Optimal Inventory for Unpredicted Production Capacity and Raw Material Supply", *Large Scale Systems,* Vol. 11, pp. 1-17, 1986.

[108]   Vanthienen L.G., " A Simultaneous Price-Production Decision-Making Model with Production Adjustment Cost", *Working Paper No 7307*, Univ. of Chicago, 1973.

[109]   Vidale M.L. and H.B. Wolfe, "An Operations Research Study of Sales Response to Advertising", *Opns. Res.*, Vol. 5, pp.370-381, 1957.

[110]   Welam U.P., "Synthesizing Short Run Production and Marketing Decisions", *AIEE Trans.*, Vol. 9, pp.53-62, 1977.

8 Flexible Manufacturing by Integrated Control and Management

R.C. Michelini, G.M. Acaccia, M. Callegari and R.M. Molfino

1 Introduction

The present chapter, which refers to the "lower" factory automation level, supervising the integrated control and management of the shop-floor operations, discusses the improvement of the manufacturing plants by information control and technology.

The discussion mainly exploits the previous work of the authors (see the listed references), and is based on the expert-simulation environment **SIFIP**, established for supporting production engineers' activity [1]:

- as an off-process consultation aid, for quality-and-efficiency upgrading of products-and-production, based on the comparative assessment of the actual performances of flexible manufacturing with competing enterprise fabrication agendas over the different (strategic, tactical and operational) horizons;
- as an on-process governing option, for enabling multiple controls: - assuring efficient parallelism of the concurrent manufacturing processes all along the steady tactical horizons; and - switching to distributed decisional manifold (under centralised supervision) during the transient (initialisation or restoring) periods.

The chapter is organised as follows: a section is devoted to the assessment of flexibility at the level of the shop-floor operations. The benefits of flexible manufacturing cannot be acknowledged, unless they can be measured; obviously, the overall efficiency depends on the ability to assure most-of-the-time optimal schedules (under fully decentralised control); this does not contradict "flexibility", rather the production engineer needs to directly investigate the *sleeping* capabilities of recovery flexibility (with minimisation of the restoring transient situations, with respect to *optimal* steady production plans). The following section considers the many requirements of the govern-for-flexibility architectures; this enabling technology benefits extensive improvements by the incorporation of artificial intelligence programming options; the field is continuously evolving, and upgraded solutions are progressively offered. The discussion will therefore be mainly focused on methodological aspects. The final section presents sample applications related to actual manufacturing plants, and discusses the typical knowledge frameworks and modelling requirements of both the physical and the logical reference environments, comparing the actual dynamical behaviour in terms of step-by-step transients and in terms of statistical performance figures.

2 Assessing Flexibility

Flexible manufacturing [2, 3] is an enabling technology aiming at the intensive exploitation of fixed assets by increasing the current net production, with versatile resources, through low inventory and with the adaptive capacity of intelligent plants. These are based on sets of interconnected manufacturing sections (FMSs) with related tool-room and fixturing facilities, having the convenient logistical supports, and are characterised by an integrated management information system, see Fig. 1, assuring the step by step govern of the progression of the production agendas. An FMS is described as a cluster of cells and workstations (fixturing units, machining centres, assembly cells, etc.) linked by parts/products and tools/fixtures, transport and handling devices, that operate under an integrated information network to perform "schedules", satisfying the productivity constraints, according to "plans" consistent with the production requirements.

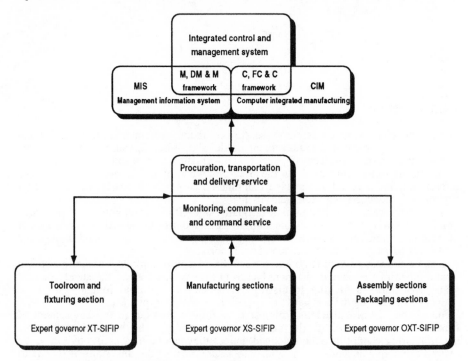

Fig. 1 - Schematic representation of the generic Management Information System for flexible manufacturing

The challenging aspect is that the FMSs can be used only after setting and upgrading. The setting is concerned with *components, facility-configuration* and *control* (C, F·C & C structural framework); the design activity also presents an everlasting activity. The selection of components, configuration and control simply provides the reference for identifying current product-and-process arrangements all along the operational life of a manufacturing enterprise. To enable the efficiency of the intelligent work-organisation, the ability to manage its flexibility is a critical request to maintain/reduce manufacturing costs, to support/improve products quality, to preserve/extend market share, etc., by the integrated exploitation of the pertinent information flow, while the principal material flow is enabled. The upgrading is grounded on a multilevel decisional logic, concerning: · the strategic fabrication agendas (inventory decision); · the tactical task programming (planning decision); · the capacity allocation updating (schedule decision); · the shop-floor logistics (dispatching decision). The efficiency upgrading is concerned with the options offered by *monitoring, decision-manifold* and *management* (M, D·M & M behavioural framework). The continuous (and incremental) acknowledgement of monitoring, decision-manifold and management framework offers the proper references for improving plant efficiency for the life-long economical exploitation of a business firm. The above mentioned (four) decisions need to be assessed and evaluated, with supports which would include the measurement of actual performances and the comparison of current figures against the expected levels of performance. In order to supply this type of decision supports, computer simulation (providing virtual reality assessments) is required. Indeed, it is not possible to manage flexibility, unless it can be evaluated [4].

2.1 Setting the Functional Descriptions

For the setting of an FMS, both static and dynamical figures need to be considered referring to a functional model, illustrating the enterprise's activities defined as ordered group of tasks and related relationships, aiming at the fulfilment of the production objectives of the business firm.

The **static characterisation** is usually expressed in terms of **production planning**. The input data are the product mix and shop productivity. The number, the type etc. of machines, of tools, etc., arranged in the fabrication lay-out, are considered the output. The solutions are weighed against certain *performance features*, such as: - *organisation requirements*, e.g.: "open shop" part-program requests, directly generated by customers' orders; "closed shop" inventory generated options for serving customers' specifications; - fabrication *scheduling requirements*, e.g.: production control on release time, completion time, etc., given as dynamic or static requests; - *complexity requirements*: from a single step, to multiple step schedules; from a single processor, to parallel processors; through flow-shop or job-shop processing; etc.; - *specification requirements*, namely: part batch figures, processing and due dates figures, etc. (expressed with deterministic or stochastic forms); - *cost/performance requirements*; typical costs refer to: production set-ups or changeovers, overtime, inventory holding, shortages, stocking out, etc.; the performances can consider: saturation/sub-utilisation of resources, tardiness (completion time minus task date), flow time (completion time minus release time) etc.

The reference description of flexible manufacturing is generally split into material and information systems, considered in a mainly CAD-CAM and, respectively, in a CAP-CAT context, as schematically shown in Fig. 2.

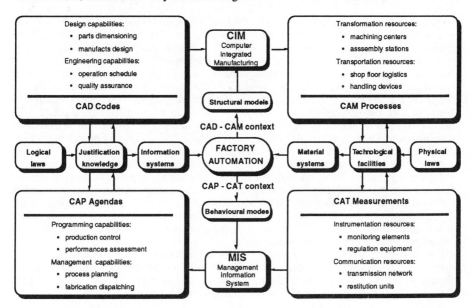

Fig. 2 - Schematic representation of the constitutive elements for factory automation

The representation suggests a vertical separation of functional tasks that actually cooperate, interacting transversally, for factory automation [5]. The vertical lines provide sample details, recalling:
· *material systems*: specified for CAM contexts (namely: transform and transfer resources) and for CAT contexts (namely: measuring and monitoring resources);
· *information systems*: specified for CAD contexts (e.g.: design and engineering capabilities) and for CAP contexts (e.g.: managing and planning capabilities).

For the individual capability and resource, typical activities and work-elements are listed in Fig. 2, within the correspondent frames (CAD codes, CAP agendas, CAM processes and CAT diagnoses). Factory automation depends critically on the functional connections. The integration of the manufacturing resources with the supervising observation schemata leads to the CAM-CAT context, providing visibility on the production activities and giving transparency to the control flow, according to the paradigms of company-wide quality-requirements, stated by the CAD-CAT context. The product design, using the manufacturing requirements as conditioning elements, assures the efficient establishment of part-programs, specified by the CAD-CAP context. The CAD-CAM context refers to a justification knowledge, mainly expressed in terms of **structural models** (with attached problem-solving methodologies based on the computers algorithmic abilities). The CAP-CAT context, based on the integration of the manufacturing technologies with the process-planning capabilities, requires a *management information system*, MIS, to enable the plant flexibility, by exploiting the technological versatility through dynamic-scheduling; then the CAP-CAT context refers to a justification knowledge, mainly, expressed in terms of **behavioural modes** (specified by the experienced production engineer, possibly helped by the computer's heuristic abilities).

The scope of industrial automation is perceived as all-inclusive, from the customer's order, to the delivery of the product; figures 1 and 2, however, suggest how to specify restricted-area developments, individually corresponding to subclasses of activities, subsequently nested to achieve factory automation. A first restriction leads to a discrete parts production model, leaving out several activities (such as: design, engineering, sales orders, finance, etc.) in order to devote attention to the fabrication shop floor operations.

When flexible automation (in view of the enabled behavioural modes) is joined with computer integrated manufacturing (granted by the implemented structural architecture), the performances are evaluated in changing environments, so that dynamic figures are needed. The **dynamic characterisation** is developed, usually, in terms of **production scheduling**, namely the operation of assigning the manufacturing resources to the (varying) batches of product mix to be processed with due regard to the programmed fabrication agendas. The solutions are weighed against some *conditioning features*, like the following: - *specification constraints*: the parts are worked after release and should not be completed after delivery dates, even if the completion times are unknown; - *capacity constraints*: queue lengths at buffering are limited; the sorting capabilities for flow balancing are fixed; etc.; - *resource constraints*: several lots may use the same resources (workstations, fixtures, etc.), therefore sequencing is required; - *maintenance constraints*: reliability data and diagnostic figures are secured against failures, by means of the trend-monitoring of the safety signatures; - *flexibility constraints*: the governing strategies will account for: real-time rescheduling, in the case of unexpected occurrences (machine breakdowns, unavailable raw material, etc.); batch rescheduling for the allocation of auxiliary resources (fixtures, pallets, etc.) to "new" products over the medium horizon range, etc.).

The individual manufacturing facility needs a functional characterisation in order to provide a time-dependent description of the current input/output flows (see the typical figures listed in Tab. 1) assuring the fulfilment of the externally driven programs, with due consideration to the internally fixed requirements.

2.2 Assessing the Decisional Manifold

For the upgrading of the FMS efficiency, a capacity-allocation decision life-cycle needs to be enabled, based on monitoring the flexibility-dependent performances. Several specifications of **flexibility** exist, namely: capability of manufacturing in different operation orders; possibility of producing different workpiece types; ability of recovering from breakdown situations; extendibility with modular additions; potentiality of jointly matching process-to-product; etc. The flexibility management is a correspondingly complex job, and reference methodologies need to be specified.

The upgrading decision life-cycle consists of three steps (Fig. 3): intelligence and design; testing and feedback; choice and redesign. - The *intelligence phase* is the process of identification of consistent alternatives; the *design phase* is the evaluation of alternatives and comparison of each. This initial step, when associated with computer-aids, requires meta-processing abilities and corresponds to the highest conceptualisation layer.- The *testing* assesses actual performances; the *feedback* is the review process of the outcome of decision and is used to influence the acknowledgement of alternatives. This step can extensively profit of computer supports; these normally split over a (relational, generative and informational) multi-layer structure. - The *choice* is the selection and implementation of a decision alternative; the *redesign* is the process of improving the selected alternative. The last step typically exploits computer aids as virtual reality references, to enhance the monitoring restitution capabilities and the acknowledging consultation options.

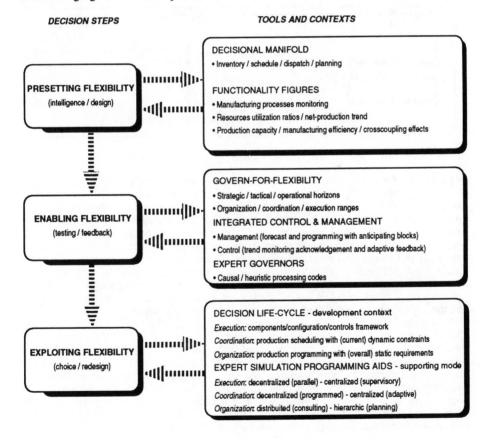

Fig. 3 - Decisional context for assessing efficient flexible manufacturing

For upgrading, the decision life-cycle begins with issues of managing flexibility, over the selected horizon (strategic, tactical, operational), according to the enabling logic: inventory decision, planning decision, schedule decision, dispatching decision. The assessment of the flexibility-dependent performances, accordingly, faces a complex decision manifold (summarised by Fig. 4); some of the characteristic features are hereafter shortly discussed.

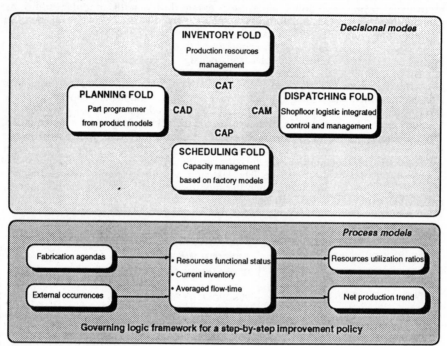

Fig. 4 - Structural context for virtual-reality computer simulation

· **Inventory fold**. The preliminary process decision setting deals with the allocation of capacity by controlling the amount of inventory in process. This decision includes the work-in-progress inventory and the (related) tools-and-fixturings inventory (assessed by tracking and locating parts). The just-in-time policy is a decision technique designed to keep a constant minimal inventory level. The capacity allocation decision is, then, concerned with the resizing of these inventory levels, with the changing work-in-progress and the related fixturing allocation. This can be obtained by considering the inventory either "active" or "idle"; idle inventory is that which is not allowed to reserve capacity. As a result, inventory decision could be carried by simply suspending the use of the pallets temporarily stored by the buffering stations. The flexibility is recovered by the switching (on and off) of inventory in-between reallocations.

· **Planning fold**. The process-plan decision deals with the choice of resources and facilities that need to be assigned to the operations to be performed. Tool capacity and pre-setting should be considered as a qualifying attribute of the on-line capability of each unit, cell, or section. This decision entails the allocation of capacity to sets of scheduled operations which define the parts program subset that needs to be accomplished. For example, a cell might have the capability to perform all the operations, ranging from deburring to finish-machining; however, the cell's capacity needs not be allocated to all these steps, rather it may be reallocated depending on the situational contexts. Flexibility, by its nature, means that the least detailed planning is needed during the setting of the strategic horizon, and the capacity can be allocated as demands for it arise. The activation of flexible capacity transfers decisions regarding

the process from the plan-management time-period, to the operations-control time-period.
· **Schedule fold**. The schedule decision deals with the choice of parts or mixes of parts which are to be allocated current capacity. The scheduling requires direct commitment and the decision tree is a straight way of filling workstations' operativity, by either pulling parts ahead of others or pushing aside parts to be processed later. The forward scheduling saturating the tactical horizons (with top-down organisation requirements) and the flexibility are opposites to one another; flexibility means that the capacity can be allocated as needed to meet current demands, and this makes inconsistent the out-of-process establishment of "optimal" schedules (that suppose steady running conditions). The dynamic scheduling (matching bottom-up coordination constraints) corresponds to the management issue of allocating capacity to the parts (or of choosing which parts to produce) in order to comply with the enterprise requested due-dates.
· **Dispatching fold**. The shop-floor logistic decision deals with the creation of a path for a part through its production program. The decision appears in the form of allocating a resource (cell, machine, fixture, pallet, tool, etc.) or a job (transfer, labour, check, assembly, etc.) to a particular part or subassembly. The dispatching becomes a needed requirement. The corresponding routing table is logically arranged as a matrix, where the columns represent the type of parts (material requirements) and the rows represent the type of jobs (production requirements). Each element of the array contains list of alternatives, consistently matching the process requirement. The decision process is then to allocate a path according to the needs of the part and the related process. In the case of an FMS, an alternative path uses equivalent-job and redundant-resource set-ups, still capable of performing the scheduled set of operations. This illustrates the difference between management and control. The management of flexibility deals with the establishing of pertinent decisions (i.e. with the consistent maintenance of the dispatching logistics); the control is the process of enabling the actually "best" alternative, within the feasible routing opportunities.

2.3 Specifying the Measurement Figures

The evaluation of the net improvements can only be metered by monitoring the consequences of flexibility, for actual running conditions, comparing alternative capacity allocation decisions, and supplying the decision manifold for the management of flexibility at the three operational, tactical and strategic ranges. At the level of setting the governing logics, the cross connection among flexibility policies and the qualitative acknowledgement of their effects can be described by decision life-cycle models. These models illustrate the coupled nature of decision-making in flexible manufacturing and specify the role that current-inventory and averaged flow-time play in assessing the effectiveness of flexibility. For a quantitative assessment, a stepwise procedure can be devised, Fig. 4, using virtual-reality computer-simulation, for implementing a reference governing logic framework based on a performance measurement scheme that exploits resource utilisation ratios and net production monitoring as critical figures.

Cross-coupling effects are a common occurrence in time-varying, unbalanced, low-inventory production surroundings. These effects have an impact upon net production in such a way that factory output is not simply related to resources efficiency. Net production is determined by a complicated relationship between many process variables ranging from production mix, resources availability, transport and machining policies, etc. up to the individual (strategic, tactical or operational) range. This relationship is so hard to fix, that computer-simulation is the only practical way to assess the efficiency of real plants, fully managing the flexibility.

The programming aids might develop: · having a modular structure, to make easy the addition of new functions or new sections; · with transparent access to any driving influence, to motivate the decision logic; · with a hierarchical architecture, for focusing the work session investigations on sub problems.

A typical simulational environment will be composed of several packages, each devoted to a specific objective for flexible-capacity management through the use of a convenient dispatching, scheduling, planning or inventory decisional logic. Furthermore, the acknowledgement of the connection between the driving influences of given decisional set-ups and the related plant exploitation figures, needs, on the one side, the explicit characterisation of the governing rules, and, on the other side, the measurement of standardised functionality figures.

Typical governing rules will be defined for each sub problem that deals with the compatibility of the requested physical resources (inventory and schedule) or with the consistency of the related logical requirements (dispatch and planning); examples of selected criteria are:

- for dispatch: avoid idle workstations due to fixturing or tooling unavailability or to the late delivering of parts; use the versatility of transportation and handling devices, exploiting operation subsidiarity for minimising the fixed investments;
- for schedule: maximise the workstations operation time for extending resources exploitation ratios; select parts mix for reducing refixturing times; exploit versatility-based recovery-flexibility to suppress set-apart resources;
- for planning: allocate functional sections of the shop floor depending on the parts programs to optimise parts tracking, transportation and positioning time; assure the continuity of strategic, tactical and recovery flexibility;
- for inventory: assure a global just-in-time policy with the matching of due-time by minimising tooling and fixturing costs; integrate the tool-room management as critical support for assuring flexibility.

For understanding the factory performance in connection to flexibility, the relations between the production variables should be stated in a structured context. The **net production**, rather than a stand-alone characteristic, needs to be related to the supporting elements, namely: production capacity, manufacturing efficiency and cross-coupling effects:

· The *production capacity* is the measure of the gross number of parts which can be produced within the planning horizon. It is not a static figure, and it changes each time the production requirements modify. Its value is computed, for a given planning horizon, according to: - production requirements; - process plans (list of acquired jobs, operation time standards); - amount of resources (number of machines, tools and operators). Any products mix (instance of production requirement) establishes a production capacity; whenever the mix changes, so will the production capacity.

· The *manufacturing efficiency* is determined by the availability of machines, labour, scrap percentage and set-up delay. These variables have an inverse impact on productivity. The efficiency is improved by appropriate techniques, such as: - trend monitoring predictive maintenance; - anthropocentric work-organisation; - total quality and on-line process control; - robotisation, quick-change fixtures. The dependent nature of all production variables makes partial figures useless; for manufacturing efficiency, their coupled effects need to be minimised. Combined figures can only be obtained from real plants or from virtual reality simulation.

· The *cross-coupling effects* are losses in production capacity due to two combined occurrences: resources shortage and continuity blocking. The figure is better explained by an example: "material shortage and bottleneck propagation", meaning that the interruption of parts feeding has downstream consequences with delay on assembly and delivering, and upstream ones with blocking formation, unless the stacking of parts between workstation is authorised.

With expert simulation, the direct experimentation of the actual dynamic behaviour of the plant assures the visibility of the governing rules, effectively exploited according to the given area context instantiation.

3 Managing Flexibility

Manufacturing flexibility is effectively exploited through dynamic scheduling. The work-in-process and the production facilities are monitored in-process. The actual situations are acknowledged and information loops are closed to update the operation sequencing (job-shop) and the part dispatching (flow-shop). The labour sequencing and the pallet forwarding, accordingly, will depend on the resources current loading (through inner-flexibility tactical decisions) and on the enterprise marketing policies (through outer-flexibility strategic decisions).

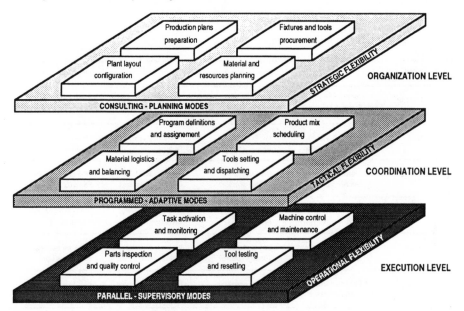

Fig. 5 - Connection between flexibility ranges and functional requirements

Governors consistent with dynamical scheduling [6] are set up, with varying ranges, namely (Fig. 5):

· *operational flexibility*, allowing the step-by-step switching between the fabrication of different part-mixes, due to the plant technological versatility governed by a centralised supervisor assuring consistent production schedules;

· *tactical flexibility*, assuring the optimal setting of the manufacturing efficiency over the programmed medium-horizon range, by enabling the parallelism of concurrent processes under decentralised control;

· *strategic flexibility*, authorising the production of varying product mixes (possibly after plant reconfiguration), according to the long-term enterprise strategies (and programmed projections).

The technological versatility is a built-in plant property and is exploited (for inner flexibility requirements) to grant the recovery ability against facility failures or functional degradation.

The "intelligent" governors, assuring the full exploitation of flexibility as it is prospected by the characterisation of Fig. 5, are different because of the hierarchical architectures, having:

· *organisational capability*, responsible for the enterprise strategic setting based on broad range information (compressed into characterising features, having comparatively high risk possibility);

· *coordination capability*, responsible for the company-wide data distribution, for commands dispatching, process diagnosis, end condition updating, etc., for tactically optimal production plans;

· *executional capability*, responsible for controlling the individual facilities, for the consistency of the running shop activity, based on process-monitoring and command updating.

The selection of the governing sequences is displayed into a hierarchical manifold, adapted to the horizon-range of the flexibility requirements and related to the functional level of the manufacturing facility areas. These are conveniently classified, from the narrowest to the largest, into seven levels: device, equipment, station, cell, section, plant, enterprise (Tab. 1).

Table 1 - Manufaturing facilities classification example (ISO TC 184) with characterising input / output figures

Level	Area	Functions	Material flow	Control flow	Information flow
6	Enterprise	Corporate policy Resources planning	Resources management Facilities allocation	Orders settings Programs authorization	Master agreements Enterprise projections
5	Plant	Production planning Activities organization	Managed resources Receivables/shipping	Production planning Resources supervising	Production & resources reports Requirement technical-data
4	Section	Scheduling coordination Shopfloor support	Resources tooling Fixtures allocation	Material flows balance Resources coordination	Production & resources status Fabrication schedules
3	Cell	Tasks coordination Activities monitoring	Work-in-process Fixtures feeding	Resources dispatching Machining scheduling	Coordinated diagnosis Process condition reports
2	Station	Sequences execution State monitoring	Parts/products handling Tools/jigs handling	Product synchronization Task sequences command	Production data report Station status report
1	Equipment	Tasks execution State measurement	Machining sequences Handling sequences	Elemental task activation Results aknowledgement	Performance monitoring Equipment conditions
0	Device	Operation execution Diagnostics setting	Commands sending Energy supply	Program feeding Sensors checking	Operation monitoring Device status

3.1 Govern-for-Flexibility Deployments

The areas of manufacturing engineering have been involved, traditionally, in a comparatively high quantity of "logic" information about the business product policies. The decision making mechanisms employed for production planning and management, have mainly been supported by appropriate heuristics, based on judgmental paradigms grounded on extensive, high-quality, specific knowledge of the manufacturing processes.

The problem of governing-for-flexibility the manufacturing facilities embeds judgmental and causal formulations [7]. The governors, developed on causal paradigms, are implemented with conventional non anticipatory blocks (e.g.: physical devices and/or algorithmic modules); the judgmental parts require inference abilities; the governors built with knowledge-based architectures, can be programmed to yield to "anticipatory" responses in the following sense: (a) the knowledge is assumed to correspond to the triple relation "facts + beliefs + heuristics"; (b) a response, accepted as successful, is a good-enough solution which is consistent with the selected hypotheses and with the available reference information. The governing facilities are, thus, based on "generative" modules, that provide the evolutionary trend of manufacturing activity; the simulated-process supplies information that corresponds to an image of what the future looks like, according to the causal paradigms, recognised as "pertinent", procedural knowledge. The recognition is obtained by means of logic-inference paradigms, formally emulating "anticipatory" answers, represented by *successful* solutions. The trustfulness of the governing strategies is checked on the (predicted) image in terms of performance figures, once the compatibility with planning requirements and the consistence with scheduling constraints are recognised. The recognition and check loops provide the updated information for dynamic scheduling, adapting the manufacturing resources for the contextual material-flow balance (modifying the part/product and the tool/fixture routing) and for the information-flow spreading (providing full mutual access to the local relevant data).

The governors are said to track *optimal* scheduling, when "just-in-time" work-organisations are approached for designed-for-assembly products. For optimality, the moving-assembly-lines of traditional automation, aiming at steady mass-production, are replaced by continuous-flow-production, in order to exploit the technological versatility to widen parts variability. The flow-shop scheme, joined to the integrated control and management of the individual manufacturing facilities, provides the flexibility (to the required horizon); the governor assures the dynamic routing of *material flow*, with balanced aggregates serving the manufacturing resources, managed by the *control flow*. The explicit visibility and convenient accessibility are basic requirements; they are obtained from a functional description of the manufacturing facilities up to the details directly involved by the governing strategies, together with the appropriate *information flow*. Example specifications, for the functional modelling of the manufacturing facilities, are shown by Fig. 6: hierarchies exist between the material, control and information flows. The control functions are present as decision operations, using current information status as input, to issue the appropriate functional commands. Two basic control functions are defined: - to request current status (situation recognition); - to change current status (situation updating). Information and material flows present similarities, showing the (four) generic functions: "storage", "transformation" and "verification", nested on "transportation".

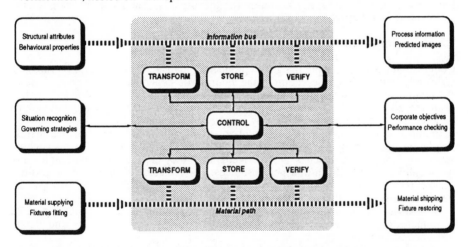

Fig. 6 - Schematic functional model of a generic manufacturing facility

The functional model of a manufacturing facility is represented by a network crossed by material transferors, serving transformation and storing devices; the links transmit information, similarly, serving processing-units and data-bases. Fig. 6 shows separate interconnections for illustration purposes; actually, the same track can be covered in the two directions by different flows, since the information about a product, the fabrication process plan, the current status of the product, the updated manufacturing situation, etc. are intrinsically interrelated entities. The governing of the "transportation functions" is clearly the critical factor in manufacturing flexibility. The model of the generic manufacturing facility of Fig. 6 shows that the govern-for-flexibility basic requirements are: **transparency** of the "modify" commands, and **visibility** on the "current" status. The management information system should cover expanded monitoring and governing capabilities of the technological and the mobility resources, for production diagnosis and process planning.

Transparency and visibility of the control functions are fixed requirements [5]; expertise and heuristics in the reference knowledge structures vary with the class level of the manufacturing facilities. Considering the example classification shown in Tab. 1, the most restricted areas (*device, equipment, station*) mainly require diagnostic abilities. The reference data are comparatively extended, presenting loose cross couplings when compared with well defined objectives (performance requests or failure figures). The diagnosis is obtained through data plausible aggregations (within some confidence bounds). On the other hand, the broadest areas (*section, plant, enterprise*) significantly require organisational abilities. The objectives are individually well specified, one at a time, and series of sub-goals could be defined with rather incomplete experimental data. The organisation typically assumes backward-chaining reasoning, detailing the elemental activities required to obtain the sub-goals and, then, the final objectives. Large intervention areas (*section, plant*) significantly require coordination abilities, conditioned by organisational ones. The data is a well acknowledged reference (the visibility on manufacturing functions is a basic requirement), but the selection of goals is not obvious (several responses, respecting the enterprise policy and the fabrication programs, may exist). Coordination of resources is obtained, by setting the task priorities and fixing the reference contexts that generate sequences of activities to direct process concurrency. Small intervention areas (*equipment, station*) require steering functions, that unite coordination and execution abilities. Expertise and heuristics in those areas are easily applied, interfaced: with diagnostic modules, for acknowledging the situations, reported from the peripheral sensing devices; and with (externally driven) agendas, for synchronisation purposes, at the same level, with the neighbouring units (*equipment, station, cell*), and/or for the organisational requirements of the surrounding facilities (*station, cell, section*).

3.2 Algorithmic-Heuristic Computer-Aids

Computer aids for supporting integrated production management and control problems, efficiently require the preservation of the algorithmic dependences that correspond to the relationships between material entities described by structured knowledge; they need the addition of the heuristic procedural abilities, enabling the relationships between the logical entities, provided by judgmental schemata.

Aiming at the development of intelligent governors, it should be remembered that, in flexible manufacturing, the processing priorities (for a given mix) and the dispatching schedule of the product batches, are not fixed, rather they depend on the current loading of the (allocated) transform/transport resources. The inclusion of inference capabilities (in the recalled sense of yielding anticipatory responses) as standard govern-for-flexibility property, is fundamental from two standpoints:
- for assessment purposes (as a CAD instrument), for expanding the expertise of manufacturing engineers, allowing for interactive enhancement of the knowledge reference environment;
- for implementation purposes (as a CAM instrument), for granting flexibility to the factory automation within the considered range, providing transparent adaptivity to the control functions.

In the restricted sense of actual unmanned exploitations, factory automation presently deals with intelligent governors, that provide all the basic steering functions only at shop floor-level (*station, cell, section*). Larger areas (*plant, enterprise*) keep on-line human commands, with, possibly, simulation programs typically as off-line consultation instruments. At the moment, however, it should be noted that such intelligent governors are still object of investigation; prototypal developments certainly exist, but the feasibility of actual deployments depends essentially on simulation results.

With such limitations in mind, a presentation of the general concepts employed for the design of the intelligent governors is considered a characteristic feature of the already existing knowledge-based computer-simulators, principally used as off-line programming instruments by the production engineer. Even within the above mentioned limitations, the development of special-purpose intelligent-governors is a noteworthy achievement of applied artificial intelligence; production engineering, for the reasons mentioned, is thus a very stimulating environment, since it requires knowledge-based computer-simulators with built-in capability for the emulation of inferential answers.

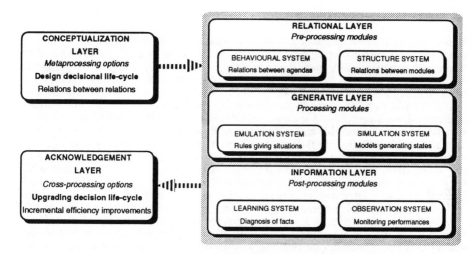

Fig. 7 - Schematics of the knowledge flow for expert computer-aids supporting intelligent manufacturing implementations

The design philosophy for an expert-simulator is presented by the Fig. 7 [8]. The programming facility develops as a multiple-layer construction, with vertical connections between the relational, the generative and the information layers. The **relational layer** embeds pre-processing modules, providing friendly interfaces for the model selection and the agenda setting, simplifying the definition of the structural attributes of the manufacturing facilities and of the behavioural properties of the related governing logic. The availability of integrated simulation environments is obtained by referring to specialised data-bases connected, through management facilities, with, both the generative and the information layers, so that the closure of the (simulated) work session is performed with full transparency of the control flow. The **information layer** performs the restitution operations, through post processing modules. The user can call for graphic presentations; the process-information is shown as sequences of relevant facts with situational specifications, or is processed to provide the performance evaluation figures as compared with competing (alternative) schedulings. The **generative layer** contains the main solving-capabilities, for propagating causal responses with algorithmic modules, and acknowledging consistent suppositions through heuristic modules.

The new design philosophies, necessary for achieving the factory automation concept, are here evident. The schematic representation of Fig. 7 shows also the two separate conceptualisation and acknowledgement layers. They are used as interfacing modules of an (outer) learning loop, closed by an intelligent governor responsible for adapting the production control and management, to the changing enterprise policy. At the **conceptualisation layer**, the user has access to the manufacturing facilities structural models and to the related control behavioural-modes; the automatic exploitation of the (built-in) technological versatility should be devised, in order to enable cooperative and coordinated processes. The **acknowledgement layer** provides the (required) full visibility; the assessed facts update the current situational characterisation of the fabrication resources and initialise the provisional analyses up to the flexibility horizon of interest. The meta-processing abilities at the conceptualisation layer grant the specification transparency and the foreknowledge opportunity to the (subsequent) relational layer. At the acknowledgement layer, specialised cross-processors, through the automatic accrediting of the hypothesised justification knowledge, might enable the expert-simulator to operate on-line as intelligent governors of the robotised manufacturing facility.

3.3 Expert-Management Supports

Understanding factory automation in the restricted sense of the integrated control and management of the running shop-floor operations, the operative environments related to medium-horizon flexibility spans (up to two or three weeks, depending on the products) are considered, oriented to small-intervention areas, that are, typically, sections of manufacturing plants. The governors perform tasks, that include *coordination abilities* [1]. These tasks enable (through the appropriate feedbacks) the (lower) *executional abilities* [4], with coordinated sequencing, to reach the expected performances within specified terminal manifolds and technological constraints. The coordination is either sought through optimisation (steady-state tactical ranges), obtaining answers in line with the static requirements of the production planning, or through expertise (recovery transients, after externally-driven occurrences), developing solutions that comply with the dynamic constraints of production scheduling. Both these results are conditioned by external requests related with the *organisational abilities*: in the first case, this applies to fabrication agendas programmed according to the enterprise objectives; in the second, the manufacturing facilities are considered in actual environmental situations. Flexibility is effectively investigated with time-varying interactive coordination schemes. The programming instruments, specifically adapted for the solutions sought, present peculiarities, summarised in Fig. 8, on which it is useful to dwell, since they are among the more representative current developments of the coming trends.

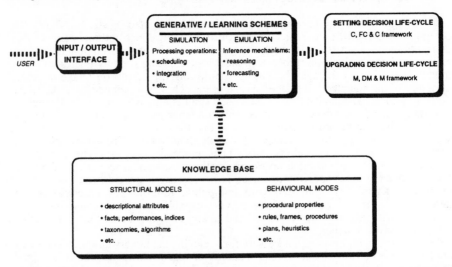

Fig. 8 - Typical arrangement for a knowledge-based computer-simulator, incorporating heuristic decisions

The control-and-management architecture for the shop-floor logistic of a flexible manufacturing facility, has to be developed in a modular form, to be easily adapted to varying functional requests and operative environments. The knowledge-based systems are a worthwhile engineering instrument; they distinguish: the **declarative knowledge,** typically the *structural attributes* of the manufacturing resources, at the input interface, and the *process information*, at the output interface; from the **procedural knowledge**, typically the *behavioural properties* of the manufacturing facilities. In general, the procedural knowledge provides the information on how to accomplish actions, or take decisions, using the declarative knowledge as data. The declarative knowledge is given at any moment in the current memory, combining the system hypotheses (problem description) with experimental results (problem status). The declarative knowledge can be retrieved and stored, but cannot provide executive instructions; to become effective it must be interpreted by procedural knowledge. Therefore, there is a connection, required with the generative programming modules, that corresponds to the learning schemata of the computer-simulators. The connection presents a separate access to the relational data-base structure to maintain visibility on the governing strategies.

An intelligent govern is based on an **expert module**; this includes declarative knowledge, corresponding to the continuously updated reference information, together with the procedural properties characterising the manufacturing facilities decisional manifold; it acknowledges the situational description in the current memory and selects the actions to activate, with an inference engine, through the recognise & act cycle. The typical components of these expert modules are: · the *conditioning logic*, encoding the procedural knowledge; · the *data-memory*, embedding the current situational characterisation; and · the *inference engine*, activating: the information-match, the conflict-resolution, and the rules-firing operations. "Expertise", in narrow domains, is exploited through the organisation of the reference information, offered with a highly structured and interconnected data-base, in the form of rules, procedures, frames, etc., associated with the situational description of the objects of investigation. Expert modules are employed for several purposes: cognitive, consultation, control, diagnosis, planning, etc. Their implementation differs, depending on the conditional operative environments of each specific application. The rules structure deals efficiently with the control network activated by interactive governors. The procedure structure is an important option for supporting mixed-mode (algorithmic-and-heuristic) programming. The frame structure assures the expandability by inheritance and is useful for providing generality to the software.

The simultaneous attainment of efficiency and flexibility [8] requires an information framework for specifying the manufacturing processes, based on functional models that combine structured analytical description with human-like judgmental abilities. The outcomes, as stated, are knowledge-based programming environments, specifically developed for production-engineering applications and related to automated manufacturing. Different manufacturing problems require "different" expert modules. Diagnosis (at the level, for instance, of equipment) is based on the acknowledgement of "facts", by performing the classification of information expressed as "attribute, value and belief"-elements; success is affected by search efficiency. Planning (at the level, for instance, of a plant) involves classificatory reasoning, singling out the goals with subgoals specification; success depends on the actions satisfying all the imposing constraints and is affected by the selection of the methods employed. In the two examples, backward chaining would be a worthwhile approach for inference, causing, however, differing implementations, since the main sources of difficulties are incorrect data and/or unsuitably interconnected data-bases, in the first case; and misleading knowledge and/or unsuitable hierarchic solving-architecture, in the second case.

The expert-modules, providing flexible control and management functions at the shop-floor operative level (typically sections supervising cells, stations, and equipment, and cooperating within the considered plants), most conveniently employ forward chaining inference. The nature of data is well understood; "optimal" goals could be devised, but they correspond to solutions too dependent on situational conditioning; "satisfying" goals provide solutions with good productivity figures (the actual throughput, for a given product mix, depends explicitly on the resource loading conditions, with time-varying information of admissible batches).

4 Example Applications

The decision life-cycles closed for the setting (intelligence and design step) and the upgrading (choice and redesign step) of the FMSs frameworks are conveniently illustrated by examples, that basically, use simulation environments for assessing the flexibility under actual running conditions (feedback and testing step). These case-investigations consider *development opportunities* (the set-up of a pilot CIM experimental plant [9] and the implementation of a multi-operational section for an established manufacturing enterprise [10]) and *governing options* (the hierarchical control of varying and sophisticated product mixes [11] and the integrated management of a market-driven diversified production [12]).

4.1 The Set-Up of a Pilot CIM Plant

To help support CIM solutions for small/medium enterprises, the availability of actual prototypal facilities is a valuable issue, aiming at: · demonstrating the effectiveness of the organisational and technologic capabilities; · supporting research and developments investigations; · providing training opportunities for personnel specialisation.

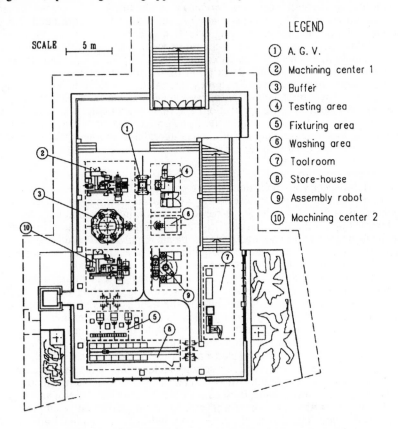

SCALE |— 5 m —|

LEGEND

① A. G. V.
② Machining center 1
③ Buffer
④ Testing area
⑤ Fixturing area
⑥ Washing area
⑦ Toolroom
⑧ Store-house
⑨ Assembly robot
⑩ Machining center 2

Fig. 9 - Design of plant lay-out for the CIM training installation

The CIM plant development (headed in Genoa by the Agency DITEL Engineering) required a preliminary feasibility study for acknowledging the characterising features that should be possessed by the components gathered to form a single pilot plant in order to:

· reproduce all the basic activities of typical mechanical manufacturing shop-floors; · support flexibility requirements with the appropriate technological versatility; · grant close connection with the evolutionary trends of innovation; · assure expandability for integrating new modules and new functions; · provide interfacing capabilities for experimenting on computer-integrated enterprise (CIE) architectures; · demonstrate the effects of flexible manufacturing and production adaptability for several (potential) industrial users.

These objectives suggested the modular lay-out of Fig. 9, having functionally independent areas, that could operate separately or mutually connected by a material transport service; for practical purposes, the manufactured parts are assumed to have "small" geometrical dimensions in order to be handled with the ISO 500 pallet. The basic manufacturing modules encompass: - a machining area, with two NC work-stations; - a testing area, with a measurement robot; - an assembly area, with a programmable manipulator. The areas are linked by an AGV operated transportation service, with intermediate storage and fixturing devices.

For the intelligence and design step, the pilot plant was simulated with the programming environment **OMX-SIFIP**, developed using the object-oriented language MODSIM II (by CACI). The objects assemble data into "attributes" with attached "methods" to describe components or activities and their related behaviour and operativity. The graphic output assures friendly restitution with icons, windows, dialog-boxes, menus, diagrams, etc. (Fig. 10).

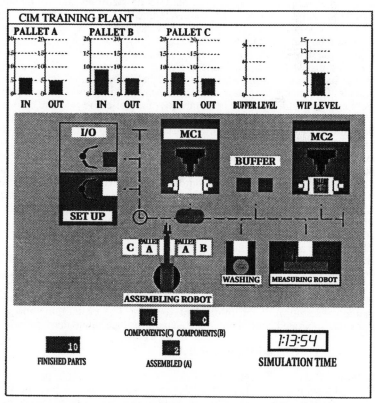

Fig. 10 - Example graphical output of the simulator

The modelling of the physical resources considers the machining, testing and assembly areas, the AGV transportation service, the loading/unloading units, the interoperational storages and the product mixes; the integration of the tool-room service has been delayed to a subsequent stage. The objects include the basic technological information (machining list and timing, assembling requirements, etc.) with the related conditions (commands, sequencing, etc.) for the synchronisation and the coordination of the concurrent processes.

The feasibility investigation has covered typical manufacturing fields, and has considered different product kinds, directly related to users potentially interested by CIM solutions, such as: - belts-meshes for tool-conveyors; - regulation valves components; - motorcycles carters; etc. The selected final products (as preliminary choice of the investigation) are obtained by assembling two parts; these could be machined by any of the (two) machining centres, and require a set of appropriate geometrical measurements for quality assessment of the (assembled) manufacts.

The investigation on the C, F·C & C structural frameworks aimed at defining convenient set-ups for: · the machining area (with the availability of two either one workstation); · the intermediate storing requirements (distribution and number of locations); · the fixturing support (number of pallets); · the assembly area (number of parts at input/output buffers); · the recovery policies; · etc.

The study of the M, D·M & M behavioural frameworks considered different situations such as: · management of parallel routing (parts assigned to individual machining centres); · management of mixed sequencing (parts machined by either one machining centre, for maximising the operative time); · adaptive scheduling for initialising "new" production programmes; · supervisory govern of transients (after station failure, part/tool unavailability, etc.) up to nominal (steady-state) production planning; · etc. The tactical horizons (standard steady-state reference) are programmed to cover a single (8 hours) work-shift, since the pilot plant is assumed to be involved into mainly demonstration business; longer horizons are (possibly) planned upon request.

Any pilot plant, certainly, will suffer of practical limitations depending on the allocated resources. It assures, indeed, physical evidence to subsets of simulation results; but its final set-up requires the correct balancing of the expected return from investments. Comments on a case example investigation provide useful hints. Let us consider the manufacturing of the belts-meshes of a tool-conveyor. The individual mesh is obtained by assembling two hollow bushes through four connecting plates. The members of the (two) part-families might differ, depending on the tool to be carried on, having, however, extended compatibility with the plant technological versatility. The simulated processes cover the manufacturing and the delivery of the meshes (bushes and plates). The first process deals with machining, washing, testing and assembling the parts; at the same time, the coordination of workstations fixturing and tooling and of material manipulation and handling is performed. The second process assures the forwarding of the assembled meshes, adapting the AGV service to comply with the delivery agendas. Bottom-up schedules are deployed, adjusted to the delivery requirements. The hypothetical plant lay-out shows the effectiveness of flexibility, demonstrating the actual manufacturing efficiency (in terms of exploitation ratios of the material resources), for the different production capacities (in terms of allocated resources availability).

The developed simulational facility **OMX-SIFIP**, having completed the support function at the conceptualisation and design stage [9], will become a standard software aid for the exploitation of the pilot plant, assuring fast prototyping for setting and upgrading the integrated control and management for the sample demonstration cases.

4.2 The Control of a Multi-Operational Area

Automation in the manufacturing of low cost thermoplastic products presents discontinuities, with local sophisticated isles (automatic moulding machinery, robotised assembling stations, etc.) and extended manual operations in between, in order to expand versatility. The example deals with the assessment of the control-and-management policies required by the multi-operational area to be added, for flexible automation, between the moulding process and the machining, assembly and packing stations. The case is discussed considering an existing manufacturing environment; the comments give details on the physical facilities, and on the modelling developments used for establishing the expert simulator **OX-SIFIP**.

The customer-oriented just-in-time planning moves products shipping and packing requirements back to parts flow (and, then, to material procurement) and to fixtures scheduling. The parts are suitably handled into containers (according to the prescribed assortments). The area to be investigated [10] superintends to the shop logistics in order to balance, both parts/products and tools/fixtures flows; it is composed by the following facilities: 1) a buffer, for storing and sorting the parts coming from the moulding shop and forwarded, into containers, to subsequent workstations; 2) a travelling lift robot, for moving the containers within and out of the buffer; 3) two machining centres, for the labours required by a given subset of components; 4) six assembly stations, for the required subset of final products (valves, etc.); 5) a bag filling machine, for packaging "small" products (valves, fittings, etc.); 6) two thermoforming machines, for packaging "bigger" manufacts.

The programming distinguishes physical resources from the logical ones. The modelling uses object-classes to exploit hierarchic inheritance. The previously listed facilities are standing physical resources not modified through manufacturing, and are associated with correspondingly stable objects. Each object belongs to a class with the hierarchical organisation shown by Fig. 11. The object definition specifies an icon and the attribute list. The basic level distinguishes between buffer shelves and delivery facilities; then details follow according to the functional requisites. A second set of objects defines the logical resources. The connections among physical resources are also specified, through links between objects that may be switched on or off by actions defined by rules or procedures. The option is useful for associating the product type-code (and manufacturing operations) timely given in charge to a given workstation, and, in general, for expanding the programming flexibility, to assure control instanciation.

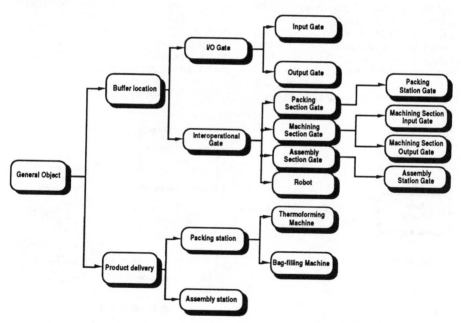

Fig. 11 - Classes hierarchy

The containers, the (element) parts and the (final) products are associated with transient objects, that are originated, modified and destroyed at given locations of the plant. The consideration of transient objects faces programming drawbacks; a model

244

based on operation agendas has thus been used. The control logic, then, exploits distributed knowledge, with combined (algorithmic and heuristic) relational structures, and concurrent processing capabilities. The inter-operational area is governed by transient logical objects, corresponding to the transportation requests generated each time a container has to be moved between two locations of the buffer, with a focus on the I/O gates. The generated requests are conveniently grouped into specialised lists (with the priority scale corresponding to the order listing), such as: - entry list, for processing the input service of the containers in the buffer; - remove list, for managing the withdrawal of empty containers, whichever location is actually occupied; - reset list, for moving back to the buffer entry the half empty containers that come from the assembling stations (typically at products changes); - forward list, for moving the containers from the machining centres or from the dedicated buffer gates, to the assembling stations; - delivery list, for planning the transport of the containers from the buffer to the machining centres and to the packaging stations.

The inter-operational area aims at establishing just-in-time policies, therefore conditioning, through the assembly lots, the moulding processes, according to the packing batches that properly match with the customers' requests. The moulding batches define the agenda of the new arrivals to the buffer, and allow for the correct feeding, through the (two) machining centres, of the (six) assembly stations and of the final packaging stations. The backward actions of the shipping requests slightly affect the packaging segment; it, rather, requires a careful analysis of the machining centres behaviour. These are, as a matter of fact, flexible cells, but appropriate refixturings are needed according to the product type to be processed; the refixturing operations are complex and time consuming, and the corresponding specialised manufacturing schedules require a management module, that is enabled to run in parallel with the main simulation program, with due conditioning effects on the assembling stations schedules.

The simulational package **OX-SIFIP** enables the concurrency of a set of mutually cooperating parallel processes: one set employs a standard programming language (actually the C language) and provides an efficient modelling of the dynamic behaviour of the plant physical resources; the second uses the expert shell G2 (by GENSYM) and assures the implementation of the control logic. The communication between the two sets of processes has been developed through specially tailored modules and appropriate interfaces. Fig. 12 schematically shows the principal features of the overall architecture, distinguishing: · the G2-block, performing the expert-control; · the interfacing protocol, assuring data communication; · the C-block, duplicating the plant behaviour.

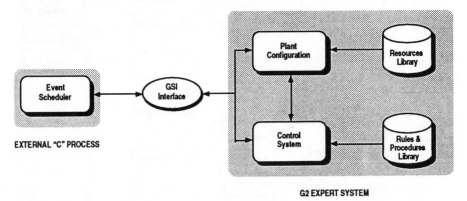

Fig. 12 - Simulator overall architecture

It should be emphasized the fact that the knowledge structure of the expert-controller is based on two different data-bases: a stationary data-base and a dynamic one. The stationary data-base is fixed for every situational updating (specifying the actual behaviour of the system, according to the driving agenda). The overall knowledge architecture is presented in the specialised workspace, corresponding to the physical frames of the display, into which the objects, connections, procedures, display-elements, etc. can be created. A knowledge architecture can contain any number of workspaces, and this is properly exploited for orderly grouping the elements into specialised layers (namely following to the definition of the related class, relation, rule, procedure, etc.). The dynamic data-base is interfaced to the governor, which appears to be a decision-keeper that runs the plant, using the established production-planning strategies, conditioned by the situational knowledge. These strategies are coded: with rules, for mainly heuristic frameworks (e.g. for the quickest selection of the appropriate container, from the sorting-and-storing system); and with procedures, for properly established algorithmic contexts (e.g. for the sequencing of operation cycles).

The simulational program **OX-SIFIP** has been tested by a number of validation runs, using different product shipping policies (according to different marketing objectives) in order to quantify the return-on-investment related to a fully automatic storage and retrieval system replacing the present requirements of manual interventions. The distributed governing structure, moreover, is based on a real-time expert shell, and can directly be transferred to the actual plant, fully exploiting the experiences previously completed through computer-simulation.

4.3 The Dynamical Scheduling for Highly Diversified Manufacts

The benefits of adaptive schedules are now considered referring to an example industrial implementation for the manufacturing of steam-turbine blades (Fig. 13 shows the lay-out).

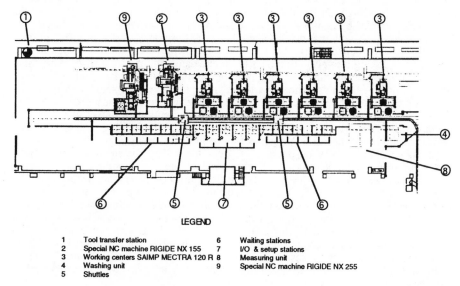

LEGEND

1	Tool transfer station	6	Waiting stations
2	Special NC machine RIGIDE NX 155	7	I/O & setup stations
3	Working centers SAIMP MECTRA 120 R	8	Measuring unit
4	Washing unit	9	Special NC machine RIGIDE NX 255
5	Shuttles		

Fig. 13 - Layout of the reference manufacturing section

The batches are highly diversified mixes (up to 78 different types of items); they require complex tasks (up to 30 machining operations for some items) and cover several part families (typically 10 to 15) with different volumes and geometry (for example: nine blade-families, the low-pressure ones longer than 1.5 m, and three-nozzle families, with internally worked sprouts, about 10 mm in length). The machining cycles require specialised set-ups, and items should be refixtured on the pallets (up to six different positions may be requested). The section contains two lines of machining centres, separately fed by appropriate tool-dispatchers and completed by washing and inspection stations. The parts are transported by two shuttles on a single track, having front palletising stations for fixturing and refixturing operations.

Efficiency is achieved with the control-for-flexibility paradigms, adapting the scheduling operations to the known running situations, with integration of tactical and operational ranges and exploitation of the decisional manifold to control the concurrency of the manufacturing processes [11].

Let us focus our attention on scheduling, assuming that the production-planning data (such as the economical size of the product batches, the management of the capacity constraints, the tooling and set-up procurement policy, the product-mix aggregation on the strategic delivery horizons, etc.) have already been established. The following six characteristic features are singled out: 1) the product batch is composed of part families A_1 ... A_k, each including m_1 ... m_k items; the due-date consistency is assessed; 2) the work cycle of the A_i family is segmented in the macro operation sequence B_i: a list of "part positions" b_{ij} ($j = 1, ... h_i$), corresponding to the items temporary palletisation for the operation sequence, is dressed; 3) the number n_{ij} of items of family A_i, fitted on a given pallet for the operation sequence b_{ij}, is known; this number is not constant for the pallet setting, and some items could be left in a store station or retrieved at refixturing; 4) a check on the "part-position" operation sequence is done in order to select the production line referring to the specialisation of the related workstations; 5) the tool sets for the specified operation sequences are handled globally (the machining times are specified individually, of course, depending on the items actually processed); 6) the workstation engagement times are known; the total figures (shuttling, loading, unloading, machining times) are listed for each "part-position" assortment.

The description leads quite naturally to the definition of a set of virtual families a_{ij}, multiplying the batch families A_i for each subsequent fixturing step "j". The pointer "j" specifies the "part positions" and the corresponding operation sequence (segment of the work cycle required by family A_i). The scheduling applies to the virtual families aij (i = 1, ... k; j = 1, ..., h_i), with due-date constraints on the (original) families A_i. Two priority patterns are established: horizontally, across the batch families (for the palletised assortments), taking into account the parts-lot extension and the total work cycle time, for a preliminary reordering; vertically, along the machining cycles, dealing with the number of part-position assortments and fulfilling the operation sequences for every item, with regard to the due date. Each virtual family a_{ik}, therefore, is reordered so that: (a) at work-step b_{ij}, the pointer "i" is dominant and items with closer due dates are processed first; and: (b) for the delivery of the family A_i, the pointer "j" is dominant and items with longer processing times before completion are machined first.

A manager should work on-process in order to modify the manufacturing plans (according to pre-established tactics) for upgrading the exploitation of the available resources. The supervisor in the example uses a twofold decisional connection with the plant-controller (Fig. 14): for task-attribution (planner) and for task-distribution (manager). The shuttle transportation, thus, would perform conditioned actions, such as: feeding the machining centres, once acknowledged the most suited station to perform the required operation sequence for given executional priorities; withdrawal of pallets from workstations, after recognition of the station-priority setting order, from "full" departure-buffer, to current destination-call, etc.; removal of pallets from fixturing-stations, using the combined assessment of machining cycle and workstation availability; etc.

At the coordination level, the updating of schedules is based on the following constraints: · 1 · batch families are reordered, with the "part-position" assignments generating the virtual families aij. · 2 · the steady-state sequencing defines the (horizontal) priorities across the families. · 3 · after refixturing, the work cycle is completed with due-date conditions along the (vertical) production flow. The supervisor operates with servicing priorities (e.g. withdrawal of full buffers, removal of pallets from intermediate stores, etc.) and instanciates the local controllers (switched to the "enabled" position). The manager then acknowledges the updated environment for the task-attribution (switching to: "selected", those tasks instantiated by the supervisors;

selecting the shuttle for servicing the station after work-completion; searching the job-consistent station for available "part positions"; etc.).

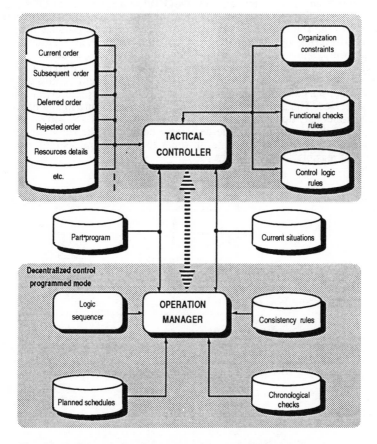

Fig. 14 - Information architecture with specialised data-bases

At the execution level, the task-activation is enabled, at the interface between scheduler and controller, by programmed occurrences (e.g. planned modifications of product mixes) or by situational occurrences (e.g. a place opens in the station output queue; the feeding station becomes temporarily empty, or a pallet leaves at the end of a work cycle). Two different operational conditions are typical of this level: *transient*, e.g.: booking the pallets sets p_{i1} for a pre-schedule depending on the set-up availability in order to dispatch the lots m_i to the initial work-steps; and *steady state*, e.g.: performing decentralised retrofitting and refixturing operations and selecting the priorities according to the horizontal/vertical patterns discussed. The rescheduling requests can easily be met if the due dates can be achieved with reallocation of available suitable resources (workstations, outfits, transporters, etc.); in other cases, the task-manager needs to modify the batch order in keeping with "recent" updating (computed as the difference between the expected due date and the remaining-operations time of the lots to be processed). A different approach is considered if the operation times can be lowered (e.g., by increasing the cutting speed); this increases the resource-availability figure, with weighted penalty indexes.

The functional conditioning aspects [11], such as the ones summarised for the example above, are fully incorporated into the **XIM-SIFIP** simulator by means of an interactive interface with the specialised module corresponding to the flexible manufacturing section; the data related to the tool/fixture-procurement operation are transmitted to the correspondingly specialised module, so that the relevant conditioning information is held in common. The **XIM-SIFIP** package uses the OPS 5 standard language for encoding the rule-based heuristical blocks, and an extended version of FORTRAN for the algorithmic blocks. The assessment of a new product mix and of the related manufacturing agenda initialises a simulation run, with the shuttles feeding the appropriate workstations according to the governing logic described above. The time evolution of the plant is continuously displayed (see Fig. 15): different colours and patterns show the actual state of the resources (buffers, machining centres and shuttles); a window specifies either the tools engaged in the section or the tools handled by the supplying service. The statistical data (utilisation ratios, failure/maintenance figures, etc.) are continuously updated, so that the actual efficiency of the activated schedules can be quantitatively compared. The decisional context of the rule-based governing logic can be backtracked at any time, in order to provide transparent access to the sequencing of feasible control regimes.

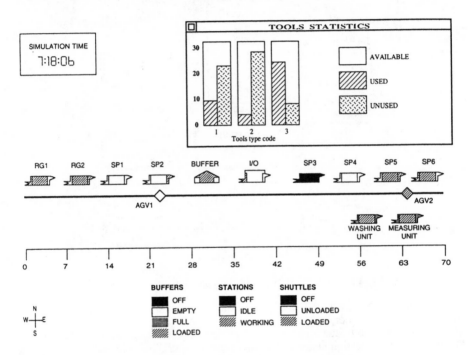

Fig. 15 - Simulator synoptics with tools monitoring window

4.4 The Dynamical Management of Time-Varying Production

The fourth case considers the re-organisation issue of the production process of a pharmaceutical enterprise [12], with focus on the on-line assessment of short-horizon adaptive scheduling. The plant develops vertically on three floors, Fig. 16, with workstations (for blending, dosing, dispensing, granulating, drying, sieving, compressing and coating) followed by the packing, printing and delivery areas. The fabrication agendas are based on batch planning, depending on the starting bins that supply the raw materials. Several bin classes are requested in relation, both to raw materials procuration and to the final products shipping plans. A bin washing area is annexed, with storing and sorting capabilities. Appropriate (and time-consuming) workstation refixturing is needed each time the production changes.

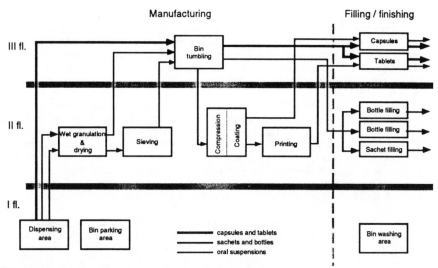

Fig. 16 - Production flow in the pharmaceutical plant

The plant integrated control and management has been given in charge to an expert-governor, programmed with the shell G2 (by GENSYM), and the assessment simulations are carried on with the programming environment **OGX-SIFIP**. The package is modularly arranged with a hierarchic architecture, which allows a double use: as a simulator for a detailed knowledge of the production process in assigned operative conditions, or as an integrated scheduler-simulator package to choose the optimal scheduling. The package has been built on an open object library that allows easy extensions to different plant set-ups and fast interfacing to other modules characterising the complex CIM pharmaceutical production systems.

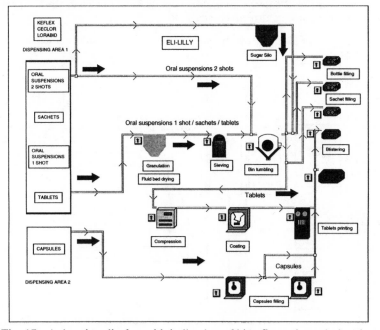

Fig. 17 - Animation display with indication of bins flows through the plant

The integrated scheduler, as usual, exploits rules and procedures. The rules are used to acknowledge the plant current situations and, with decision-and-check functions, to set the procedures required for activating the production cycles. Consequently, the procedures operate at the execution-level, to perform the actions selected by the rules that operate at the coordination level. The simulator assures high details degree, providing every relevant information that might modify the production programmes at the organisational level.

The production manager can be provided with different scheduling policies on the tactical horizon. It is interfaced to the user who, for any production programme, can require the indication of "optimal" schedules in relation to typical operative or functional figures, such as: - queue priorities resetting at the blending workstation (bottleneck of the plant); - queue balancing among the compressing or the packing centres; - priority modification between products types (capsules, tablets, liquid suspensions); - dispensing work-centre conduction, with alteration of the minimal amount of delivered raw materials; - control on the bin availability/unavailability policy; - etc.

The **OGX-SIFIP** package supplies several outputs and windows, that provide in real time every relevant characteristic of the physical resources (work-centres, bins, etc.) by means of graphics, diagrams, read-out tables, messages, etc. The restitution exploits animation displays [12] that are available on graphical workspaces, Fig. 17: they assure quick understanding of plant current status and its related time evolution; optional data are displayable with indication of the net cumulated production by means of Gantt diagrams or through resources utilisation ratios by means of histograms.

A schematic presentation of the integrated scheduling/simulation environment is shown in Fig. 18. The descriptional data are conveniently stored into specialised libraries, so that virtual reality generation of alternative plant behaviours can be instanciated. Through dynamic scheduling and simulation, anticipatory images of the production programmes are made available. The plant manager, presently written with the G2 shell, can be readily replaced by different governing blocks and interconnected to internal or external modules in order to modify the scheduling policies. The results, compared with pre-established production programmes, have fully satisfied the industry (a primary multi-national enterprise) and the simulational environment has proved its validity for the integrated control and management.

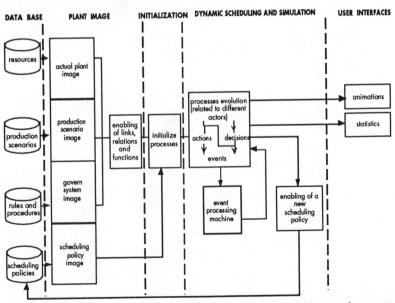

Fig. 18 - General view of the integrated scheduling / simulation environment

5 Concluding Comments

For most of the existing manufacturing enterprises, flexibility is a novel technological issue which uses, most of the time, existing facilities, after (limited) reorganisation, (small) modifications, and (some) additions. The options of integrated control and management present a critical opportunity for further improving the plant efficiency with a correct exploitation of flexibility without additional costs. Managing flexibility, moreover, essentially entails decisions involving dynamical schedules with an allocation of capacity to meet current production needs. The upgrading is based on a multi-level decisional logic, concerning: the shop floor logistics (routing tables decisions); the fabrication agendas updating (schedules decisions); the tactical production-planning (planning decisions); and the strategic organisation policy (inventory decisions). These four decisions need to be assessed and evaluated with supports granted by an information architecture, which include the monitoring of the actual performances and the comparison of current figures against a standard or an expected level of performance. In order to provide this type of decision supports, computer simulation (through virtual reality duplication) of the actual plant dynamical behaviour will be required; the assessment of the return from investments in flexible manufacturing, indeed, can only be reached after conveniently extended simulation work sessions.

In the present chapter the discussion is general, with due consideration to the basic characteristics of computer-simulation programmes having in parallel heuristic and algorithmic modules. Practical references and example illustrations are provided, using specialised programming supports established by the authors over the past years. These specialised packages have been tailored on the individual application case.

The first example faces the challenging problem of assessing the effectiveness of a prototypal CIM lay-out to be used for demonstration and training. Critical schedules are considered referring, for instance, to the need of the occurrences synchronisation along the machining and testing processes, with due regard to the concurrent management of the transportation service up to the assembly area requirements.

The second example refers to very cheap manufacts and their efficiency is related to the ability of enabling just-in-time production-programs strictly dependent on customers' orders. The need of continuously resetting the components, facility-configuration and control framework has been solved, until now, with extended manual operations linking the (eventually existing) automated isles. The switching to factory automation solutions requires an accurate assessment of the return from investments.

The third example deals with the manufacturing of very expensive mechanical components, that require comparatively long machining operations and need to be delivered with fixed assortment batches for assembly purposes. The efficiency through flexibility is sought by means of the rescheduling (when appropriate), managed on-process by a supervisor. Factory automation, thus, requires combined control and management; the critical feature is the dynamical allocation of inventory across the part-families and along the labour-sequences; the problem is solved, giving transparent access to the control and management functions.

The fourth example takes into consideration a mixed-mode (continuous-to-discrete) production plant, that needs important resetting operations each time the delivered products are modified. The economical exploitation of just-in-time policies requires preliminary verifications of the fabrication agendas, depending on the delivery dates and on the batches sizes, that are practically enabled only if off-process simulation is available, since over than 400 different final products are given in charge to be produced by the plant.

Only the first case was concerned with the implementation of a "totally" new manufacturing facility; the practically conditioning factors, however, make the lay-out of a pilot plant extensively defined. The OMX-SIFIP package was then essentially aiming at demonstrating the options of flexibility for very simple manufacturing processes for the direct use of small-to-medium enterprises, that still need to accept the "idea" of computer-integration. The second case originated, as well, in the same range of industries; the investments in automation need a pace-wise planning and the support programming instruments, such as the OX-SIFIP package, was similarly established for demonstrating the advantages of technologic innovation, obtained with the addition of a "new" multi-operational area, which will assure enough flexibility for factory automation.

The industrial concern changes with the subsequent cases, related to the range of large enterprises. The advantages of intelligent manufacturing are already recognised; still the full benefits of flexible automation need to be enabled. The capabilities of integrated control and management are, perhaps, better accepted; the critical hindrance is represented by the difficulties in the measurement of the actual improvements that flexible manufacturing assures, if properly exploited. The XIM-SIFIP package was developed for investigating the opportunities offered by dynamical scheduling for the manufacturing of sophisticated mechanical parts, requiring very complicated machining cycles; the separation of transients (based on recovery-flexibility) and steady-state (exploiting tactical adaptability) govern has been demonstrated to be the winning option for economical return from investments. Similarly, the OGX-SIFIP package has shown the advantages of re-scheduling based on the back-tracking, along comparatively simple production cycles, of the delivery requests, in front, in this case, of a very large number of diversified final products. Object-oriented modelling is a powerful aid for production engineer, facing "open" manufacturing cases, since this helps splitting the control logic level from the management of individual customer-oriented production programmes.

References

[1] - Michelini RC, Acaccia GM, Callegari M, Molfino RM. *Integrated management of concurrent shop floor operations*. Intl. J. Computer Integrated Manufacturing Systems 1990; 3:27-37
[2] - Acaccia GM, Michelini RC, Molfino RM. *Automazione della fabbrica: Sistemi integrati di lavorazione*. Il Progettista Industriale 1984; 7:68-79
[3] - Michelini RC, Molfino RM, Acaccia GM. *Automazione della fabbrica: Robotizzazione dei processi manifatturieri*. L'Industria Meccanica 1984. 387:589-600
[4] - Michelini RC. *Decision anthropocentric manifold for flexible-specialization manufacturing*. In: Ming Leu (ed) Proc. of the 4th Japan-USA Intl. Symposium on Flexible Automation, vol. 1. ASME, USA, 1992, pp.467-474
[5] - Acaccia GM, Michelini RC, Molfino RM. *Design of intelligent governors for the material-handling equipment of automated factories*. In: Sata T, Olling G (eds) Software for Factory Automation. North-Holland, Amsterdam, 1989, pp.297-312
[6] - Acaccia GM, Michelini RC, Molfino RM, Piaggio P. *X-SIFIP: a knowledge-based special-purpose simulator for the development of flexible manufacturing cells*. In: Proc. of IEEE Intl. Conf. Robotics & Automation, vol. 1, IEEE Computer Society Press, Washington, 1986, pp.645-653
[7] - Michelini RC, Acaccia GM, Callegari M, Molfino RM. *Simulation facilities for the development of computer-integrated manufacturing*. Int. J. of Advanced Manufacturing Technology 1992; 7:238-250.
[8] - Acaccia GM, Michelini RC, Molfino RM. *Knowledge-based simulators in manufacturing engineering*. In: Sriram D, Adey RA (eds) Proc. of the 2nd Intl. Conf. on Applications of Artificial Intelligence in Engineering, Computational Mechanics Publ, Unwin Brothers, Southampton, 1987, pp.327-344
[9] - Milanesio R, Rossi A. *Progettazione di un impianto manifatturiero per scopi dimostrativi e di formazione*. Tesi di Laurea, Univ. Genova, 1993
[10] - Acaccia GM, Callegari M, Michelini RC, Molfino RM, Peri P, Ricci A. *Control automation of a multioperational section for the flexible manufacturing of highly-diversified products*. In: Mezgár I, Bertók P (eds) Proc. of KNOWHSEM '93. North Holland Elsevier Science Publ, Copenaghen, 1993, pp.187-195
[11] - Acaccia GM, Michelini RC, Molfino RM, Piaggio M. *Govern-for-flexibility of manufacturing facilities: an explanatory example*. Intl. J. Computer Integrated Manufacturing Systems 1993; 6:149-160
[12] - Firenze G, Firenze G. *Sistema esperto in tempo reale per il controllo e la supervisione di un impianto farmaceutico*. Tesi di Laurea, Univ. Genova, 1993

9   Modeling of Manufacturing Systems

Zbigniew Banaszak

1   Introduction

The modern manufacturing systems are characterized by automation of material, energetic, and information flows. A set of computer numerically controlled (CNC) facilities (e.g. machine tools, robots, and measuring machines) interconnected by an automated material handling and storage system, and supervised by a distributed computer control system, shows the background of manufacturing processes automation. Its main attributes, i.e. structural and functional flexibility, concurrency of variety material and information processes, and hierarchism of control organizations, allow the simultaneously processing of variety part types.

This ability however, develops many different decisions and optimization problems including design problems, planning problems, scheduling problems, and control problems [3], [17], [37].

The methods offered by the theories of operational research, stochastic processes, computer simulation, and artificial intelligence provide versatile conceptual models. Their usage, however, is usually limited, and no single method can guarantee an efficient solution adequate for all tasks in both technical and organizational aspects of manufacturing systems design and operation. The lack of a unified theory which describes and evaluates the entire manufacturing system provides a strong motivation for a better understanding of this class of systems.

In recent years, the usefulness of the Petri net models to efficiently model and analyze manufacturing systems has been widely recognized [3], [14], [30], [31], [33], [36], [42]. They have been used in the study of parallel computations [18], multi-processing [41], computer and flexible manufacturing systems modeling [1], [10], [16], [37], [40]. Thus, the Petri nets theory seems to offer a unified framework for modeling a class of systems as large as that offered by the discrete dynamic systems [21], [39].

To make the section self-contained, an overview-like part has been developed for the discussion of the main modeling techniques used for the performance evaluation. The other three parts are based mostly on the research work that the author has done in this field in recent years. The second part contributes to the problem of automated synthesis of the performance-oriented Petri net models. The third part focuses on the specific topics of the performance optimization including deadlock handling and efficiency predicting techniques. Directions for further development and application of the methods presented are discussed in the concluding part.

2 Modeling and Performance Evaluation

In this section, we discuss the need for and methods of evaluating the performance of the manufacturing systems. Therefore, we provide a brief introduction to the modeling with the Petri nets.

2.1 Logistics Systems for Manufacturing

The rapidly evolving market environment requires manufacturing systems which are capable of treating a wide variety of products. The response to this demand is the rapid development over the last two decades of a new automation technology for discrete-parts manufacturing.

2.1.1 Modern Manufacturing Systems

In the early seventies, a new generation of CNC machine tools and industrial robots developed. This emergence resulted in the computer-aided design and computer-aided manufacturing systems (CAD/CAM), and flexible manufacturing systems (FMS) applications [3], [17], [24], [37], [42]. The systems mentioned above have been introduced to provide:

- rapid response to manufacturing and market demands,
- batch processing with mass-production efficiency,
- mass-production with flexibility of batch processing, and
- reduction of manufacturing costs.

These objectives address many issues concerning the problems of FMSs design and operation, and their interaction with the environment. The first group of issues regards the structure design, processes and production planning, as well as scheduling and dispatching problems. The problems that an enterprise encounters in order to satisfy environment requirements (including the pollution protection and natural resources exploitation), and the ability to respond to customers' demands (concerning costs, quality, throughput, time and flexibility in the ordered products delivery) belong to the second group.

The problems of both groups are tightly related. Their solutions require an application of a new unified approach which allows the integration of different material, energetic and information processes as follows:

- product, part, fixture, and jigs design,
- technological processes and production planning,

- manufacturing , including machining, assembling, inspection and quality control,
- human resources management,
- receiving, storage and shipping, including material, components, products, and resources.

The process execution is supported by a variety of computer-aided tools which provide knowledge and information basis for the integration of the flows of goods and services with the relevant monitoring and control processes. The integrated use of computers in all the sections of an enterprise ranging from the planning of production, and the design and manufacture of products, to the supply services, production, and stock holding, supports the concept of computer-integrated manufacturing (CIM) [17], [37]. Its objective is to integrate the functions of all the particular computer-aided and decision supporting systems into one information processing system [19].

2.1.2 Logistic Systems Management

A typical logistic system consists of the flow of goods and services, and its monitoring and control [27]. Some of the most characteristic activities of the system are: transportation, inventory management, order processing, warehousing, distribution, and production. The main objective of the management of the logistic systems is the coordination of these processes and activities; in other words, its goal is to achieve a well-synchronized behavior of the dynamically interacting components, where the right quantity of the right material is provided in the right place, and at the right time [27].

To reach this goal, a system decision maker has to deal with a complex multidimensional optimum control problem, due to the fact that the logistics management includes a variety of the decision making processes. These processes cover different areas of an enterprise activity, ranging from the control of the flow of materials, and the flow of goods between the stages of manufacturing, to the distribution of products to the customers. These areas have usually conflicting interests within the same enterprise. For example, a large batch production that leads to the reduction of manufacturing costs may also result in high stock levels and lead to the increase of the inventory costs, which proves that a solution to one particular problem in one area has an impact on all the other areas.

2.2 Techniques of Performance Evaluation

Performance evaluation techniques can be classified into two main areas of measuring and modeling, respectively. While the measurements are performed in the real system under real operation conditions, the modeling involves constructing a model of the system that is not physically available [23], [28], [29].

2.2.1 Queuing Networks

Among the other analytical models of stochastic analysis, the queueing networks have become quite popular in the field of performance evaluation since the early sixties [10]. A typical queueing network is a system of interconnected queues in which customers circulate. In turn, the queues are the systems to which customers arrive to receive service, wait for their turn when the system is busy serving other customers, and leave the system after having been served.

Queueing models provide information about the average system behavior observed over a long period of time, and are useful for quantitative answers to a number of design and exploitation problems, including machine grouping and loading, determination of overall production capacity of the system affected by the mix of part types, the selection of the number of pallets, the type of material handling facilities and so on.

The models are very useful at the preliminary design stage when it is desirable to determine such performance measures as the mean throughput, the inventory level, the machine utilization level, etc. [37]. In spite of the simple and inexpensive models offered by queueing networks, several practically important features such as synchronization, blocking and splitting of customers, cannot usually be modeled. This means that these models can be treated as a first approximation of system performance at its early design stages.

2.2.2 Operations Research

Due to the discrete character of processes in the manufacturing systems, many of the optimization problems are strictly combinatorial. Among such problems as machine layout planning, job routing, tool path sequencing, assignment of tools to machines, etc., the job scheduling problem seems to be the most representative [4], [28].

The operational approach deals with behavior sequence rather than random processes. The nature of this approach lies in the modeling of the goal seeking procedure which searches a predefined space of feasible solutions. This is reflected by the models of combinatorial programming supporting such classical methods as mathematical programming (including integer, linear, non-linear, and dynamic programming), transportation algorithms, and branch and bound algorithms [12].

Typical performance measures are costs (including transportation, inventory, machining, etc.), times (including completion time, mean flow and mean weighted time, total tardiness and lateness, and work-in-progress), and design and operation characteristics (including buffers capacity, batch size, throughput, number of set-ups, rate of machine utilization). However, the important quality aspects of system performance such as deadlock avoidance, transient stages of production processes, flexibility and reliability are beyond the scope of the models employed so far.

2.2.3 Computer Simulation

The most detailed representation of the manufacturing system functioning can be achieved by using computer simulation methods. The major advantage of simulation is its generality and flexibility. It enables the detailed representation of characteristics of jobs, facilities design and operation, work flow and job routing, complex control rules and examination of system behavior in both steady and transient states.

Simulation modeling has generally been applied in problem areas where the variables and their relationships are clearly understood, but where there exists no efficient analysis solution method. In the domain of manufacturing systems, the simulation techniques provide answers to the resource capacity, layout structure, scheduling strategy, etc., which are of primary importance in the course of designing and exploitation of real life systems [27], [28].

Compared to the queueing networks and models of operations research, the computer simulation provides a tool for system behavior examination in both steady and transient states.

Usually, the simulation is a costly technique that requires a large amount of computer time for a reliable estimation of performance indices. This is a result of sampling nature of simulation modeling. It can easily be seen that confidence intervals computed for each performance measure may be difficult to apply by using the actually available methods.

2.3 Modeling with Petri Nets

Although the Petri nets have been introduced to cope with modeling and performance evaluation problems arising from the area of computer science, their graphical nature, firm mathematical foundation, analysis methods, and the availability of computer support, they have also proved their usefulness in the domain of the modeling of the discrete manufacturing systems.

The modeling power of Petri nets derives from their inherent capability of modeling causal relationships between events. Thus, by offering the representation of a parallel occurrence of different events, they provide a framework which allows the modeling of different material and information flows in a unifying way. Moreover, Petri nets can represent the system in a top-down fashion where an event at one level of abstraction may be represented by a series of events at a lower level of abstraction.

2.3.1 Basic Definitions

In the graphical representation, the structure of a standard PN [29] (sometimes called classical PN [23] or basic PN [36]) is a bipartite graph that comprises a set of places P (drawn as circles), a set of transitions T (drawn as bars), and a set of directed arcs E. Places may contain tokens which are drawn as black dots.

The state of PN is defined by the number of tokens in each place and is denoted by vector $M = (M(p_1),M(p_2),...,M(p_i),...,M(p_n))$, where the i-th component represents the number of tokens in the i-th place. The state of a PN is usually called its marking.

A formal definition of a frequently used class of standard PNs (so called Place/Transition Systems [33]) is the following: $PN = (P,T,E,K,W,N,M_o)$, where $P = \{p_1,p_2,...,p_n\}$, $T = \{t_1,t_2,...,t_m\}$ are the finite and non-empty sets of places and transitions,

$E \subset (P \times T) \cup (T \times P)$ is a flow relation,

$W: E \longrightarrow N$ is a weight arc function,

where N_o is the set of natural numbers including zero, $N = N_o \setminus \{0\}$,

$K: P \longrightarrow N$ is a place capacity function,

$M_o: P \longrightarrow N_o$ is an initial marking such that $M_o(p) \leq K(p)$ for all $p \in P$.

In the modeling of the manufacturing systems places represent conditions and transitions represent operations.

The dynamic behavior of a PN is described by the changes of the net markings, i.e. by a sequence of transition firings. The marking may change as a result of the execution of a PN. A PN executes according to the following rules:

1. A transition t is enabled when the following conditions hold:
   (i) $M(p) \geq W(p,t)$, $\forall p \in {}^\circ t$,
   (ii) $M(p) \leq K(p) - W(t,p) + W(p,t)$, $\forall p \in t^\circ \cup \{{}^\circ t \cap t^\circ\}$, (2.1)
   where ${}^\circ t = \{p \mid (p,t) \in E\}$, $t^\circ = \{p \mid (t,p) \in E\}$.

2. An enabled transition can fire, thus removing $W(p,t)$ tokens from each input place $p \in t^\circ$ and placing $W(t,p) - W(p,t)$ tokens in each output place $p \in t^\circ$.

3. Each firing of a transition modifies the distribution of tokens in places and thus produces a new marking for the PN.

In the PN shown in Fig. 2.1 the initial marking $M_o = (1,0,0,0,1,0,2,0,0)$ enables only transition t_6. After the firing t_6 the marking becomes $M = (1,0,0,0,1,0,0,0,3)$. In this case, transition t_5 is enabled and can fire.

2.3.2 Properties

This is a matter of compromise observed between the PN modeling power and decision power. Besides the standard PNs, their subclasses, e.g. Marked Graphs, Free Choice Petri nets, State Machines, and their extensions, e.g. Inhibitor Petri nets, Timed-, Coloured-, Stochastic Petri nets are frequently used [14], [31], [33]. It is worth noticing that the extension of a PN model usually increases the modeling power of Petri nets, however it may decrease their decision power. Also, the subclasses might have good decision properties, however, their modeling power may be decreased.

The decision properties commonly addressed in the modeling of manufacturing systems are liveness, reachability, conservativeness, boudness, and persistency. Their detailed description can be found in [31], [33], [36].

The time-independent properties of a PN can be studied with the help of techniques based on the reachability graph (tree) analysis, the matrix-based analysis and the reduction (or decomposition) analysis [26], [30], [31], [33], [34].

2.3.3   Algebraic representation

Since our further considerations will focus on the matrix-based analysis, let us consider the following algebraic representation of the PN shown in Fig. 2.1 (a):

$$PN = (C, K, M_o) \tag{2.2}$$

where K and M_o denote the row vectors, of size $1 \times n$, corresponding to the capacity function and an initial marking, respectively.

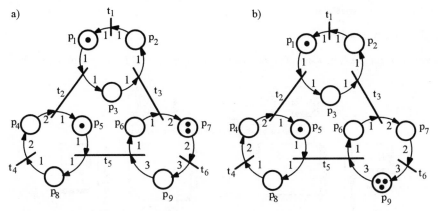

Figure 2.1. An example of graphical representation of a Petri net (a) before firing and (b) after firing. The numbers labeling the arcs denote values of the weight arc function W. The corresponding place capacity function K is defined as follows: $K(p_i) = 1$ for all $i \in \{1,2,3,5,6,8\}$, $K(p_4) = K(p_7) = 2$, and $K(p_9) = 3$

The incidence matrix of the PN $= (P,T,E,W,K,M_o)$ without loops, denoted C, is defined as the $m \times n$ matrix of c_{ij}'s, where $m = \|T\|$ and $n = \|P\|$ denote the cardinality of sets T and P, respectively, and

$$c_{ij} = \begin{cases} W(t_i,p_j) & \text{for } (t_i,p_j) \in E \\ -W(p_j,t_i) & \text{for } (p_j,t_i) \in E \\ 0 & \text{for } \{ (p_j,t_i) \cup (t_i,p_j) \} \notin E \end{cases} \tag{2.3}$$

The matrix representation provides a well suited form useful for the simulation of a net execution. Thus, the dynamic of the PN is described by the following equation:

$$M' = M + e[i] \, C \tag{2.4}$$

where e[i] is a unit row-vector of size $1 \times m$, which is zero everywhere, except the i-th component corresponding to the transition t_i enabled at the marking M.

3   Modeling Automation

In this subchapter, we shall formulate the problem of design, as opposed to analysis, where the objective is to find optimum values of the input parameters that yield the desired values of the performance measures of interest. Next, we shall provide a standard form specification of concurrent processes, and finally, discuss a Petri net-based approach to the automated synthesis of control flow models.

3.1   Problem Statement

The objective of modeling is to provide both a representation which allows an insight analysis of the system behavior, and a tool for the design of a system with presumed behavioral characteristics.

3.1.1   An Illustrative Example

Let us consider the following flexible manufacturing module (FMM) shown in Fig. 3.1.

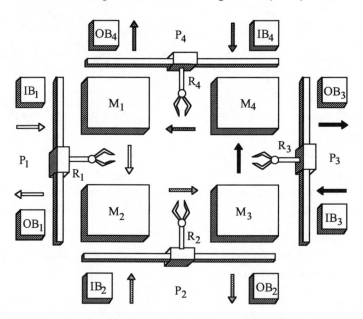

Figure 3.1. The flexible manufacturing module; M_i is the i-th workstation, R_i is the i-th robot, P_i is the manufacturing process of the i-th type product, IB_i (OB_i) is the i-th input (output) storage

The module consists of : four workstations (M_1, M_2, M_3, and M_4), four inputs (IB_1, IB_2, IB_3, and IB_4) and four output (OB_1, OB_2, OB_3, and OB_4) storages of limited capacities, and four robots (R_1, R_2, R_3, and R_4) responsible for the transport operations.

The four product types are manufactured by the module. Every product is associated with its production route, i.e. a sequence of transport and technological operations runs in a specified time on the relevant module components.

The production steps are described as follows: a row part of type 1 (of type 2, 3, and 4, respectively) is delivered to input storage IB_1 (IB_2, IB_3, and IB_4 respectively), then the robot R_1 (R_2, R_3, and R_4 respectively) takes it and loads workstation M_1 (M_2, M_3, and M_4 respectively) if it is available. Next, the robot R_1 (R_2, R_3, and R_4 respectively) takes a processed part from M_1 (M_2, M_3, and M_4, respectively) and loads M_2 (M_3, M_4 and M_1, respectively). Finally, R_1 (R_2, R_3, and R_4) takes a finished product from M_2 (M_3, M_4, M_1) and places it in OB_1 (OB_2, OB_3, and OB_4).

It is assumed that raw parts of every type are always available in IB_i , $i \in \{1,2,3,4\}$, and finished products of every type are carried out so that OB_i, $i \in \{1,2,3,4\}$, cannot be overflowed. Besides these assumptions, the following constraints should be respected:

- A resource (e.g. a robot or a workstation) can be allocated to execute only one operation at a time.
- A process can release a resource only if an operation is completed and all the resources requested by the following operation in the production route are available.
- There may be different part types processed (manufactured) in the module at a time.
- Operation and storage time of every operation are finite.

3.1.2 Problem Description

We assume that the functioning of a typical FMS can be seen as a result of the interaction of different material processes. Thus, the problem considered is a task of finding out a procedure A which provides a given process specification PS with a dynamic model of the multiple interactions DM, i.e. the following transformation:

$$A: PS \longrightarrow DM \qquad\qquad (3.1)$$

The process specification captures the information regarding the precedence relation which defines the execution of component manufacturing operations as well as their technical parameters and characteristics. Such information mainly describes all the potentially possible realizations of the material flows. However, for the purpose of performance evaluation, a subset of being feasible, e.g. flexible, deadlock-free, and finite is of primary importance. This observation implies the division of our main problem (3.1) into the following two subproblems:

- Automatic design of a Petri net model PM encompassing all possible material flows for a given production process specification PS, i.e. development of transformation:

$$A1: PS \xrightarrow{\hspace{2cm}} PM \qquad\qquad (3.2)$$

- Modification of the **PM** into a modified Petri net model **DM** encompassing only admissible realizations of the material flows modeled, i.e. the development of transformation:

$$A2: PM \xrightarrow{\hspace{2cm}} DM \qquad\qquad (3.3)$$

3.2   Process specification

The structure of the parameters included in a process specification depends on the performance measures used in the course of manufacturing systems evaluation. Some illustrative examples can be found in [3], [6].

Our further considerations will be restricted to the following structure:

$$PS = (PR, RC, OT) \qquad\qquad (3.4)$$

where PR is the description of the production routes which define process operation order and resources required in the course of operation execution, RC is a set of resource capacities associated with the components of the modeled system, and OT is the set of operation times associated with pairs (operation, resource) which contain technological operations and resources required for their execution.

3.2.1   Production Rule-based Specification

One of the possible specifications of production routes is a representation that employs the production rules. To illustrate this possibility, let us consider the FMM shown in Fig. 3.1. Supposing that the resultant model should provide a possibility to study the material flows with respect to the specific machine operations, processed on a given set of workstations; the relevant model of production process involved in the course of manufacturing of type 1 product, could be described by the following set of production rules:

R1: IF $\boxed{w \text{ on } IB_1}$ AND M_1 *available* THEN s *on* M_1,

R2: IF s *on* M_1 AND M_2 *available* THEN s *on* M_2 AND M_1 *available*,

R3: IF s *on* M_2 THEN M_2 *available* AND $\boxed{s \text{ on } OB_1}$.

where w and s correspond to a part stored on IB_1, and processed on workstations M_1, M_2, then placed on OB_1, respectively.

According to the assumptions of the FMM functioning, the conditions put in a frame are always fulfilled. The notation p_i, when $i = \{1,2,9,10\}$, stands for the abbreviation of the distinguished conditions. With this notation and without the conditions which are always fulfilled (in the frames), the set of production rules has the following form:

t_1 : IF p_9 THEN p_1,

t_2 : IF p_1 AND p_{10} THEN p_2 AND p_9, $\qquad$ (3.5)

t_3 : IF p_2 THEN p_{10}.

The concept employed assumes that each rule of the following form:

$$t_i : \text{IF } p_a \text{ AND } \ldots \text{ AND } p_k \text{ THEN } p_l \text{ AND } \ldots \text{ AND } p_r \qquad (3.6)$$

corresponds to transition t_i and the set of places $\{p_a,...,p_k, p_l,...,p_r\}$ such that $p_i \in {}^\circ t$ for all $p_i = \{p_a,...,p_k\}$ and $p_i \in t^\circ$ for all $p_i = \{p_l,...,p_r\}$.

3.2.2 Production Route-based Specification

An alternative approach to the specification of production routes uses the structure:

$$PR = \{PR_j \mid j = \{1,2,...,v\}\} \qquad (3.7)$$

where $PR_j = (O_{j,1},RE_{j,1}),...,(O_{j,r},RE_{j,r})$ is the j-th production route encompassing the technological order of operations $O_{j,k}$ performed on the system resources $RE_{j,k}$ used for the completion of the j-th type of product. In the case of the considered FMM, the process specification corresponding to the production route of type 1 product has the following form:

$$PR_1 = (O_{1,1},M_1),(O_{1,2},M_2). \qquad (3.8)$$

Using the above representation, the whole structure of the process specification PS of the considered FMM has the following structure, which generates the Petri net model as shown in Fig. 3.2.

$$PS = \{PR_i \mid i = \{1,2,3,4\}, \qquad (3.9)$$

$PR_1 = (O_{1,1},M_1),(O_{1,2},M_2), \qquad PR_2 = (O_{2,1},M_2),(O_{2,2},M_3),$

$PR_3 = (O_{3,1},M_3),(O_{3,2},M_4) , \qquad PR_4 = (O_{4,1},M_4),(O_{4,2},M_1).$

Consider now the subnet composed of $\{t_1, t_2, t_3\}$ and $\{p_1, p_2, p_9, p_{10}\}$ modeling the production process of manufacturing the type 1 product. The initial marking corresponds to a situation in which all the system resources are available, i.e. $M_o(p_9) = M_o(p_{10}) = 1$, hence no technological operation is being executed, i.e. $M_o(p_1) = M_o(p_2) = 0$. The places p_1, and p_2 correspond to machining operations executed on M_1 and M_2, respectively. In turn, places p_9, and p_{10} correspond to conditions expressing the availability of M_1, and M_2, respectively.

Transitions t_1, t_2, and t_3 encompass events along the production route. Transitions are associated with taking a raw part from IB_1 by R_1, fixing it on M_1 and starting the machining, then finishing and taking the part from M_1 by R_1, transporting to M_2, fixing on M_2 and starting the machining, then taking the part from M_2 by R_1, and placing it finally on OB_1, respectively.

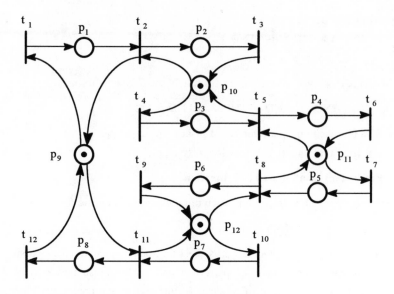

Figure 3.2. The Petri net model of the FMM specified by (2.13)

3.2.3 Model Extension - Material Flows Level

The net model from Fig. 3.2 can be easily extended by introducing the additional subnets which capture the level of robots and operation. The relevant extension of the process specified by (3.8) has the following form:

$$PR_1 = (O_{1,1},R_1),(O_{1,2},M_1),(O_{1,3},R_1),(O_{1,4},M_2),(O_{1,5},R_1) \qquad (3.10)$$

The graphical representation of the resultant net model is presented in Fig. 3.3. The places p_{13}, p_1, p_{14}, p_2, p_{15} correspond to the consequent process operations and places p_9, p_{10}, p_{16} represent the respective system resources.

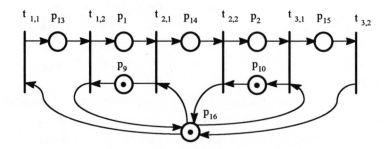

Figure 3.3. The extended Petri net model specified by (3.10)

3.2.4   Model Extension - Robot Operation Level

Besides the representations encompassing material flows, we can also consider the other specifications which describe the functioning of the system components such as: robots, AGVs, and warehouses. In order to illustrate this possibility, let us return to the FMM from Fig. 3.1.

Consider now the production process of type 1 product. It can be easily noted that the employed robot R_1 has a finite number of positioning points at which it waits for the execution of the consequent transport operations (resulting from the considered production route). In the previous example, there are four such points associated with IB_1, M_1, M_2, and OB_1 respectively.

Using the production rule-based specification, it is possible to create the relevant net model which is shown in Fig. 3.4 (a).

The places of the model correspond to the robot positioning points. Transitions represent the operations of the robot moves between the respective points. The transitions distinguished by t_{12}^*, t_{13}^* and t_{14}^* correspond to robot moves while caring a part. All the other transitions concern returning moves (made in order to take another part, or moves from/to some positioning places) including the initial position IP.

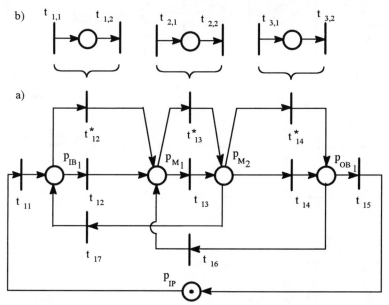

Figure 3.4. The Petri net model; (a) encompassing robot operation; (b) encompassing extension of events modeled by t_{12}^*, t_{13}^* and t_{14}^* into subnets including elementary events $t_{1,1}$, $t_{1,2}$, and $t_{2,1}$, $t_{2,2}$ and $t_{3,1}$, $t_{3,2}$ respectively

A token located in a certain spot of the net model indicates the current position of the robot whose description is provided by an interpretation of that particular spot.

The graphical representation of the resultant Petri net model is shown in Fig. 3.5. In this model each compound event represented by t_{12}^*, t_{13}^* and t_{14}^* is replaced by a pair of transitions $t_{1,1}$, $t_{1,2}$, and $t_{2,1}$, $t_{2,2}$, and $t_{3,1}$, $t_{3,2}$, respectively. This solution fits into the frames of mathematical formalism, event modeling, material flows and robot movements.

3.2.5 Model Extension - Robot Path Level

The other extension of the Petri net models which could be considered, regards their places. To illustrate this possibility, let us consider the PN from Fig. 3.5. The places p_{13}, p_{14}, and p_{15} correspond to the conditions describing the execution of the relevant transport operations. Assuming that the robot work envelope is composed of a set of elementary subspaces, it is evident that each robot path can be described by the sequence of subspaces passed on through the robot gripper. It means that the robot gripper path can be specified by a sequence of subspaces quite similar to original (3.5).

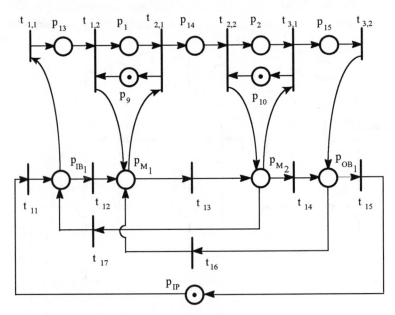

Figure 3.5. The integrated Petri net model of the parts flow and the robot operation

In order to illustrate this case, let us consider the operation corresponding to the transport of a part of type 1 between IB_1 and M_1, i.e. the place p_{13} from the PN shown in Fig. 3.5. Supposing the robot working envelope is arbitrary, subdivided into a set of elementary subspaces SU_k, $k \in \{1,2,...,h\}$, the typical specification of the i-th robot path matches the sequence $SQ_i = SU_{i_1}, SU_{i_2}, ..., SU_{i_v}$. Using the production rule-based approach, the resultant model is shown in Fig. 3.6.

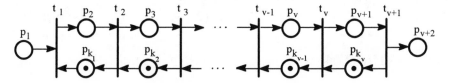

Figure 3.6. The Petri net model of the robot path

The places p_1 and p_{v+2} encompass the conditions related to the state preceding the first shift of the already grasped part, and to the state in which the part is released, respectively. In turn, places p_2, p_3,..., p_{v+1} correspond to the conditions encompassing the execution of the elementary moves of the robot gripper. Each elementary move results in passing through the relevant subspace. The subspaces occurring in the robot path can be treated as resources. Their availability (i.e. the state describing whether the subspace is occupied or not) is modeled by places p_{k1}, p_{k2},..., p_{kv}. Transitions $t_1, t_2,..., t_{v+1}$ encompass events which occur when the surfaces separating subsequent subspaces of the sequence SQ_i are passed on by the robot gripper.

Finally, the model extension results in the replacement of the place p_1 by the PN shown in Fig. 3.6. The obtained net provides the more detailed insight into the robot operation, and enables its more accurate evaluation.

3.3 Model Generation

In order to present our contribution to the problem of the automated generation of the performance oriented Petri net-based models, let us focus on its first subproblem (3.2), i.e. the synthesis of algorithm transforming a given process specification PS into a relevant Petri net model.

3.3.1 Algorithm Description - Material Flows Level

Due to the assumptions imposed on the problem (3.2) the time dependent parameters are omitted, which means that the transformation we are looking for, has the following form:

$$A1: PS = (PR, RC, OT) \longrightarrow PM = (C, K, M_o) \tag{3.11}$$

where OT is the set of operation times in which each element is equal to zero.

Thus, the algorithm considered should consist of three basic steps aimed at the calculation of elements C, K, and M_o.

In general, the production routes specification, PR, can be expressed in many different ways that depend on a particular case model, as well as on the applied description [3], [6]. To illustrate this, it is sufficient to concentrate on the representation defined by (3.7). This means that PR has the following form:

$PR = \{ PR_i \mid i \in \{1,2,..., \upsilon \}$, where

$PR_i = (O_{i,1},RE_{i,1}),(O_{i,2},RE_{i,2}),...,(O_{i,k},RE_{i,k}),...,(O_{i,I_i},RE_{i,I_i})$ is a specification of

the i-th production route,

$$RE = \bigcup_{i=1}^{\upsilon} \{crd^2(crd^k(PR_i)) \mid k = \{1,2,...,I_i\}, \ i = \{1,2,...,\upsilon\}\} \text{ is a set of all the}$$

resources specified in PR,

$crd^i(S) = s_i$, $S = (s_1,s_2,...,s_i,...,s_q)$, is the i-th element in the sequence S.

Let us suppose, that the workpieces are processed part by part, i.e. the weight arc function W of the resultant Petri net model is equal to one, for all the elements of E. Taking this into account, the considered **NetSynthesis** algorithm consists of the following three steps:

step 1

According to a given process specification PS set up the following incidence matrix

$C = [C'|C'']$ of size $m \times n$, where

$$\sum_{k=1}^{\upsilon}|PR_k|+\upsilon \ , \quad n = m - \upsilon + \|RE\|. \text{ The matrix C' of size } m \times (m - \upsilon) \text{ has the}$$

following structure and the elements of each submatrix C^k of size

$$C' = \begin{bmatrix} [C^1] & & & & 0 \\ & \ddots & & & \\ & & [C^k] & & \\ & & & \ddots & \\ 0 & & & & [C^{\upsilon}] \end{bmatrix} \quad , \qquad C^k = \begin{bmatrix} 1 & & & & \\ -1 & 1 & & 0 & \\ & -1 & \ddots & & \\ & & & 1 & \\ 0 & & & -1 & 1 \\ & & & & -1 \end{bmatrix} .$$

$m_k \times n_k$, where $m_k = |PR_k| + 1$, $n_k = m_k - 1$ are determined as follows:

$$C^k_{rq} = \begin{cases} 1 & \text{for } q = r, \quad r = \{1,2,...,m_k - 1\} \\ -1 & \text{for } q = r-1, \quad r = \{1,2,...,m_k\} \\ 0 & \text{otherwise} \end{cases}$$

In turn, the matrix C" of size $m \times \|RE\|$ has the following structure, and the elements of each submatrix $\tilde{C}^k$, of size $m_k \times \|RE\|$ are determined as follows:

$$
C" = \begin{bmatrix} \left[\tilde{C}^1\right] \\ \vdots \\ \left[\tilde{C}^k\right] \\ \vdots \\ \left[\tilde{C}^\upsilon\right] \end{bmatrix}, \quad \tilde{c}^k_{rq} = \begin{cases} 1 & \text{for } r = u, \quad q = m - \upsilon + j, \quad RE_j = crd^2(crd^u(PR_k)) \\ -1 & \text{for } r = u - 1, \quad q = m - \upsilon + j, \quad RE_j = crd^2(crd^u(PR_k)) \\ 0 & \text{otherwise} \end{cases}
$$

step 2

The capacity function K is determined as follows:

$$
K(j) = \begin{cases} RC_i & \text{for } j = m - \upsilon + i, \quad i = \{1, 2, ..., \|RE\|\} \\ 1 & \text{for } j = \{1, 2, ..., m - \upsilon\} \end{cases} .
$$

step 3

The initial marking M_o is determined according to the following formula:

$$
M_0(j) = \begin{cases} 1 & \text{for } j = m - \upsilon + i, \quad i = \{1, 2, ..., \|RE\|\} \\ 0 & \text{for } j = \{1, 2, ..., m - \upsilon\} \end{cases} .
$$

3.3.2 Algorithm Operation

Applying the algorithm to the process specification $PS = (PR, RC, OT)$ where PR is the process specification determined by (3.9), RC (OT) is the set of resource capacities (operation times) so that each element is equal to one (zero), the resultant model has the following form:

$$
PM = (C, K, M_o) \tag{3.12}
$$

where

$$
C = \begin{bmatrix}
1 & & & & & & & & & & & -1 & & & & & & & & \\
-1 & 1 & & & & & & & & & & 1 & -1 & & & & & & & \\
& -1 & & & & & & & & & & & 1 & & & & & & & \\
& & 1 & & & & & & & & & & & -1 & & & & & & \\
& & -1 & 1 & & & & & & & & & & 1 & -1 & & & & & \\
& & & -1 & & & & & & & & & & & 1 & & & & & \\
& & & & 1 & & & & & & & & & & & -1 & & & & \\
& & & & -1 & 1 & & & & & & & & & & 1 & -1 & & & \\
& & & & & -1 & & & & & & & & & & & 1 & & & \\
& & & & & & 1 & & & & & & & & & & & -1 & & \\
& & & & & & -1 & 1 & -1 & & & & & & & & & 1 & & \\
& & & & & & & -1 & 1 & & & & & & & & & & &
\end{bmatrix}
$$

$K = (1,1,1,1,1,1,1,1,1,1,1,1)$, and $M_o = (0,0,0,0,0,0,0,0,0,1,1,1,1)$. Blank spaces in matrix C correspond to zero. The graphic representation of the net model (3.12) is shown in Fig. 3.2.

3.3.3 Integrated Modeling

In case the process specification has a production rule-based form, the relevant transformation algorithm consists of two stages. At the first stage, the incidence matrix is designed so that its rows correspond to rules (transitions), and the columns correspond to conditions (places), (see 3.6). The capacity and initial marking functions are determined at the second stage according to the specification based interpretation of the net places. Therefore, the integrated models combining the material flows and the robot operations can also be considered. This means that the models of the form (3.12) provided for different levels of system detail (and/or different process specification) can be compounded into one Petri net model.

The integration process of models compounding as well as the mathematical foundations of the approach presented can be found in [3], [4].

4 Performance Optimization

The approaches discussed here apply to both quantitative and qualitative performance characteristics. Qualitative characteristics are discussed in an example envolving a deadlock handling problem. For the quantitative measures we consider techniques that provide the optimal solutions to some cases of the optimal control of manufacturing systems.

4.1 Petri Net-based Performance Modeling

The major advantage of the Petri net-based modeling is that the resultant model can be directly transformed (converted) into a computer code, and then simulated.

4.1.1 Petri Net-based Tools

A typical Petri net-based computer-aided tool consists of a general purpose specification language, and its software implementation supporting the process of a system modeling [3]. Let us consider ExSpect (EXecutabe SPECification Tool) [36] aimed at the specification, modeling, and performance evaluation of discrete dynamic systems. A typed functional language provides the basis for a set of tools including the design interface (graphical editor), the interpreter, and the analysis tool. The window oriented graphical editor has the ability to represent the systems formed of graphical objects such as channels, stores, and processes. Moreover, it allows the user to build a system definition from the other, already existing, definitions.

Besides the standard well-known techniques based on simulation, reachability analysis, continuous time Markov chain, and P and T-invariants analysis, the ExSpect analysis tool offers some new Interval Timed Coloured Petri Net based techniques. The application domain of ExSpect, which covers the issues of the design and operation of logistic systems, is guided by the domain oriented library supporting the user in the analysis and validation of systems supply and demand, stock points and transport systems, etc.

The Grafcet that is designed for the specification, modeling and programming of the discrete event systems plays a dominant role among the industrial application oriented packages [14]. The Grafcet is a Petri net-based tool whose descriptive power is the same as that of the Moore and Mealy machines. Because of this property it is mainly used in the description of the input/output behaviors, i.e. the description of the logic controllers and real time systems.

Moreover, its software implementation supported by a graphic editor enables the introduction of data in the form of drawings as well as the elimination of certain description errors. In turn, the behavioral characteristics of the designed model (e.g. programming hazards) are evaluated with a Grafcet interpreter. Because the Grafcet software is directly linked to that of the programmable logic controllers, further conversion of the resultant model into a relevant control code can be executed automatically.

Besides the standard Petri nets which are used mainly for the qualitative validation of a modeled system, a variety of their extensions are applied in order to evaluate the system performances. The first category of nets provides the description of the system the system whose functioning is time-independent (transitions fire in zero time), i.e. provides an opportunity to explain "how it could work". The extensions introducing the notion of time can be used for the quantitative performance evaluation of systems, i.e. can be used to explain "how it will work".

According to either the constant (transition fire after a presumed delay time) or the stochastic (transition fire after a random, exponentially distributed, enabling time) character of the time dependent parameters, two classes of models can be considered: Timed Petri Nets (TPN) [14], [36], and Stochastic Petri Nets (SPN) [22], [29], respectively.

The generalization of the above mentioned class leads to so called Generalized Stochastic Petri Nets [1], [23], [29], [37] whose transitions belong to two different subsets of firing in zero time, and firing after a random enabling time transition, respectively. Of all the other extensions, the following two are used most frequently: the timed Continuous Petri Nets [14], and the timed Coloured Petri Nets that combine the properties of TPN (or SPN) and Coloured Petri Nets (CPNs) [36].

4.1.2 Areas of Control Applications

Besides the software modeling (including compilers and operating systems) where Petri nets based validation methods are mainly used to direct the programmer to rethink the program design, the applications of fast prototyping of the PLC [13], [38], verification of data bases [2], [19], specification and validation of communication protocols [35], monitoring and fault detection [32], still remain their major domain.

Among many qualitative properties (such as bottlenecks and starvation) that determine the quality of system behavior, the process blockings (deadlocks) play the most important role. Thus, the models searching for a good deadlock avoidance policy are of crucial importance. Their objective is to develop a control that achieves high resource utilization while avoiding deadlocks. In real-life applications, the relevant deadlock avoidance test-procedure is combined with a given dispatching policy to select valid and utilization maximizing control operations.

Unlike the case of qualitative properties, the performance evaluation of the quantitative characteristics is mainly based on computer simulation techniques. The results of the recent investigations suggest the possibility of the implementation of two alternative techniques employing the max-algebra [9], [21] and Timed Event-Graph [20], [25] formalism, respectively. These techniques allow the calculation of the dynamic parameters of the manufacturing systems under different assumptions, irrespective of the initial state and the types of jobs flow.

Their main shortcomings, however, stem from the deterministic condition of their functioning. In order to overcome these disadvantages, we shell consider a synthesis of systems possessing the self-synchronized property. The main feature of such systems is their robustness for small random fluctuations of certain time-dependent parameters.

4.2   Deadlock Avoidance

The phenomenon of deadlocks has been extensively studied in the context of computer operating systems and data communication systems [3], [29]. However, the objective of the investigations is still pursued [5], [6], [41], [42].

4.2.1   Deadlock in a FMC

A state of mutual process deadlock denotes the distribution of non-pre-emptive resources in the system, where the completion of certain processes is impossible, since they request non-pre-emptive resources allocated to other processes (tasks).

The existing methods of deadlock handling can be classified according to the following approaches:

- deadlock detection and recovery: allows deadlock detection, i.e. determines which processes and resources cause blocking, removes it by the successive reallocation of resource units from the blocked processes, and checks whether that helps to recover the blocking,
- deadlock prevention: uses protocols specifying ways of requesting resources which make blocking impossible (do not satisfy one of the conditions necessary for blocking),

- deadlock avoidance: based on additional a priori information, describes ways of using the resources in each process, and allows a selection of resource requests which ensures the system transition from one safe state to another.

In general, the methods of deadlock avoidance implemented in computer operating systems assume nothing about the order in which resources are requested or released, and differ in ways of testing algorithms with different computational complexity. An alternative approach takes advantage of the information on the resource use order, which is available in the manufacturing processes.

To illustrate the method proposed [3], [6], let us consider the Flexible Machining Cell (FMC) shown in Fig. 4.1. The FMC consists of machine tools M_1 and M_2, industrial robots R_1 and R_2, a gripper changing station equipped with grippers G_1 and G_2, input storage IB_1 and IB_2 output storage OB_1 and OB_2. A machining cycle is initiated when a raw part is available at the input storage IB_1 (IB_2). The robot R_1 (R_2) equipped with gripper G_1 (G_2) picks up a raw part from IB_1 (IB_2) and loads it on the machine tool M_1 (M_2) to carry out the machining operation. The finished part is unloaded from the machine M_1(M_2) by the robot R_2 (R_1) equipped with gripper G_2 (G_1) and placed in the output storage OB_1 (OB_2).

Assuming that the capacity of each resource is equal to one, the process specification *PS* has the following form

$$PS = (PR, RC, OT) \tag{4.1}$$

where $PR=\{PR_1,PR_2\}$, $PR_1 = (O_{1,1},\{R_1,G_1\}),(O_{1,2},\{M_1\}),(O_{1,3},\{R_2,G_2\})$,
$PR_2 = (O_{2,1},\{R_2,G_2\}),(O_{2,2},\{M_2\}),(O_{2,3},\{R_1,G_1\})$,

$(O_{i,u},\{RE_j \mid j \in J_{i,u}\})$ means the u-th operation in the i-th production route completion which requires the subset $\{RE_j \mid j \in J_{i,u}\}$ of the FMC resources, RC (OT) is the set of resource capacity (operation times) so that each element is equal to one (zero).

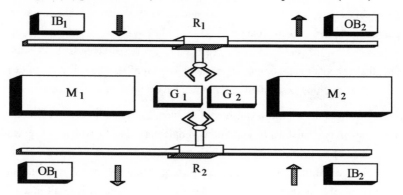

Figure 4.1. A Flexible Machining Cell

274

The specification (4.1) can be treated as a direct extension of the already considered specification for production routes defined in (3.7). In order to cope with such a case, the generalized version of the **NetSynthesis** algorithm has been developed in [6]. According to this algorithm, the Petri net model corresponding to the specification (4.1) has the form shown in Fig. 4.2 (a).

a)   b)

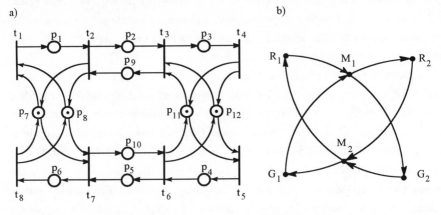

Figure 4.2. Graphical representation of: (a) Petri net model and; (b) resource precedence graph determined by (4.1)

As one can see the asynchronous execution of processes can result in their mutual blocking. The possible deadlock situation corresponds to the marking $M = (1,1,0,1,1,0,1,1,1,1,1,1)$. Moreover, from the analysis of the reachability graph it follows, that either the transition t_1 at the marking $M' = (0,1,0,1,1,0,0,0,1,1,1,1)$ or the transition t_5 at the marking $M'' = (1,1,0,0,1,0,1,1,1,1,0,0)$ has to be inhibited from firing in order to avoid the deadlock marking M occurrence. The above observations lead to the following conclusions:

- the enableness rules (2.1) should be extended to take into account the conditions that guarantee the deadlock avoidance,
- because one of the problems of the reachability graph searching is the NP-complete one, the computational complexity of an algorithm of checking if a given deadlock avoidance condition is satisfied or not, must have a polynomial character.

Taking this into account, the considered problem (3.3) can be seen as a task of the development of the satisfactory conditions, implementation of which will extend the firing rules (2.1) to guarantee the deadlock-free execution of the resultant models. In other words, we are looking for some satisfactory conditions whose examination will require a reasonable amount of time and whose implementation will result in avoidance of firings, leading to deadlocks.

4.2.2 Deadlock Avoidance Conditions

In order to develop such conditions, let us consider the following definition of the resource request graph G:

$$G = (RE, \prec) \ , \qquad \prec \subset RE \times RE \ , \tag{4.2}$$

where

$(RE_i, RE_j) \in \prec$ if and only if RE_i and RE_j belong to the subsets of the two subsequent elements in a production route, i.e. $RE_i \in crd^2(crd^k(PR_n))$ and $RE_j \in crd^2(crd^{k+1}(PR_n))$ hold for $i \neq j$ and $n \in \{1,2,...,\upsilon\}$, $k \in \{1,2,...,l_{n-1}\}$. The resource request graph obtained for the process specification (4.1) is shown in Fig. 4.2 (b).

From the analysis of deadlocks we can conclude that the system state S is a state of process blocking if and only if there exists a cycle in the resource request graph G so that all its resources are allocated (possessed) at this state. Its implementation to the systems where every resource capacity is equal to one, leads to the following rule: the state S is safe if the following condition holds:

$$N(S,G) \leq L(G) - 1 \ , \tag{4.3}$$

where $N(S,G)$ is the number of parts processed along the production routes PR at a state S which encompasses the current resources allocation, $L(G)$ is the length of the shortest elementary cycle in the graph G.

Therefore to avoid deadlocks, no more than $L(G) - 1$ processes (parts) can be simultaneously processed in the system. The disadvantage of this solution is the low level of system resources utilization observed, for instance, in processes where two or more disjoin cycles exist.

The new less restrictive conditions sufficient for deadlock avoidance which take into account the information on resources use order have the following form:

(i) $N(S,G_{i,k}) \leq L(G_{i,k}) - 1$ for $L(G_{i,k}) \neq 1$ or $N(S,G_{i,k}) \leq 1$ for $L(G_{i,k}) = 1$, and

(ii) $N(S,Z_{i,k+1}) \leq L(Z_{i,k+1})$, and $\qquad\qquad\qquad\qquad\qquad\qquad$ (4.4)

(iii) $N(S,Z_{i,k+1}) + L(G_{i,k}) < L(G) - 1$ for all zones $\{ Z_{i,k} \mid i \in \{1,2,...,\upsilon\}$ &
$$k \in \{1,2,...,l_i\}\},$$

where $G_{i,k}$ is the subgraph composed of resources corresponding to the shared zone $Z_{i,k}$ which precedes the non-shared zone $Z_{i,k+1}$, $N(S,G_{i,k})$ is the number of parts processed by shared operations at a state S, the shared (non-shared) zone is a sequence of resources that multiply occur (uniquely occur) in production routes of process specification, $N(S,Z_{i,k})$ is the number of parts processed by the operations in the $Z_{i,k}$ -th non-shared zone at a state S, $L(Z_{i,k})$ is the number of operations in the non-shared zone $Z_{i,k}$.

In the general case, the condition (4.4) is less restrictive than (4.3) which leads to the better utilization of systems resources. This is because the resources determined in the non-shared zones (see $N(S,Z_{i,k+1}) \leq L(Z_{i,k+1})$ and a number of non-shared resources (see $N(S,G_{i,k}) \leq L(G_{i,k})$ can be used simultaneously.

To summarize our considerations, let us point out the following:

- while taking advantage of the information on resource requirements of each particular operation of a manufacturing process, it is possible to develop more efficient deadlock avoidance methods than the standards proposed for computer operating systems,

- since the Petri net models of possible (including deadlocks) realizations of the modeled processes can be automatically generated, and then modified to avoid transitions firing that would lead to deadlock occurrence, it is possible to realize an automated conversion of an input specification of processes into the Petri net models encompassing only admissible (deadlock-free) realizations of the competing processes,

- the modified Petri net models are suitable for being implemented as real-time control programs and provide a workbench supporting the operation of the Petri net-based computer-aided performance evaluation tools.

4.3   Modeling and Control

PNs have long been used for the modeling and verification of the correctness of the synchronized behavior of concurrent systems. With the concept of time, however, they are suitable for performance studies as well.

4.3.1   Timed Petri Nets

In essence, there are two ways to introduce the concept of time into a standard PN; either to associate time with the places or to associate time with the transitions. A TPN can be defined as

$$TPN = (P,T,E,W,K,\Theta,M_o) \ , \tag{4.5}$$

where P and T is a finite and non-empty set of places and transitions, respectively, E is a flow relation, W and K is a weight and a place capacity function, respectively, $\Theta = \{ \theta_1, \theta_2,..., \theta_m\}$ is the set of delays associated with transitions T, and $M_o : P \longrightarrow N_o \times N_o$ is an initial marking such that $pr + pu \leq K(p)$ for all $p \in P$, and where $M_o = (M_o(p_1),M_o(p_2),...,M_o(p_i),...,M_o(p_n))$, $M_o(p_i) = (pn_i,pr_i)$ represents numbers of reserved tokens pr_i and non-reserved tokens pn_i in the i-th place, respectively.

It is assumed that a reserved token has been generated by transition t_i firing being unavailable during a delay θ_i. Then, deposited over θ_i it becomes available, i.e changes its state to non-reserved. Because only non-reserved tokens enable a transition, the case (i) of condition (2.1) has to change as follows:

$$pn \geq W(p,t) , \forall p \in P , M(p) = (pn,pr) \tag{4.6}$$

After transition t_i is enabled, it fires removing a non-reserved token from all its input places, and deposits reserved tokens in all its output places in a single atomic operation. The reserved token then remains unavailable for a time interval θ_i thus, modeling the state of the system when the relevant (modeled) operation is performed.

This mechanism is illustrated in Fig. 4.3. Transition t_1 is enabled by the non-reserved token in p_3, i.e. at the state $M = ((0,0),(0,0),(1,0))$. After the firing of t_1, the TPN reaches marking $M_1 = ((0,1),(0,0),(0,0))$ corresponding to the reserved token in p_1. Then, after delay θ_1 results in marking $M_2 = ((1,0),(0,0),(0,0))$ it associates with the non-reserved token in p_1. The TPN then stays in M_2 until the firing of t_2. The TPN will return to its initial state M_o after firing t_3.

Of all the possible executions, the one corresponding to the case in which the time intervals of non-reserved tokens are equal to zero is of primary importance. This is because such an execution (known as functioning at maximal speed [4]) encompasses the case where the transitions are fired as soon as they are enabled, i.e. enables to validate system performance as a function of its presumed time dependent parameters. This restriction provides a basis for investigating a system at its stationary behavior.

4.3.2 Modeling

Let us consider some conditions sufficient for periodical stationary behavior of systems modeled. Their development provides the restrictions the system should follow so that the presumed performance measure could be calculated. Consider some extensions of the PN models introduced in Section 2.3 on the basis of the manufacturing process corresponding to the production route $PR_1 = (O_{1,1},M_1),(O_{1,2},M_2)$ processed in the FMM shown in Fig. 3.1.

The different versions of the PN models are shown in Fig. 4.4. The Petri net from Fig. 4.4 (a) is a simple extension of the subnet from Fig. 3.2, where the added place $\tilde{p}$ corresponds to an input/output buffer limiting the number of workpieces that can be simultaneously processed. Its alternative solution is the PN from Fig. 4.4 (b) where the transition $t_{1,3}$ models the event composed of two elementary events of the PN from Fig. 4.4 (a), i.e. t_1 and t_3, associated with the loading of the machine M_1 and unloading of the machine M_2, respectively.

The timed version of the steady-state reduced PN models, encompassing the following markings $M_o=((0,0),(0,1),(0,0),(0,1))$, $M=((1,0),(0,0),(1,0),(0,0))$, and $M'=((0,1),(0,0),(0,1),(0,0))$ are shown in Fig. 4.4 (c), (d), and (e), respectively. It can be noted that the transition t_2 is enabled at M_o. Its firing results in the marking M which after delay θ_2, results in the marking M'.

It is easy to note, that the steady-state reduced model of the PN from Fig. 3.2 provides a closed-chain structure of cyclic processes that leads to the deadlock. Thus, the only possible structures that can be taken into account, regard the open chains of processes.

4.3.3 Performance measure

In order to discuss performance measures and the sufficient conditions that guarantee their calculation, let us focus on the possible open-chain structure of the cyclic processes shown in Fig. 4.5. The initial state

$$M_o=((0,0),(0,1),(0,0),(0,0),(0,0),(0,1),(0,0),(1,0),(0,0),(1,0)) \tag{4.7}$$

of the considered net can be presented in the following simplified form $M_o=(*,*,1,0)$.

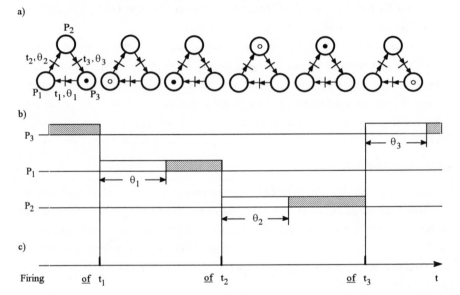

a)
b)
c)

Figure 4.3. A timed Petri net (a) phases of transitions firing and token reservation (b) time intervals of reserved and non-reserved tokens placed in p_i (● non-reserved token, O reserved token, ▨ a time interval of non-reserved token placed in p_i, ▢ a time interval θ_i of reserved token in p_i) (c) instants of transitions firing

This is because the set of all possible markings of each elementary (component) net (see Fig. 4.4 (c)) contains the following elements $M_1=((0,0),(1,0),(0,0),(1,0))$, $M_2=((0,0),(0,1),(0,0),(0,1))$, $M_3=((0,1),(0,0),(0,1),(0,0))$, $M_4=((1,0),(0,0),(1,0),(0,0))$. It is worth reminding the interpretation of the pairs that compose the above markings.

Let us assume the following conversion code $M_1=(0,1)$, $M_2=(0,*)$, $M_3=(*,0)$, $M_4=(1,0)$, where "1" means the relevant operations processed, "*" means the process execution is suspended after the relevant operation was completed, "0" means the relevant operation is waiting for its processing. Thus, the elementary model (see Fig. 4.4 (c)) can be in one of the following two states, it is either processed (on M_1, M_4), or its execution is suspended (on M_2, M_3).

In the case of the chain composed of two processes, the marking can be treated as a result of the overlap of its component markings. For instance, the marking $M=(1,1,0)$ is an overlap of the following two markings $M'=(1,0)$ and $M'=(1,0)$, i.e.

$$M = \frac{(\ 1,\ 0\ )}{(\ 1\ ,0\ )} = (\ 1,\ 1,\ 0)$$

Thus, finally the marking (4.7) can be treated as a result of the subsequent overlap of the following markings: $M_1=(*,0)$, $M_2=(*,0)$, and $M_3=(1,0)$.

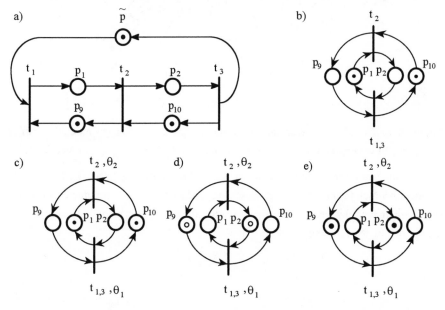

Figure 4.4. Modeling of the manufacturing process specified by $PR_1=(O_{1,1},M_1),(O_{1,2},M_2)$, (a) the PN model of single workpiece flow realization, (b) steady-state reduced version of the PN model, (c) (d), and (e) the TPN version of the steady-state reduced PN models corresponding to $M_o=((0,0),(0,1),(0,0),(0,1))$, $M=((1,0),(0,0),(1,0),(0,0))$, and $M'=((0,1),(0,0),(0,1),(0,0))$, respectively

From our definition of TPN we can conclude, that each state has its duration time as a function of delays Θ. According to the principle of the maximal speed behavior, the duration time is associated with markings such as $M=(1,1,*,0)$, i.e. corresponding to the situations where the suspended processes do not perform their next operation, which is waiting for the processing. Thus, the periodicity of the resultant behavior can be calculated as a sum of duration times associated with markings in a sequence SQ.

The considered net model (see Fig. 4.5) is composed of the elementary nets modeling component cyclic processes. The nominal value of each elementary cycle T_i is determined by a sum of delays. This means that the cycle time of each separately acting component process is optimal in the sense that it is equal to the nominal value. However, the allowance of the interaction of the component processes may lead to the suspension of some of them, and this results in the decrease of resources utilization.

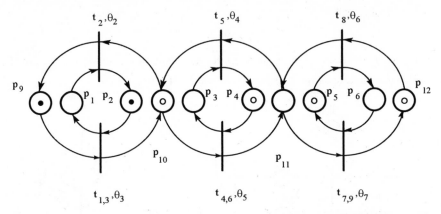

Figure 4.5. The TPN representation of the steady-state reduced model of single workpiece flow version of the PN corresponding to PR_1, PR_2, PR_3 determined in (3.9)

Thus, the natural measure of the efficiency of the cooperating processes is a ratio of the time in which the component processes are executed and the cycle time of the whole system of interacting processes. Therefore, the relevant performance measure can be determined by the following formula [7]:

$$\eta = \frac{\sum_{i=1}^{n} n_i T_i}{nT} . \tag{4.8}$$

where n is a number of all the component processes, T is a cycle time of the system of the interacting processes, T_i is a nominal cycle time of the i-th component process, n_i is a number of times the i-th process occurs during the period T.

4.3.4 Control

The satisfactory conditions that guarantee the cyclic, steady-state behavior can be formulated using a concept of the Flow Graph (FG) [7]. The considered graph can be treated as a time-dependent extension of the reachability graph of a TPN model. The condition resolves itself to the examination of the cycles occurrence. In case a FG contains a unique cycle, a chain of interacting processes possesses cyclic steady-state behavior.

Let us consider the TPN from Fig. 4.5 assuming the following set of delays $\theta_2 = 7$, $\theta_3 = 2$, $\theta_4 = 7$, $\theta_5 = 3$, $\theta_6 = 2$, and $\theta_7 = 2$. The nominal values of the component processes cycle times are as follows $T_1 = 9$, $T_2 = 10$, and $T_3 = 4$. The reachability graph obtained for the marking $M_o = (*,1,0,1)$ contains two cycles. However, the corresponding FG $\Gamma = (S,\alpha)$, $S = \{S_i \mid i \in \{1,2,...,n\}\}$, $\alpha \subset S \times S$, such that $S_i \, \alpha \, S_j$ if and only if $crd_1(S_i)$ follows $crd_1(S_j)$ in the relevant reachability graph, contains a unique elementary cycle. This is because each node $S_i = (M,\tau)$, $i \in \{0,2,...,15\}$, where τ is a duration time of the marking M, can be treated as an element of a equivalence class defined as follows

$$S_i \sim S_j \overset{def}{\Leftrightarrow} crd^1(S_i) = crd^1(S_j) \text{ and } \sum_{k=1}^{k=i} crd^2(S_k) = \sum_{k=1}^{k=j} crd^2(S_j) + nT \qquad (4.9)$$

where

$\sim$ is the equivalence relation, T is the cycle time, n is the natural number $n \in N_o$.

Note that values $crd^2 S_i$, $i \in \{0,2,...,15\}$, follow from the Gantt's chart representation of steady-state behavior of the concurrently interacting processes (see figure 4.6).

As during the period T, the first and the second component process repeats $n_1 = n_2 = 1$ times, while the third component process repeats $n_3 = 2$ times, the value of the considered performance measure η is equal to 9/10. Therefore, the cycle time T is different from a number of nodes in the elementary cycle in FG, i.e. it is not equal to the number of states corresponding to the sequence SQ.

In order to present the conditions that enable us to calculate the performance measure η, let us introduce the concept of the critical component process (CCP - for short) [7]. The k-th component process in a chain of cyclic processes is said to be critical if and only if it never suspends. Thus the sufficient conditions that allow the calculation of the cycle time of the whole system are the following:

(i) if the k-th component process is a unique CCP in a chain of n cyclic processes and such that $T_k > T_i$ for all $i \in \{1,2,...,n\} \setminus \{k\}$, then T_k is the cycle time of the whole system,

$$\qquad (4.10)$$

(ii) if the k-th component process is a unique CCP in a chain of n cyclic processes and there is the j-th process of n processes such that $T_j > T_i$, $i \in \{1,2,...,n\} \setminus \{k\}$ where $j \in \{k-1,k+1\}$, then there is m such that $(m - 1)T_k < T_j < mT_k$ holds and mT_k determines the cycle time of a whole system.

In turn, in case the chains of n component processes satisfy the following conditions,

(i) $T_i > T_{i+1}$ for all $i \in \{1,2,...,n-1\}$,

(ii) $\theta_i < T_{i+1}$ for all $i \in \{2,3,...,n-1\}$, $\qquad (4.11)$

(iii) $\theta_i + \theta_{i+1} < \theta_{i-1} + \theta_{i+2}$ for all $i \in \{2,3,...,i-2\}$,

and its first component process is a CCP, then T_1 is the cycle time of the whole system and the n_i can be calculated according to the following formula:

$$n_i = n_{i-1} \, n'_i, \quad \text{where} \quad n'_i = T_{i-1} \text{div } T_i \quad \text{for all} \quad i \in \{1,2,...,n\} \tag{4.12}$$

where the result of a div b is an integer part of division a by b. Thus, in case the chain of processes satisfies the conditions (4.10), and (4.11), then the performance measure η can be determined analytically.

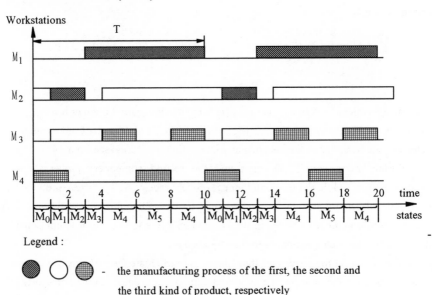

Figure 4.6 Gantt's chart representation of the steady-state behavior modeled by the TPN from figure 4.5

The above discussion can also be used to illustrate an approach to a certain control problem of the distributed systems. In order to state this problem, let us consider a chain of n cyclic processes following the conditions (4.11). Assuming that the first process is CCP and the cycle time T is equal to T_1, find out the cycle times T_i , $i \in \{1,2,...,n\}$. The solution provided under assumption $n_1 < n_2 <...< n_n$ enables us to calculate the required nominal cycle times from the set of equations [7]:

$$\begin{cases} n'_i = T_{i-1} \text{ div } T_i \\ \qquad\qquad\qquad\qquad \text{for all} \quad i \in \{2,3,...,n\} \\ n_i = n_{i-1} \, n'_i \end{cases} \tag{4.13}$$

The solutions to the different variants of the considered problem can then be evaluated by the presumed performance measure η. Thus, the relevant control problem can be considered as a problem of the search for the two vectors $(n_1,n_2,...,n_n)$ and

$(T_1,T_2,...,T_n)$ that extremize the criterion function η. Following the task oriented assumptions, the problem can be finally specified either as a satisfaction or an optimization problem.

The above mentioned results prove the possibility of a design of the analytical models in the case of chain of cyclic processes. Furthermore the values of n_i and T_i, $i \in \{1,2,...,n\}$, once adjusted (for instance in the course of a solution to a control problem) they can stay unchanged during the system operation time. Also, the small disturbances in the nominal cycle times (excluding a CCP) do not change the cycle time T of the whole system, so the resulting changes in η value can be ignored.

5   Conclusions

The common features of the considered systems are their complexity and the lack of adequate performance evaluation tools. Because the number of the discrete physical states explodes in a combinatorial fashion, the existing approaches to the performance evaluation are either inadequate, because they are based on the unrealistic assumptions (e.g. queuing networks, Markov chains), or are numerically impossible to evaluate, because of a combinatorial explosion (e.g. computer simulation).

The mathematical problems associated with the modeling and the performance evaluation of the discrete manufacturing systems can be easily expressed in terms of the Petri nets theory. This theory offers a unified framework which allows the study of the interconnections existing among a variety of different energy, material, and data flows observed on different organizational and functional levels of the system design. Its main advantage is a workbench which enables to develop methods aimed at the evaluation of quantitative and qualitative characteristics of system functioning. In this context, their application to the development of methods aimed at the automated Petri net models synthesis, seems to be the most important issue.

The results provided in this study illustrate our approach, namely, the conditions leading to the procedures that allow the achievement of the required performance of the system behavior, while preserving some of its presumed qualitative constraints. Although the models are simplified in scope, they provide an insight into the benefits of the implementation of the proposed approach. The Petri net-based tools and the analytical formulas that allow us to predict a system behavior according to its component characteristics, are its main applications.

References

1. Al-Jaar R.Y., Desrochers A.A., Performance evaluation of automated manufacturing systems using generalized stochastic Petri nets. IEEE Trans. on Robotics and Automation (1990) Vol.6, No.6., pp. 621 - 639.

2. Agrawal R., Tanniru M., A Petri-net based approach for verifying the integrity of production systems. Int. J. Man-Machine Studies (1992) No.4, pp. 447 - 468.

3. Banaszak Z. (Ed.) Modeling and control of FMS (Petri net approach), Wroclaw Technical University Press, Wroclaw 1991.

4. Banaszak Z., et al. Computer-aided control of processes in group technology systems. Wroclaw Technical University Press, Wroclaw 1990.

6. Banaszak Z., Krogh B., Deadlock avoidance in flexible manufacturing systems with concurrently competing process flows. IEEE Trans. on Robotics and Automation (1990) Vol.6, No.6, pp. 724-734.

6. Banaszak Z., Wojcik R., D'Souza K.,A., Petri net approach to the synthesis of self-reprogramming control software. Appl. Math. and Comp. Sci. (1992) Vol.2, No.2, pp. 65 - 86.

7. Banaszak Z., Jedrzejek K., On self-synchronization of cascade-like coupled cyclic processes. Appl. Math. and Comp. Sci. (1993) Vol. 3, No 4., pp. 101 - 127.

8. Banaszak Z., Control oriented models of interprocess cooperation. Systems Science (1988) Vol.14, No.2, pp.31-59.

9. Braker J.G., Max-algebra modeling and analysis of time-table dependent transportations networks. Proc. of the ECC'91 European Control Conference, July 2 - 5, 1991, Grenoble, France, pp. 1831 - 1836.

10. Buzacott J.A., Yao D.D., Flexible manufacturing systems: a review of analytical models. Management Science (1986) Vol.32, No.7, pp. 890 - 905.

11. Capkovic F., Modeling and justifying discrete production processes by Petri nets. Computer Integrated Manufacturing Systems (1993) Vol.6, No.1, pp. 27 - 35.

12. Chow W.S., Heragu S., Kusiak A., Operations research models and techniques. Asbojorn Rostadas (Ed.) Computer-Aided production Management, Springer-Verlag, London, Paris, Tokyo 1988, pp. 135 - 148.

13. Cutts G., Rattigan S., Using Petri nets to develop programs for PLC systems. K. Jensen (Ed.) Application and Theory of Petri Nets 1992, Springer-Verlag, Berlin 1992, pp. 368 - 378.

14. David R., Alla H., Petri nets and Grafcet (Tools for modeling discrete event systems), Prentice Hall, New York 1992.

15. Feldbrugge F., Petri net tool overview 1989. In: Lecture Notes in Computer Science No. 424, G. Rozenberg (Ed.) Advances in Petri nets 1989, Springer-Verlag, Berlin 1990, pp. 150 - 178.

16. Freedman P., Time, Petri nets, and robotics. IEEE Transactions on Robotics and Automation, vol. 7, No.4, 1991 pp.417-433.

17. Gunasekaran A., Martikainen T., Yli-Olli P., Flexible manufacturing systems: and investigation for research and applications. European Journal of Operational Research No.66, 1993, pp.1-26.

18. Heiner M., Petri net based software validation (Prospects and limits). Technical Report No. TR-92-022, International Computer Science Institute, Berkeley 1992.

19. Heinschmidt K.F., Finn G.A., Integration of expert systems, databases and computer-aided design. A. Kusiak (Ed.) Intelligent design and manufacturing, John Wiley and Sons, New York 1992, pp. 157 -178.

20. Hillion H.P., Proth J.-M., Performance evaluation of job-shop systems using timed event-graphs. IEEE Trans. on Automatic Control (1989), Vol.34, No.1, pp. 3-9.

21. Ho Y-C., Dynamics of discrete event systems. Proc of the IEEE (1989) Vol.77, No. 1, pp. 3 - 6.

22. Jothishankar M.C., Wang H.P., Determination of optimal number of Kanbans using stochastic Petri nets. Journal of Manufacturing Systems (1992), Vol.11, No.6, pp.449-461.

23. Kant K., with contributions by M.M. Srinivasan, Introduction to computer system performance evaluation. McGrow-Hill, Inc., N.Y. 1992.

24. Kusiak A. (Ed.), Intelligent design and manufacturing. John Wiley and Sons, N.Y. 1992.

25. Laftit S., Proth J-M., Optimization of invariant criteria for event graphs. IEEE Trans. on Automatic Control (1992), Vol.37, No.5, pp. 547-555.

26. Lee-Kwang H., Favrel J., Baptiste P., Generalized Petri net reduction method. IEEE Trans. on Systems, Man and Cybernetics, (1987), Vol. 17, No.2, pp. 297 - 303.

27. Lemmer K., Schnieder E., Modeling and control of complex logistic systems for manufacturing. K. Jansen (Ed.) Application and Theory of Petri Nets 1992, Springer-Verlag, Berlin 1992, pp. 373 - 378.

28. Leung Ying-Tat, Rajan S. Performance evaluation of discrete manufacturing systems. IEEE Control Systems Magazine , June 1990, pp. 77 - 86.

29. Marsan M.A., Balbo G., Conte G., Performance models of multiprocessor systems. The MIT Press, Massachusetts 1990.

30. Murata T., Petri nets: properties, analysis and applications. Proc. of the IEEE (1989) , Vol.77, No.4, pp.541 - 579.

31. Peterson J., Petri net theory and modeling of systems. Prentice Hall Int., Englewood Cliffs, New York 1981.

32. Prock J., A new technique for fault detection using Petri nets. Automatica (1991) Vol.27, No.2, pp. 239 - 245.

33. Reisig W., Petri nets. Springer-Verlag, Berlin 1982.

34. Reutenauer Ch., The mathematics of Petri nets. Prentice Hall Int. Englewood Clifs, New York 1988.

35. Strayer W.T., Weaver A.C., Performance measurement of data transfer services in MAP. IEEE Network (1988), Vol.2, No.3, pp. 75 - 81.

36. Van der Aalst W., M.,P., Timed coloured Petri nets and their application to logistics. Technische Universiteit Eindhoven, Eindhoven 1992.

37. Viswandham N., Narahari Y., Performance modeling of automated manufacturing systems. Prentice Hall, N.Y. 1992.

38. Wilson R.G., Krogh B.H., Petri net tools for the specification and analysis of discrete controllers. IEEE Transactions on Software Engineering (1990), Vol.16, No.1, pp. 39 - 50.

39. Zeigler B.P., DEVS representation of dynamical systems: event-based intelligent control. Proc. of the IEEE, Vol.77, No. 1, January 1989, pp. 72-80.

40. Zhang W., Representation of assembly and automatic robot planning by Petri net. IEEE Trans. on Systems, Man and Cybernetics, (1989), Vol. 19, No.2, pp. 418 - 422.

41. Zhou MengChu, DiCesare F., Parallel and sequential mutual exclusions for Petri net modeling of manufacturing systems with shared resources, IEEE Trans. on Robotics and Automation (1991) Vol. 7, No. 4, pp. 515 - 523.

42. Zhou MengChu, DiCesare F., Petri net synthesis for discrete event control of manufacturing systems. Kluwer Academic Publishers, Massachusetts, 1993.

10 Fault Tolerance in Flexible Manufacturing Systems Parameters

Anders Adlemo and Sven-Arne Andréasson

---

1 Introduction

The primary objective in this chapter is to describe a systematization of aspects and terminology of fault tolerance applied to distributed systems and, more specifically, to distributed, computerized manufacturing systems. The purpose of this is to facilitate the transition from manual manufacturing systems to computerized manufacturing systems.Until recently a system designer has considered such aspects as:

1. how to obtain good economy,

2. how to obtain the correct product quality,

3. how to obtain the maximum throughput,

4. how to obtain the maximum flexibility, and

5. how to minimize the amount of time during which a system does not produce.

All these aspects are very important in establishing a "good" system. However, most of the aspects mentioned implicitly take into consideration fault tolerance; for example, in order to minimize the amount of time a system is inactive, some kind of fault tolerance must be introduced. Until recently, the introduction of fault tolerance has been made more or less *ad hoc* and has not been considered at the design stage. We therefore suggest the explicit introduction of fault tolerance at the specification and design stage and we propose and demonstrate some different forms of achieving fault tolerance that are feasible to implement in a modern, computerized manufacturing system.

The chapter can be looked upon as consisting of two parts. The first part, consisting of section 2 to 4, is an introduction to the area of dependability and fault tolerance in computer systems and manufacturing systems, the purpose of which is to briefly introduce the reader to this field of investigation.

The second part, consisting of section 5, deals with some applications of fault tolerance to a case study. This part can be further divided into two directions.

1. The first direction deals with configurations and reconfigurations of a manufacturing system in the case of errors. The original idea is to make it possible for a manufacturing system to reconfigure before the complete stoppage of production. We suggest three different types of (re)configuration that can be applied in a manufacturing system. The ways in which the (re)configurations can be implemented in a manufacturing system are shown through examples.

2. The second direction deals with information flows and information placement to obtain fault tolerance in a manufacturing system. The intention is to make it possible for a manufacturing system to survive data network partitions. How these aspects of information can be implemented in a manufacturing system is shown through examples.

The case study is a part of the Volvo Uddevalla plant in Sweden that assembles kits used in production of cars.

2   Aspects of dependability

The terms dependability and, to a greater extent, fault tolerance, have existed for several years in the computer world. In the 1950's, von Neumann [28] presented his original work on improving system reliability, which has since developed into the area of fault tolerance and dependability. The field of incorporating means for tolerating faults in order to improve the reliability of a computing system has experienced a staggering increase in terms of the definition of new terminologies for fault tolerance and identifying new areas of application.

Most fault tolerant computer systems are constructed to provide protection against errors in the hardware caused by physical phenomena. Over the course of time, systems have grown larger and more complex, which has resulted in an increased amount of errors in the construction stage. Such errors in hardware and software were once considered easy to handle for the reason that structured programming would reduce the amount of errors, and the remaining errors in software and hardware could the be removed at the testing stage. This approach is possible for small computer systems but not for large ones. Large, distributed computer systems are especially difficult to fully test, as they are not deterministic.

The first area in which fault tolerance was introduced was computer hardware. Various approaches were used, such as introducing replicated hardware units. Over time, computer software has also adopted fault tolerance aspects. Various aspects of hardware and software fault tolerance techniques are described in [15]. A similar approach can be found in [27]. Some of the techniques mentioned in the articles are also described in the following text. A good reference text on reliability techniques is found in [34].

The descriptions in the following sections are by no means complete, and thus should be regarded only as examples of fault tolerance approaches in the computer world, some of which have been applied to the research presented in this chapter.

2.1 Basic dependability concepts

Laprie was one of the first to systematize fault tolerance in the area of reliable computing, e.g., [22]. His view of fault tolerance is illustrated in figure 1.

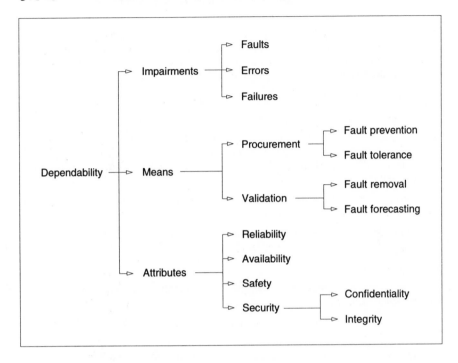

Figure 1: The dependability tree, *Laprie* [22]

To include various aspects of fault tolerance into one single concept a unifying concept was introduced called *dependability*. Dependability is defined as follows:

Dependability is the quality (or trustworthiness) of the delivered service such that reliance can justifiably be placed on this service.

The dependability concept comprises three aspects.

1. *Impairments*, which are undesired, but not unexpected, circumstances.

2. *Means*, which are the methods, tools and solutions that enable

   — the ability to deliver a service on which reliance can be placed, and

   — the possibility to reach confidence in this ability.

3. *Attributes*, that enable

   — the quantification of service quality resulting from the impairments and the means, and

   — the properties that are expected from the system to be expressed.

The impairments are further divided into faults, errors and failures.

1. A *fault* is the original cause of any impairment, such as a defective component or a programmer's mistake. A fault causes a latent error.

2. An *error* is that part of the system which is liable to lead to a failure, i.e., an impairment that has taken effect, such as using a defective component. An activated error causes a failure.

3. A *failure* occurs when the delivered service deviates from the specified service, i.e., an error produces erroneous data which affects the delivered service. A failure on one level in a system can be viewed as a fault on the next higher level. This means that there is a chain that might appear in the following way:

$$\text{fault}_{\text{level } n+1} \rightarrow \text{error} \rightarrow \text{failure}_{\text{level } n+1} \rightarrow \text{fault}_{\text{level } n} \rightarrow \text{error} \rightarrow \text{failure}_{\text{level } n} \rightarrow$$

$$\text{fault}_{\text{level } n-1} \rightarrow \text{error} \rightarrow \text{failure}_{\text{level } n-1} \rightarrow \text{fault}_{\text{level } n-2} \rightarrow \text{error} \rightarrow \text{failure}_{\text{level } n-2}$$

This way of viewing faults, errors and failures is not the only existing form [8], but does occur most frequently among those who work in the area of dependability [27].

The dependability means are divided into *procurement* and *validation*. Procurement is further divided into:

1. *fault prevention*, i.e., how to prevent, through construction, the occurrence of a fault (*fault prevention* is also called *fault avoidance* further on), and

2. *fault tolerance*, i.e., how to provide, through redundancy, a service that complies with the specification in the presence of faults.

Validation is divided into:

1. *fault removal*, i.e., how to minimize, through verification, diagnosis and correction, the presence of latent errors, and

2. *fault forecasting*, i.e., how to estimate, through evaluation, the presence, creation and consequence of an error.

Finally, there are four attributes: reliability, availability, safety and security.

1. *Reliability* is a measure of the continuous delivery of correct service or, equivalently, of the time period of correct service until the occurrence of a failure.

2. *Availability* is a measure of the delivery of correct service with respect to the time required to repair a fault.

3. *Safety* is the probability that a system either performs its intended functions correctly, or that the system has failed in such a manner that no catastrophic consequences occur.

4. *Security* is defined as the possibility for a system to protect itself with respect to confidentiality and integrity, where

   — *confidentiality* is defined as how to protect a system from the unauthorized release of information or from the unauthorized use of system resources, and

   — *integrity* is defined as how to avoid the unauthorized modification of a system.

2.2   Stages in response to a failure

In a redundant system, it can be necessary to go through a number of stages to act on a failure. Designing a reliable system involves the selection of a coordinated failure response that combines several reliability techniques. Some of these techniques are described in the next section. Siewiorek [32] divided the actions into ten different stages. The ordering of these stages in the following paragraphs corresponds roughly to the normal chronology of a fault occurrence, although the actual timing may be different in some instances.

1. *Fault confinement* is achieved by limiting the spread of fault effects to only one area of the system, thereby preventing the contamination of other areas.

2. *Fault detection* can be divided into two major classes:

   — *On-line detection* provides real-time detection capabilities.

   — *Off-line detection* means that the device cannot be used during the test.

3. *Fault masking* is a technique for hiding the effects of a failure. Fault masking can be divided into:

   — *Hierarchical fault masking,* in which higher levels mask the further propagation of a failure from lower levels.

   — *Group fault masking,* in which a group of devices together mask the further propagation of a failure.

4. *Retry,* i.e., a second attempt of an operation, may be successful, particularly if the cause of a first-try failure is transient and causes no damage.

5. *Diagnosis* can be required if the fault detection technique does not provide information about the failure location and/or its properties.

6. *Reconfiguration* can be used to replace a failed component with a spare or to isolate the failed component.

7. *Recovery* must be done after the detection of a failure (and the possible reconfiguration) to eliminate the effects of the failure. A recovery can be performed in two ways:

   — *Backward recovery.* A system is backed up to some point in its processing that preceded the fault detection, and the operation is then restarted from this point.

   — *Forward recovery.* A system is moved forward to a fault-free state and then restarted from this point.

8. *Restart* of a system can be necessary if too much of the information is damaged by a failure or if a system is not designed to handle a recovery.

9. *Repair* of a failed and replaced device can be either

   — *on-line repair,* or

   — *off-line repair.*

10. *Reintegration* of a failed device can be done after its repair.

2.3 Reliability techniques

The reliability techniques spectrum can be divided into four major classes, as defined in [34]:

1. *Fault avoidance* reduces the possibility of a failure through, for example, highly reliable components.

2. *Fault detection* provides no tolerance to failures but gives a warning when they occur.

3. *Masking redundancy,* also called static redundancy, provides tolerance to failures but gives no warning when they occur.

4. *Dynamic redundancy* covers those systems whose configuration can be dynamically

changed in response to a fault, or in which masking redundancy, supplemented by on-line fault detection, allows on-line repair.

The range in the cost of fault tolerance techniques is almost a continuum in terms of percentage of redundancy. Figure 2 depicts four regions of hardware redundancy, each corresponding to one of the four major areas of the fault tolerance technique spectrum. Even though most techniques in each area fit within these regions, individual techniques may fall well outside them.

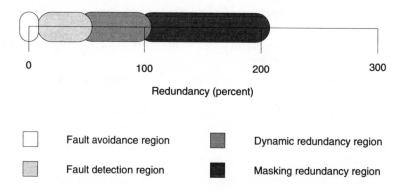

Figure 2:   Cost range of redundancy techniques, *Siewiorek and Swarz* [34]

2.4   Hardware fault tolerance

One common approach of enhancing fault tolerance in computer hardware is to replicate sensitive parts.

The AXE telephone exchange, e.g., reaches an acceptable level of fault tolerance and system availability by duplicating the main computer [29]. The computer pair then works actively in parallel, with one as the executive and the other as the standby. This approach is also known as a duplex system. The results from the pair are continuously compared to detect any errors.

Another approach that has been used is TMR (Triple Modular Redundancy). The results from three parallel components are checked. If one component is erroneous, work can still proceed, as the remaining two reach the same result.

Many other variants based on the replicating theme have also been developed. One project based on the idea of replicated components was the FTMP (Fault Tolerant Multi Processor) project, described in [18]. The object of this project was to fly an active-control transport aircraft. The FTMP consisted of ten identical processor/cache modules, ten shared memory modules, ten I/O ports and ten clock generators. Each one of the parts were packed into a module. The ten modules were interconnected by a total of six serial busses that were quintuple-redundant.

Self-checking circuits provide another way to achieve fault tolerance. Such circuits are based on error-control coding, which means that the information that is passed in a system must be coded in some form. This coding then permits one or several errors to appear in the information without altering the correct function of the system. An excellent reference material to this area is [31].

An article describes various fault tolerance techniques used in commercial computers, some of which are hardware error detection, duplication and matching, and checkpointing

[32]. However, the penalty for non fault tolerant users ranges from ten percent to 300 percent when considering processor logic alone, as compared with the use of a system without any fault tolerance. This is an important factor: improved fault tolerance almost always implies an increased cost, either in hardware or software components or in the extra computing time imposed by the fault tolerance. However, it must be remembered that the key issue is to obtain the best possible *overall* cost/performance/dependability result.

2.5 Software fault tolerance

A project aimed at achieving fault tolerant software in the area of active-control transport aircrafts (as the FTMP project) was the SIFT project (Software Implemented Fault Tolerance), described in [26]. The main design objective of the SIFT system was to implement the fault tolerance techniques as much as possible in software with a minimum development of specialized hardware. A task can be executed with one, three or five identical copies, depending on how critical the task is. The three (or five) copies are extracted from a data file, voted on and, finally, stored in a buffer.

A way of obtaining fault tolerance in software is similar to that of TMR for hardware. A computer system uses replicated blocks of a program that run in parallel and the result is then voted upon (so called *N-version programming*, see, e.g., [1] and [9]). If a result from one of the blocks deviates from the others, it is assumed that the block is erroneous. Figure 3 is an illustration of the N-version programming mechanism.

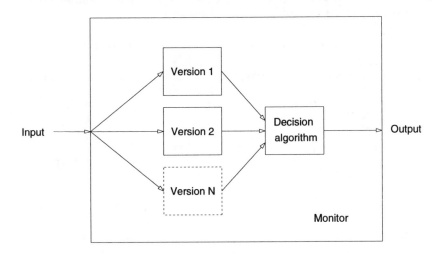

Figure 3: N-version programming

Another software fault tolerance approach is known as *recovery blocks* (see, e.g., [1] and [30]), in which one version of a program block is assumed to be the main version. If the result from this version does not pass a check of its result, another version is started. If the new version also fails, another version is started and so on. Figure 4 is an illustration of the recovery block mechanism.

294

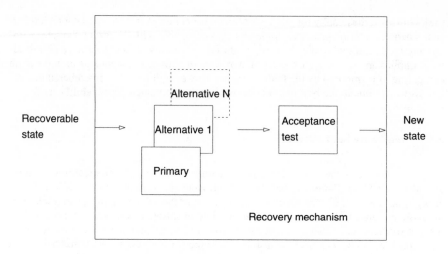

Figure 4: Recovery block

Table 1 describes the costs for fault tolerant software, C_{FT}, as compared to non fault tolerant software, C_{NFT}, for the N-version programming, recovery block and self-checking programming techniques. The table gives the ranges for the ratio C_{FT}/C_{NFT}, as well as the average values and the average values per variant (NC_{NFT}, where N is the number of variants). C_{FT}/C_{NFT} consists of the cost distribution percentages for requirements, specification, design, implementation, verification and validation.

Table 1: Cost of fault tolerant software versus non fault tolerant software, *Laprie et al.* [23]

Faults tolerated	Fault tolerance method	N	CFT/CNFT minimum	CFT/CNFT maximum	CFT/CNFT average	CFT/NCNFT average
1	N-version programming	3	1.78	2.71	2.25	0.75
1	Recovery blocks	2	1.33	2.17	1.75	0.88
1	N self-checking programming					
	Acceptance test	2	1.33	2.17	1.75	0.88
	Comparison	4	2.24	3.77	3.01	0.75
2	N-version programming	4	2.24	3.77	3.01	0.75
2	Recovery blocks	3	1.78	2.96	2.37	0.79
2	N self-checking programming					
	Acceptance test	3	1.78	2.96	2.37	0.79
	Comparison	6	3.71	5.54	4.63	0.77

The exact definition of C_{FT}/C_{NFT} can be found in [23]. One observation that can be made from the table is that N-variant software is less costly than N times a non fault tolerant software (i.e., the last column).

3   Dependability in computer systems

Dependability involves a number of aspects, as shown in section 2.1. Some of these will be discussed in the next section, e.g., fault tolerance, reliability and availability. Fault tolerance for a distributed computing system can be divided into fault tolerant hardware, software and communication. Some methods for dealing with fault tolerant hardware and software were described in sections 2.3, 2.4 and 2.5. Approaches on how to obtain fault tolerant communication has been described in, for instance, [24] where a low-quality backup link is used in cases of data network partitionings in distributed database systems.

The computers described in this section have all been developed to respond to failures. However, various aspects may influence the type of fault tolerance required, e.g., whether the system should be highly reliable or highly available. The cost of having fault tolerance must furthermore be weighed against the cost of suffering an error or a failure.

The simplest type of computer consists of a uniprocessor that includes a processor, a memory and input/output devices. Components can be replicated to enhance performance and fault tolerance. When the entire uniprocessor is replicated, the result is a multicomputer. Multicomputer systems consist of several autonomous computers which may be geographically dispersed, and are therefore sometimes also referred to as *loosely coupled systems*. If only the processor is replicated, the result is a multiprocessor. Multiprocessor systems usually have a common memory for all processors, with a single, system-wide address space available to all processors. Owing to their close connection, these systems are also called *tightly coupled systems*.

A simple illustration of a multicomputer system and a multiprocessor system is shown in figure 5.

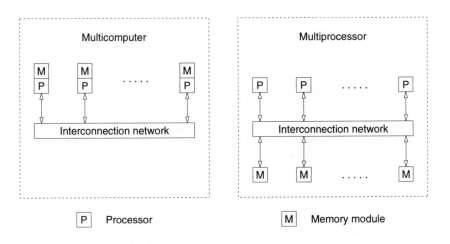

Figure 5: Multicomputer and multiprocessor systems

Seven different commercial computers are described from a fault tolerance point of view in [33]. They consist of uniprocessors, multicomputers and multiprocessors. The fault tolerance techniques used by the different computer systems are briefly described in table 2.

All the systems focus on high availability, correct storage of data, unique designs and configuration into multiprocessor systems. The VAX 8600, the IBM 3090 and the VAXft 3000 use general purpose operating systems, while Tandem, Stratus, Teradata and Sequoia systems use operating systems optimized for transaction processing.

Table 2: Fault tolerance techniques in commercial computing systems, *Siewiorek* [33]

Structure	Detection	Recovery	Sources of failures tolerated	Techniques
Uniprocessor				
VAX 8600	Hardware	Software	Hardware	Hardware error detection
IBM 3090	Hardware	Hardware/ software	Hardware	Hardware error detection, retry, work-around
Multicomputer				
Tandem	Hardware/ software	Software	Hardware, design, environment	Checkpointing, "I'm alive" messages
Stratus	Hardware	Hardware	Hardware, environment	Duplication and matching
VAXft 3000	Hardware	Hardware	Hardware, environment	Duplication and matching
Multiprocessor				
Teradata	Hardware	Software	Hardware, environment	Duplication
Sequoia	Hardware	Software	Hardware, environment	Duplication and matching

4 Dependability in manufacturing systems

Dependability and fault tolerance are two concepts that have been thoroughly investigated over the last decades. Some examples of research performed in the area of fault tolerant manufacturing are described in this section.

4.1 Distributed fault tolerant real-time systems

Kopetz pointed out in several articles that the real-time control of a manufacturing system must be present on time and must also be correct in order for the system to function according to specifications (see, for example, [20]). One system that incorporates these aspects is Mars (Maintainable Real-Time System). A sketch of the system is shown in figure 6.

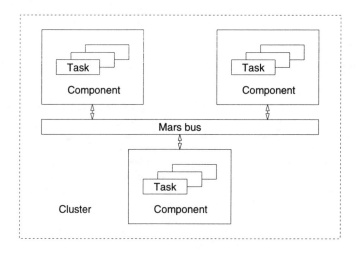

Figure 6: A Mars cluster, *Kopetz et al.* [20]

A Mars cluster is a subset of the network with a high functional connectivity. Clusters are the basic elements of the system architecture. Each cluster consists of several components interconnected by a synchronous, real-time Mars bus. The concept of clustering helps to manage the complexity of the large network of components.

Other key points addressed in the Mars project are:

1. limited time validity of real-time data,

2. predictable performance under peak load,

3. maintainability and extensibility, and

4. fault tolerance.

Fault tolerance is achieved through active redundancy. The fault hypothesis in the Mars design covers permanent and transient physical faults in the components and on the real-time bus. The fault tolerance relies on self-checking components that run with active redundancy. Fault tolerance also relies on multiple transmissions of messages on the real-time bus. Active redundancy is used because of its superior timing properties as compared with passive redundancy. As the components are self-checking, they are fail-stop, i.e., they either operate correctly or they do not produce any results at all. As long as at least one component in a redundant set of components works, the required service can be maintained.

The messages between different components are transmitted n times, either in parallel over n busses or sequentially over a single bus. That means that n-1 messages can be lost without causing problems. Mars messages are sent as periodic real-time datagrams, each containing a validity time. Each message is sent twice or more, depending on the fault hypothesis and the transient-failure probability of the bus. As regards the components, redundant components can be inserted into an operational system without notification to or reconfiguration or modification of the running components.

The mechanisms in the Mars operating system check both the correctness of the information in the value domain as well as its correctness in the time domain. Checks in the value domain include plausibility tests and time-redundant execution of tasks. In the time domain, these mechanisms check runtime limits, global time limits and the timing behavior of the tasks with respect to the timing requirements of the controlled system. When the operating system detects an error, it attempts to logically turn off the component, regardless of whether the fault is transient or permanent or has occurred in hardware or software.

4.2   Reliability and availability in automatic manufacturing

In [17] is shown that a continuous model for the analysis of reliability and availability based on the concept of a specific failure rate is a useful tool for analyzing the performance of automatic manufacturing systems and for formulating a planning strategy. The specific failure rate is a fundamental limitation which determines the maximum size of a manufacturing cell for a certain availability to be reached.

However, the continuous model cannot be considered complete if higher control and administration levels are not considered. These higher levels are subject to a much more coarse segmenting. To achieve this, a model was introduced which has hierarchical levels that handle the same product flow but have separate resource consumptions and separate specific failure rates.

The total system availability for a system consisting of n levels is given by the product of the availabilities in the different levels:

$$\text{Availability}_{\text{system}} = \text{Availability}_{\text{level 1}} * \text{Availability}_{\text{level 2}} * \ldots * \text{Availability}_{\text{level n}}$$

The equation assumes that the higher levels directly control the lower levels in real-time and, consequently, that a failure in a higher level leads to a standstill at lower levels. This means that if the availability at higher levels is unsatisfactory, the advantages obtained by segmenting the manufacturing system cannot be realized. The remedy is to design a certain autonomy into the manufacturing cells, so that each cell can function autonomously for a certain amount of time which exceeds the MTTR (Mean Time To Repair) required to correct failures at higher levels.

The use of redundant cells always increases availability without the need to increase product buffer storage. From an availability point of view, the optimum cell configuration consists of many, physically long, parallel manufacturing cells.

4.3   Modeling of fault tolerant distributed manufacturing systems

The importance of fault tolerance in distributed computerized manufacturing systems has been demonstrated in, e.g., [11]. A manufacturing system is viewed as a large, distributed system consisting of interconnected networks of computing nodes and automated devices. Functionally, a manufacturing system is involved in a variety of activities, including planning, design, process scheduling, tool and materials handling, product assembly, process quality control and inventory control.

Some typical features of a manufacturing system are:

1. distributed nature,

2. hierarchical control structure,

3. real-time constraints,

4. heterogeneous nodes,

5. varying consistency requirements at different production levels,

6. heterogeneous traffic patterns, and

7. hostile environments.

All these different features must be considered when designing a fault tolerant manufacturing system. The primary types of failures include:

1. node failures and software failures of the computing nodes, and

2. communication failures and timing failures on the interconnected network.

Examples of ways to deal with node failures are the following:

1. The primary site approach, in which one site is designated as primary and some others as backup. If the primary site fails, a backup takes over as primary.

2. Multiple executions of identical processes. As long as one of the processes is still alive, the task can be completed on time.

Ways of dealing with software failures are:

1. recovery blocks,

2. N-version programming, and

3. exception handling.

Communication failures can be dealt with by using specialized protocols supporting multiple channels. Another possibility is to retransmit messages.

Timing failures, finally, can be handled by the use of a scheme similar to recovery blocks.

As mentioned earlier, the introduction of fault tolerance also implies an overhead cost. In the case of manufacturing systems, these costs can be classified as being caused by:

1. duplicated resources,

2. communication overheads, and

3. time overheads.

4.4   Causes of operational interruptions

The following section illustrates the distribution of operational outages in automatic manufacturing systems. This gives an indication of the possible sources of faults which, in turn, indicates where fault tolerance should be introduced, if possible.

Little is actually known about the distribution and causes of failures in automatic manufacturing systems, but some results indicate that longer stops in production are caused by disturbances that affect the cooperation of machining centers. Other failures are related to the positioning and tolerance of workpieces, tool changing and transport disturbances.

An investigation of 25 automatic assembly systems showed an average availability of 80% [37]. This is unacceptable in modern production where the goal is to obtain systems with *high-* or even *very-high-availability*, i.e., an average availability of 95% or 99.5%, respectively [15]. The investigation distinguished between rotary indexing systems, linear indexing systems and linear free systems. The causes of operational interruptions can be related to positioning and tolerance of workpieces (29%), malfunction in assembly devices (18%) and disturbances in buffer storage (19%) (see table 3).

Table 3: Percentage of operational interruptions for various sources, *Wiendahl and Winkelhake* [37]

Source of failure	Assembly plants
Component	29
Defective part	6
Contamination	1
Assembly device	18
Maintenance	3
Replenishing	13
Buffer	19
Other causes	11

5 Dependability applied to a case study

The Volvo Uddevalla plant described in this case study is a non-traditional assembly plant. Two of the main differences, as compared to traditional plants, are the number of parallel assembly teams resulting in very long cycle times and the material feeding system which is a pure kitting system. The plant has been studied extensively and the layout and function of the plant is described in [19]. The main purpose of this section, however, is to describe some fault tolerance strategies that can be applied to the plant in order to improve productivity, throughput etc.

5.1 The case study

The objective of the system is to produce kits, which are plastic bags containing small size parts (e.g., screws, nuts, and plugs) to be used in the assembly department of a company. Each kit contains the small parts needed for a specific portion of the assembly of one specific object.

There are a large variety of kits (several hundred), each containing between 2 to 20 different parts, i.e., parts with different part numbers, and between 10 to 40 single parts. A typical value is 8 part numbers corresponding to 25 single parts. Kits are produced according to a fixed production program (on order), which means that each kit can be produced once a week. This results in small lot sizes with a maximum of 1000 kits per lot and a typical lot size of a few hundred. The total production volume is about four million kits/year.

The production of kits is divided into three main activities corresponding to physical areas in the system; automatic counting, manual counting and packaging (bagging).

Counting is done into sectioned trays (0.6 x 0.4 m) where each tray has 18 sections. The purpose of the trays is to hold the parts in a kit apart from each other until they reach the bagging equipment. One such tray never contains more than one type of kit. At the counting and bagging operations, piles of trays are loaded into the equipment, processed one by one and automatically stacked up again after the operation. Between these main activities, the piles of trays are buffered and handled by manual lifters.

The automatic counting is done by 18 identical vibratory bowl feeders that work in four so called counter-groups. There are three counter-groups with four feeders each and one group with six feeders. A feeder can only handle one part number at a time, which means that four or six part numbers per kit is the maximum number that can be counted at one time. For this reason many of the kits have to pass the counter-group more than once. One such run, as well as the manual counting and bagging, is called an operation. Normally, one kit is not processed in more than one counter-group.

The manual counting work area consists of two identical work stations at which the trays are exposed to the operator one by one. After the parts have been placed manually into a tray, the tray is piled up and buffered before the bagging operation.

Bagging is done by means of two identical bagging machines. In a special machine, the tray is turned and the kits are placed into a bucket elevator which leads to the bagging machine. A printer that is attached to the bagging machine prints the internal part number of the kit on the bag.

The quality of the kits, with regard to part numbers and quantities of each part number, is controlled by means of a check weigher connected to each bagging machine. The production system is shown in figure 7.

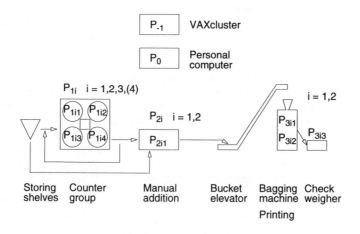

Figure 7: Layout of the production units in the production system

The small lot sizes and the large number of different kits means that measures have to be taken to keep set-up times low. This is done in two ways. Firstly, all the non-mechanical settings are done from a personal computer (PC). Secondly, the production planning for the kitting process makes use of the similarities between kits, forming sequences of operations that are similar, or identical, to each other.

The task of the PC system is to maintain a database containing the equipment settings with regard to part numbers (sub-parts and complete kits). Furthermore, manufacturing orders created in the central computer are downloaded to the PC, from which the operators control equipment settings.

5.2   Dynamic configuration

In this section a GRS description is given of the manufacturing system described in the previous section. GRS stands for General Recursive System and it is a useful tool for constructing abstract, hierarchical models of manufacturing systems into which different types of fault tolerance can be introduced [3], [4]. The various parts of the system are identified with so called GRS atomic entities and these are grouped together into entities at different levels to form a GRS hierarchy. The work that is going to be performed by the GRS is described in missions which in detail specifies the execution of each entity in the GRS.

The three different forms of possible system configuration for an intelligent manufacturing system [3], as we see it, are:

1. Dynamic configuration of the manufacturing system, in which units are moved to replace faulty units. This type of configuration is called Hardware Configuration. Hardware Configuration can be further divided into Physical Hardware Configuration and Logical Hardware Configuration.

   — The actual physical layout of a manufacturing system is important to describe suitable places to store information and possible redundant ways of information transportation. The graphical representation of a physical system is called a Physical Hardware Configuration, or PHC.

   — A graphical illustration of a GRS representation of a manufacturing system is called a Logical Hardware Configuration, or LHC. An LHC can be represented as a tree showing the recursive levels of a GRS. The leaves in an LHC tree represent atomic entities. An LHC can be changed by adding, moving or deleting any node.

2. Dynamic configuration of the missions, in which alternative production is prepared in the CAD/CAM process. This type of configuration is called Mission Configuration. A Mission Configuration can be represented by a tree, indicating how the missions must be divided into entity missions at each hierarchical level within the GRS. Each mission is associated with a unique tree; it is not possible to remove a sub-tree (i.e., an entity mission) and place it in another sub-tree. The missions are stored in a mission pool, MP. An MP can be considered to be an abstract data structure in which missions are inserted in order. An MP is used in such a way that missions are entered into it and entity missions are chosen from it [7].

3. Dynamic configuration of the work distribution, in which faulty units are avoided during work distribution and work is transferred to redundant units when a unit fails. This type of configuration is called Work Configuration. A Work Configuration can be viewed as "how to program" a GRS on a macro level. The program is achieved by associating each node in a Mission Configuration tree with a node in a Logical Hardware Configuration tree, i.e., the Mission Configuration tree is mapped on the Logical Hardware Configuration tree.

5.2.1 Hardware configuration

The first step is to find a Logical Hardware Configuration of the system where the different parts of the Physical Hardware Configuration, called production units and labeled P, are identified with the leaves and nodes in a Logical Hardware Configuration, or GRS, called entities and labeled G.

An example of a PHC of the case study is shown in figure 8. However, only the physical data network and the manual data transportation are shown; the material network has been left out. The nodes in the PHC-graph represent production units, and the lines represent physical data networks or manual data transportation.

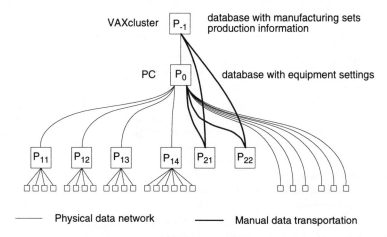

Figure 8: The Physical Hardware Configuration describing production units with storage capacity and paths of information distribution

The LHC of the production system is illustrated in figure 9. The LHC only considers the hierarchical structure of the GRS-abstraction of the system.

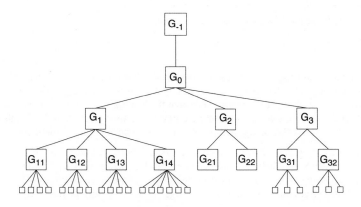

Figure 9: The Logical Hardware Configuration describing the logical dependencies

A description of the different nodes in figure 9 can be found in table 4. From now on we will only write Hardware Configuration when we mean Logical Hardware Configuration except when it is not clear from the text what we mean.

Each vibratory bowl feeder P_{1ij} are identified with a leaf (atomic entity) in the GRS hierarchy. These are designated G_{1ij} where i designates the counter-group that the feeder belongs to and j gives a unique number to the feeders in each counter-group. The counter-groups P_{1i} are designated G_{1i}. The first three of these (P_{11}, P_{12}, P_{13}) consist of four feeders and are designated G_{11}, G_{12}, G_{13}, while the fourth (P_{14}) consists of six feeders, and is designated G_{14}. The four counter-groups are in turn grouped together to form an automatic counting area, G_1. However, there exists no physical equivalence.

Table 4:   Functions for the GRS entities

G_{111} :	automatic counting	G_{112} :	automatic counting	G_{113} :	automatic counting
G_{114} :	automatic counting	G_{121} :	automatic counting	G_{122} :	automatic counting
G_{123} :	automatic counting	G_{124} :	automatic counting	G_{131} :	automatic counting
G_{132} :	automatic counting	G_{133} :	automatic counting	G_{134} :	automatic counting
G_{141} :	automatic counting	G_{142} :	automatic counting	G_{143} :	automatic counting
G_{144} :	automatic counting	G_{145} :	automatic counting	G_{146} :	automatic counting

G_{11} :   counter group (4) with automatic internal transport
G_{12} :   counter group (4) with automatic internal transport
G_{13} :   counter group (4) with automatic internal transport
G_{14} :   counter group (6) with automatic internal transport
G_1 :   4 counter groups with manual internal transport
G_{21} :   manual counting G_{22} :   manual counting
G_2 :   2 manual counting stations with manual internal transport
G_{311} :   text printing G_{312} :   bagging G_{313} :   check weighing
G_{321} :   text printing G_{322} :   bagging G_{323} :   check weighing
G_{31} :   one bagging station with automatic internal transport
G_{32} :   one bagging station with automatic internal transport
G_3 :   2 bagging stations with manual internal transport
G_0 :   production system with internal transport (using a PC for the system control)
G_{-1} :   central control using a VAXcluster for the database
            (the VAXcluster is shared with the rest of the production plant

The manual counting area consists of two work stations P_{21} and P_{22}, which are designated G_{21} and G_{22}. Together they form the manual counting area, G_2, which have no physical equivalence. Note that the atomic entities at the manual counting area are situated one level closer to the root than the automatic counting. This reflects that the location of the different parts within G_{21} or G_{22} is not modeled while performing manual counting.

For the bagging part of the manufacturing system three types of atomic entities are identified, i.e., printing (P_{3i1} designated G_{3i1}), bagging (P_{3i2} designated G_{3i2}) and check weighing (P_{3i3} designated G_{3i3}). The work is done at two stations (G_{31} and G_{32}) which together form the bagging area, G_3. Neither G_{31} and G_{32} nor G_3 have any physical equivalences.

The three areas, together with the PC, form the production system P_0 designated G_0. At the highest level, P_{-1} designated G_{-1}, there is a computer center with a large database.

The transportation of material is done automatically at the $P_{k,i}$ level, while the transportation is done manually at the P_k level.

5.2.2   Hardware reconfiguration

In this case study Hardware Reconfiguration will be the replacement of parts from one entity to another in order to repair a defect entity. An example is the replacement of a faulty feeder in one counter-group with a correct feeder from another partly faulty counter-group. In this way it is possible to have at least one working counter-group instead of none. However, this type of repair is unlikely because of the amount of time and work required compared to the benefits from the use of other solutions.

Other Hardware Reconfigurations to consider are the different faulty states. When an entity fails, and subsequently leaves the configuration of working entities, this is in fact a Hardware Reconfiguration, albeit an undesirable one.

To demonstrate an example of a possible Hardware Reconfiguration a case in which all 4-groups of counters are defunct while the 6-group counter is still working is chosen. This is illustrated in figure 10 with the resulting Hardware Reconfiguration $\overline{G}$.

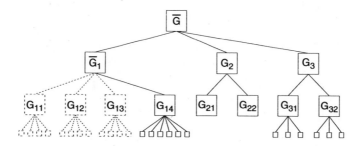

Figure 10: A Hardware Reconfiguration

5.2.3 Mission configuration

A real system can be exemplified with some simple missions labeled A, B and C as in figure 11. The entity missions are more closely described in table 5.

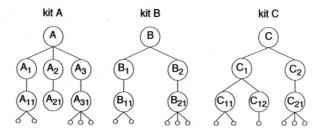

Figure 11: Mission Configurations

Mission A describes how to assemble a kit with three different pieces. From the structure it is clear which part number is to be counted manually (A_{21}). The kit will then be bagged which is preceded by text printing on the bag and followed by check weighing. These precedences are given in figure 12.

Mission B is similar to A but it has no manual counting. Mission C has no manual counting either, but the automated counting is divided into two steps since some of the parts are the same as in mission A and subsequently can be optimized to be counted together on the same setup.

The requirements are indicated in each node in each Mission Configuration tree. On the lowest level, M_{kij}, the identity of the part to be counted is given. From this it is possible to extract all the necessary requirements about weight, size and so on from the database. On the next level, M_{ki}, how the counting should be grouped together is given. On the next

Table 5: Entity missions

A	≡	kit #1299				
A_{111}	≡	5 of part #22746		A_{112}	≡	3 of part #4657
A_{21}	≡	1 of part #132 (manually)				
A_{311}	≡	text printing		A_{312}	≡	bagging
A_{313}	≡	check weighing				
B	≡	kit #999				
B_{111}	≡	2 of part #27321		B_{112}	≡	6 of part #8102
B_{211}	≡	text printing		B_{212}	≡	bagging
B_{213}	≡	check weighing				
C	≡	kit #3422				
C_{111}	≡	5 of part #22746		C_{112}	≡	3 of part #4657
C_{113}	≡	7 of part #11345				
C_{211}	≡	text printing		C_{212}	≡	bagging

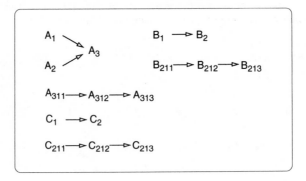

Figure 12: Mission precedences

highest level, M_k, the requirements for the mode of counting, i.e., whether the counting should be automatic, manual, or if it does not make a difference is indicated. On the highest level, M, the requirements about material transportation between the areas are given.

On each level the requirements also include settings and programs that may possibly be needed during production. Indications about time limits for different phases of production are also included in the requirements.

5.2.4 Mission reconfiguration

One way to achieve fault tolerance according to the GRS model is to prepare for an alternative operation order (structure) for a mission that can be used when it is not possible to continue using the given structure. This is called a Mission Reconfiguration. For each of the three missions A, B and C are identified Mission Reconfigurations labeled $\overline{A}$, $\overline{B}$ and $\overline{C}$. The production problem that is taken care of is a situation where all automated counting is defunct. Subsequently is described the production of the three kits with only manual counting. The alternative mission structures are given in figure 13.

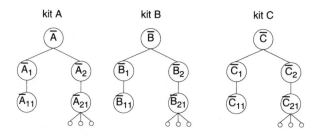

Figure 13: Alternative Mission Configurations

The corresponding mission precedences are given in figure 14 and the entity missions are described in table 6.

$$\overline{A}_1 \longrightarrow \overline{A}_2 \qquad\qquad \overline{B}_1 \longrightarrow \overline{B}_2$$
$$\overline{A}_{211} \longrightarrow \overline{A}_{212} \longrightarrow \overline{A}_{213} \qquad \overline{B}_{211} \longrightarrow \overline{B}_{212} \longrightarrow \overline{B}_{213}$$
$$\overline{C}_1 \longrightarrow \overline{C}_2 \qquad\qquad \overline{C}_{211} \longrightarrow \overline{C}_{212} \longrightarrow \overline{C}_{213}$$

Figure 14: Mission precedences

Table 6: Alternative entity missions

$\overline{A}$	≡	kit #1299			
$\overline{A}_{11}$	≡	5 of part #22746, 3 of part #4657, 1 of part #132 (manually)			
$\overline{A}_{211}$	≡	text printing	$\overline{A}_{212}$	≡	bagging
$\overline{A}_{213}$	≡	check weighing			
$\overline{B}$	≡	kit #999			
$\overline{B}_{11}$	≡	2 of part #27321, 6 of part #8102 (manually)			
$\overline{B}_{211}$	≡	text printing	$\overline{B}_{212}$	≡	bagging
$\overline{B}_{213}$	≡	check weighing			
$\overline{C}$	≡	kit #3422			
$\overline{C}_{11}$	≡	5 of part #22746, 3 of part #4657, 7 of part #11345 (manually)			
$\overline{C}_{211}$	≡	text printing	$\overline{C}_{212}$	≡	bagging
$\overline{C}_{213}$	≡	check weighing			

5.2.5   Work configuration

In this section is given an example of a possible Work Configuration. A configuration on how to produce a kit of type A (as in table 5) is given in figure 15. This implies a Work Configuration where the entity missions are associated to corresponding GRS entities.

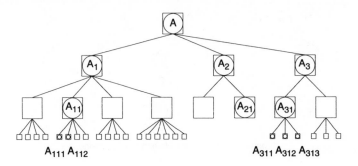

Figure 15:   A Work Configuration

When performing Work Configuration the requirements at each node in the Mission Configuration must be matched by the operations offered at the corresponding nodes in the Hardware Configuration. This can be done according to two different principles. Which of the principles depends on the type of manufacturing system that the model is aimed at [3]. In addition there are variants of the two principles.

The first principle does not take into account the tree structure and consequently only matches operations on the leaf level. It is called *unstructured matching*. This principle does not necessarily take replicated entities into account. Thus, redundant entities are not chosen when performing a reconfiguration. The principle might lead to the selection of one entity for more than one task, i.e., time sharing of the entities might occur. This is the desired approach in some types of manufacturing system, especially those that deal with many different missions in a short period of time.

The second principle is based on knowledge of the structure of a complete sub-tree at each level in the Work Configuration. It is called *structured matching*. This principle makes it possible to optimize the configuration in advance, which is an important consideration when the configuration is to be used over a long period of time.

First Principle

Using the first principle for matching nodes in performing a Work Configuration will provide a simple configuration algorithm. When a given Hardware Configuration sub-tree is capable of performing the production according to a given Mission Configuration sub-tree, it is sufficient to check that the required operations of a mission can be satisfied by the Hardware Configuration sub-tree.

We define the offered services of a GRS, O_{GRS}, and the required services of a mission, R_M, in the following way:

$$GRS \equiv \{e_1, e_2, ..., e_f\}$$

$$M \equiv \{m_1, m_2, ..., m_n\}$$

$$O_{GRS} = \bigcup_{e_i \in GRS} O_{e_i}$$

$$O_{AE} = \{\text{offered operations}\}$$

$$R_M = \bigcup_{m_i \in M} R_{m_i}$$

R_{AM} = {required operations}

A mission can be fulfilled by a GRS if

$R_M \subseteq O_{GRS}$

A GRS, G, that fulfills a mission, M's, required services is, for example:

$R_M = \{a, b, c, d\}$

$O_G = \{a, b, c, d, e\}$

Second Principle

The second principle takes into account the structure of the offered services. To do so, a new operator for comparing required services with offered services must be defined. The offered services of a GRS, O_{GRS}, and the required services of a mission, R_M, are defined in the following way:

GRS $\equiv \{e_1, e_2,..., e_f\}$

M $\equiv \{m_1, m_2,..., m_n\}$

$O_{GRS} = \{ O_{e_i} \mid e_i \in GRS\}$

$R_M = \{ R_{m_i} \mid m_i \in M\}$

The symbol $\sqsubseteq$ is used for the new operator, defined in the following way:

$R_{AM} \sqsubseteq O_{AE}$ iff $R_{AM} \subseteq O_{AE}$

$R_M \sqsubseteq O_{GRS}$ iff $\exists i_1, i_2,..., i_n: R_{m_1} \sqsubseteq O_{e_{i_1}} \wedge R_{m_2} \sqsubseteq O_{e_{i_2}} \wedge ... \wedge R_{m_n} \sqsubseteq O_{e_{i_n}}$,

$\qquad i_1 \neq i_2 \neq ... \neq i_n \wedge e_{i_j} \in GRS, j = 1..n$

A GRS, G, that fulfills a mission, M's, requirements is, for example:

$R_M = \{\{\{a\},\{b\}\},\{\{a\}\},\{\{c\},\{d\}\}\}$

$O_G = \{\{\{a\},\{b\},\{b\}\},\{\{a\},\{b\}\},\{\{c\},\{d\}\}\}$

In this example $R_M \sqsubseteq O_G$. An example for which it is **not** true that a GRS fulfills the required services owing to the structure is

$R_M = \{\{\{a\},\{b\}\},\{\{a\}\},\{\{c\},\{d\}\}\}$

$O_G = \{\{\{a\},\{b\},\{b\}\},\{\{a\},\{b\},\{c\}\},\{\{d\}\}\}$

In [7] a basic algorithm is given for obtaining Work Configuration. This algorithm uses the data structure called mission pool. Before the missions can be distributed, the structure of the Hardware Configuration must be collected. It must be known what each sub-tree (entity) can perform in order for missions to be assigned to them.

Because of the combinatorial explosion in computing an optimal configuration, such a computation cannot be done completely automatically. Instead, the possibility must exist for an operator to interact with the system. This means that an interface between the human operator and the system must be included, e.g., by using a scheduling editor. Two examples are the editors developed at the Mitsubishi Electric Corporation in Japan [13] and the Imperial College in London [16].

In order to have a flexible and fault tolerant system two different algorithms for Work Configuration are used. The first algorithm, ConfigureA (figure 16), uses *structured match-*

ing. This optimizes production in advance. The Mission Configuration tree structure must follow the Hardware Configuration tree structure.

```
ConfigureA(G:Hardware Configuration; M:Mission Configuration);
begin
    for all entity missions (em) in M do
        begin
            E := {eᵢ | eᵢ ∈ G ∧ Rₑₘ ⊑ Oₑᵢ };
            e := choose(E, ... );
            ConfigureA(e, em);
        end;
end;

procedure choose(E, em, ... );
begin
    if lower requirements of em already in earlier chosen eᵢ then
        choose := eᵢ;
    else if free eᵢ choose := first free eᵢ;
    else choose any eᵢ;
end;
```

Figure 16: Algorithm ConfigureA

The second algorithm, ConfigureB (figure 17), uses *structured matching* as long as it is feasible but it also allows for the use of *unstructured matching* when *structured matching* is no longer feasible. Both algorithms have an extra condition which says that identical entity missions should be distributed to the same entity. This is an application condition.

A manufacturing order contains, except from the settings above, all necessary production information for a corresponding "manufacturing set". One could say that a manufacturing order is created when the orders on hand for the kits in the manufacturing set are added to the set. Such a manufacturing set (identified by a name and a number) is built up of one or several (*no* restrictions) complete kits divided into operations (identified by an operation number), where one operation refers only to a part of one of the kits in the set. For example, an operation can be the automatic counting of four different part numbers in a kit in one of the counter-groups, the manual adding of parts that are not suitable for automatic counting, or the bagging of one of the complete kits in the set. A manufacturing set is equivalent to a Work Configuration.

The manufacturing set (Work Configuration) contains the operations, the type of each operation, the equipment to use (e.g. automatic or manual counting), the part numbers involved in the operation (sub-parts for the counting and kit part number for the bagging operations) and the amount of each part number.

Now the evaluation of a Work Configuration for a specific set of kits is shown. For the normal, "default", Work Configuration algorithm ConfigureA is used. The symbol ⇨ is used as a priority ordering within the mission pool. It indicates the sequence in which the missions are inserted without requiring that this sequence be followed. It is only a sequence of fairness.

```
ConfigureB(G:Hardware Configuration; M:Mission Configuration);
begin
  for all entity missions (em) in M do
    begin
      E := {ei | ei ∈ G ∧ Rem ⊏ Oei };
      if E ≠ ∅ then
        begin
          e := choose(E, ... );
          ConfigureB(e, em);
        end;
      else
        begin
          E := {ei | ei ∈ G ∧ Rem ⊆ Oei };
          if E ≠ ∅ then
            begin
              e := choose(E, ... );
              ConfigureB(e, em);
            end;
          else configuration impossible;
        end;
    end;
end;
```

Figure 17: Algorithm ConfigureB

As an example is chosen the following manufacturing sequence:

100 A ➪ 500 B ➪ 50 C

One hundred kits of type A, 500 kits of type B and 50 kits of type C should be made. The sequence is just a priority sequence and the system is allowed to alter this sequence if it is convenient. To evaluate the Work Configuration, missions are inserted into the mission pool called G. Subsequently 100 A, 500 B and 50 C missions are inserted into G's mission pool. The missions A, B and C are the same as described earlier. The mission pool will then have the following sequence:

MP_G : A ➪ A ➪ ... ➪ A ➪ B ➪ B ➪ ... ➪

➪ B ➪ C ➪ C ➪ ... ➪ C

These missions are then expanded to show their sub-missions in order to do the configuration:

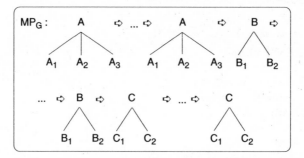

Now the sub-missions are distributed among the entities of G according to algorithm ConfigureA. If G_1 is operating correctly this distribution can only be done in one way. For all automatic counting G_1 is chosen, for manual counting G_2 is chosen and for bagging G_3 is chosen. This implies that the sub-missions are inserted into the three subsequent mission pools as follows. Into G_1's mission pool the automatic part of the counting of A, B and C; all A_1's, all B_1's and all C_1's are inserted. Into G_2's mission pool the manual part of the counting of A, B and C; all A_2's are inserted. Into G_3's mission pool the bagging part of A, B and C; all A_3's, all B_2's and all C_2's are inserted.

This distribution follows the *structured matching* principle since algorithm ConfigureA is used. If G_1 is not operable, a Work Configuration using ConfigureA is not achieved.

After performing ConfigureA at each hierarchical level, the following Work Configuration at the lowest (atomic) level is obtained.

MP_{G111} : $(100\ A_{111} \Rightarrow 50\ C_{111}) \Rightarrow 50\ C_{121}$

MP_{G112} : $(100\ A_{112} \Rightarrow 50\ C_{112})$

MP_{G121} : $500\ B_{111}$

MP_{G122} : $500\ B_{112}$

MP_{G22} : $100\ A_{21}$

MP_{G311} : $100\ A_{311} \Rightarrow 50\ C_{211}$ MP_{G321} : $500\ B_{211}$

MP_{G312} : $100\ A_{312} \Rightarrow 50\ C_{212}$ MP_{G322} : $500\ B_{212}$

MP_{G313} : $100\ A_{313} \Rightarrow 50\ C_{213}$ MP_{G323} : $500\ B_{213}$

5.2.6   Graceful degradation

To achieve necessary availability in production, graceful degradation is introduced. Some parts of a mission may be compulsory while others may only be desirable. In our case study it is essential that the parts be sorted together and put in plastic bags while check weighing and text printing are dispensable. A failure in either step should not be enough to halt production entirely. In case that there is no possibility for printing or weighing this will be neglected by the system. In order to perform a Work Configuration in this case it is necessary to have some information included in the requirements concerning the compulsion of a mission. In its simplest form there are only two values, *required* and *desired* (table 7).

Table 7: Entity missions

A	$\equiv$	kit #1299, R				
A_{111}	$\equiv$	5 of part #22746, R		A_{112}	$\equiv$	3 of part #4657, R
A_{21}	$\equiv$	1 of part #132 (manually), R				
A_{311}	$\equiv$	text printing, D		A_{312}	$\equiv$	bagging, R
A_{313}	$\equiv$	check weighing, D				
B	$\equiv$	kit #999, R				
B_{111}	$\equiv$	2 of part #27321, R		B_{112}	$\equiv$	6 of part #8102, R
B_{211}	$\equiv$	text printing, D		B_{212}	$\equiv$	bagging, R
B_{213}	$\equiv$	check weighing, D				
C	$\equiv$	kit #3422, R				
C_{111}	$\equiv$	5 of part #22746, R		C_{112}	$\equiv$	3 of part #4657, R
C_{113}	$\equiv$	7 of part #11345, R				
C_{211}	$\equiv$	text printing, D		C_{212}	$\equiv$	bagging, R
C_{213}	$\equiv$	check weighing, D				
R	:	required		D	:	desired

Graceful degradation is obtained with an algorithm called ConfigureC that takes into account the values included in the production allocation requirements (see figure 18).

5.2.7 *Fault tolerance through different distribution principles*

To achieve fault tolerant production one can choose between different methods, such as different distribution principles, Mission Reconfiguration or Work Reconfiguration.

In this section is described how fault tolerance can be achieved in the production process using different distribution principles. How to obtain fault tolerance through Mission Reconfiguration and Work Reconfiguration are described in the following two sections. It is assumed that the automatic counting area of the manufacturing system is out of order. This means that the Hardware Configuration, called $\underline{G}$, lacks G_1. If the principle of *structured matching* were to be used, it would have been impossible to find a configuration for the missions. If the *unstructured matching* distribution principle is used instead it is still possible to find a configuration by doing all counting manually. As a result, algorithm

ConfigureC(G:Hardware Configuration; M:Mission Configuration)
begin
    for all entity missions (em) **in** M **do**
      begin
        $E := \{e_i \mid e_i \in G \wedge R_{em} \sqsubseteq O_{e_i}\}$;

        if $E \neq \emptyset$ **then**
          begin
            e := choose(E, ...);
            ConfigureC(e, em);
          end;
        else
          begin
            $E := \{e_i \mid e_i \in G \wedge R_{em,min} \sqsubseteq O_{e_i}\}$;

            if $E \neq \emptyset$ **then**
              begin
                e := choose(E, ...);
                ConfigureC(e, em);
              end;
            else configuration impossible;
          end;
      end;
end;

Figure 18: Algorithm ConfigureC

ConfigureB is used and since G_1 is missing the following mission pools are obtained for G_2 and G_3:

$$MP_{G2}: A_1 \Rightarrow ... \Rightarrow A_1 \Rightarrow A_2 \Rightarrow ... \Rightarrow A_2 \Rightarrow B_1 \Rightarrow$$
$$... \Rightarrow B_1 \Rightarrow C_1 \Rightarrow ... \Rightarrow C_1$$
$$MP_{G3}: A_3 \Rightarrow ... \Rightarrow A_3 \Rightarrow B_2 \Rightarrow ... \Rightarrow B_2 \Rightarrow C_2 \Rightarrow$$
$$... \Rightarrow C_2$$

For the G_3 sub-tree there will be no difference from the former example. To fulfill the distribution for the G_2 sub-tree the missions in its mission pool are expanded:

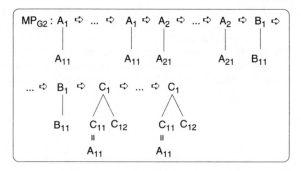

The *unstructured matching* distribution principle has to be followed for the next distribution too, to get a configuration. It is possible to chose between the two manual counting areas.

How the counting is performed at the manual counting area is not given here since each manual counting area is regarded as an atomic entity. The counting is scheduled manually within the area. The following production is obtained:

$$MP_{G21} : 100 \ A_{11} \Rightarrow 100 \ A_{21}$$
$$MP_{G22} : 500 \ B_{11} \Rightarrow 50 \ C_{11} \Rightarrow 50 \ C_{12}$$

5.2.8   Fault tolerance through mission reconfiguration

Another way to achieve fault tolerance is through Mission Reconfiguration instead. This method is used when it is impossible to get a functioning Work Configuration with the original Mission Configuration. Since it is a cumbersome process to create a Mission Configuration it is assumed that it has been created in advance and not after discovering that a Mission Reconfiguration is needed. To get a Work Configuration for the Hardware Configuration $\underline{G}$, the alternative Mission Configurations that were described in table 6 are used. $\underline{G}$ is the Logical Hardware Configuration when all automatic counting is defunct in G, (i.e., the G_1 sub-tree is missing in figure 9). The Work Configuration algorithm ConfigureA can now be used. However, there still exists the possibility to use algorithm ConfigureB as a backup.

As an example is used the same production as in the first example:

$$100 \ \overline{A} \Rightarrow 500 \ \overline{B} \Rightarrow 50 \ \overline{C}$$

The difference is the alternative missions $\overline{A}$, $\overline{B}$ and $\overline{C}$ instead of A, B and C for the three different kits. The contents in the root mission pool is:

$$MP_{\underline{G}} : A \Rightarrow A \Rightarrow ... \Rightarrow A \Rightarrow B \Rightarrow B \Rightarrow ... \Rightarrow$$
$$\Rightarrow B \Rightarrow C \Rightarrow C \Rightarrow ... \Rightarrow C$$

The missions are then expanded and at the atomic level is obtained the following production for manual counting:

$$MP_{G21} : 100 \ \overline{A}_{11} \Rightarrow 50 \ \overline{C}_{11} \qquad MP_{G22} : 500 \ \overline{B}_{11}$$

Then the Mission Reconfiguration provides a Work Configuration for $\overline{A}$ as shown in figure 19. The same can be done for $\overline{B}$ and $\overline{C}$ as well.

5.2.9   Fault tolerance through work reconfiguration

Fault tolerance can be achieved using algorithm ConfigureA and the primary Mission Configurations A, B and C. This is possible if, for instance, all three 4-counter-groups are defunct but the 6-counter-group still is working. This implies the Hardware Configuration $\overline{G}$ that was given in figure 10.

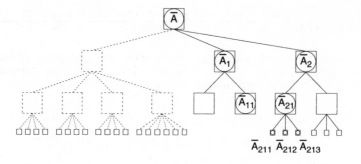

Figure 19: A Work Configuration for A on $\underline{G}$

The same work order as in the first example in section 5.2.5 is used. $\overline{G}$'s mission pool will then have the following sequence:

$$MP_{\overline{G}}: A \Rightarrow A \Rightarrow ... \Rightarrow A \Rightarrow B \Rightarrow B \Rightarrow ... \Rightarrow$$
$$\Rightarrow B \Rightarrow C \Rightarrow C \Rightarrow ... \Rightarrow C$$

These missions are then expanded to their sub-missions in order to do the configuration and the following Work Configuration is finally obtained for the manufacturing set:

$$MP_{G141}: (100\,A_{111} \Rightarrow 50\,C_{111}) \Rightarrow 500\,B_{111} \Rightarrow 50\,C_{121}$$
$$MP_{G142}: (100\,A_{112} \Rightarrow 50\,C_{112}) \Rightarrow 500\,B_{112}$$

This gives a production that is unnecessarily complicated since it takes two rounds to make C. If instead an additional rule is included in the distribution algorithm, that missions on a higher level in the Mission Configuration tree have priority over lower levels, the termination of C will overtake B. That gives the following mission pools for G_{141}, G_{142}, and G_{143}:

$$MP_{G141}: 100\ A_{111} \Rightarrow 50\ C_{111} \Rightarrow 500\ B_{111}$$
$$MP_{G142}: 100\ A_{112} \Rightarrow 50\ C_{112} \Rightarrow 500\ B_{112}$$
$$MP_{G143}: \qquad\qquad 50\ C_{121}$$

The synchronization rules for the counter-group force the following precedence rules on the production:

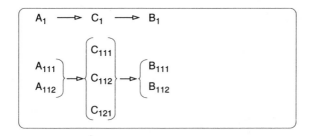

The resulting Work Reconfiguration for A is shown in figure 20. Similar results can be obtained for B and C.

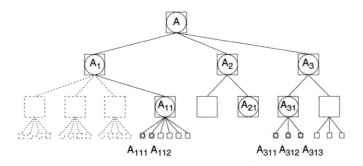

Figure 20: A Work Reconfiguration

5.2.10   *Fault tolerance through graceful degradation*

To achieve production also under circumstances with difficulties it is allowed to skip some of the operations in the missions. In this case study we reflect that the check weighing and the text printing are not regarded as indispensable as to stop production entirely. In case that there is no possibility for printing or weighing this will be neglected by the system.

In order to perform Work Reconfiguration in this case we need to have weights on the requirements. In the simplest case, which we will follow, there are only two values on the weights: required or desired. We use the same missions as in figure11 but will need additional information regarding the requirements as in table 7.

318

In the following example we assume a Logical Hardware Configuration as in figure 21, i.e., without printing and weighing.

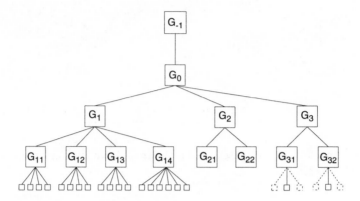

Figure 21: The Logical Hardware Configuration without printing and weighing

We also need a new Work Configuration algorithm which uses the weights on the requirements. The new algorithm, ConfigureC, was given in figure 18.

We use the same production as in the original example, i.e.:

$$100\ A\ \Rightarrow\quad 500\ B\ \Rightarrow\quad 50\ C$$

Since the counting part of the Logical Hardware Configuration is the same as in the original example, the solution will be identical and we will not show it here. We will instead show the bagging part G_3. Its mission pool, MP_{G3}, will have the following content:

$$MP_{G3} : A_3 \Rightarrow A_3 \Rightarrow \dots \Rightarrow A_3 \Rightarrow B_2 \Rightarrow B_2 \Rightarrow \dots \Rightarrow$$
$$\Rightarrow B_2 \Rightarrow C_2 \Rightarrow C_2 \Rightarrow \dots \Rightarrow C_2$$

which expanded becomes:

$$MP_{G3} : A_3 \Rightarrow \dots \Rightarrow A_3 \Rightarrow B_2 \Rightarrow \dots \Rightarrow B_2 \Rightarrow$$
$$\qquad\quad A_{31} \qquad\quad A_{31} \quad B_{21} \qquad\quad B_{21}$$
$$\Rightarrow C_2 \Rightarrow \dots \Rightarrow C_2$$
$$\qquad C_{21} \qquad\quad C_{21}$$

The corresponding entity missions are distributed among the two bagging units as:

$$MP_{G31} : A_{31} \Rightarrow ... \Rightarrow A_{31} \Rightarrow C_{21} \Rightarrow ... \Rightarrow C_{21}$$

$$MP_{G32} : B_{21} \Rightarrow ... \Rightarrow B_{21}$$

and is expanded to:

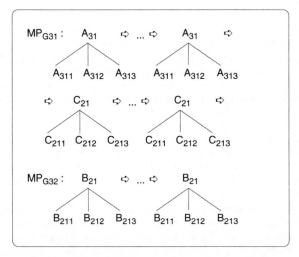

Since text printing and check weighing is missing in the Logical Hardware Configuration for both bagging stations we have to take away the entity missions which are only desired. Subsequently we get the following situation:

$$MP_{G31} : A_{31} \Rightarrow ... \Rightarrow A_{31} \Rightarrow C_{21} \Rightarrow ... \Rightarrow C_{21}$$
$$A_{312} \qquad A_{312} \quad C_{212} \qquad C_{212}$$

$$MP_{G32} : B_{21} \Rightarrow ... \Rightarrow B_{21}$$
$$B_{212} \qquad B_{212}$$

The entity missions are distributed which give:

$$MP_{G312} : A_{312} \Rightarrow ... \Rightarrow A_{312} \Rightarrow C_{212} \Rightarrow ... \Rightarrow C_{212}$$

$$MP_{G322} : B_{212} \Rightarrow ... \Rightarrow B_{212}$$

and the following production at the bagging stations:

$$MP_{G312}: \ 100 \ A_{311} \Rightarrow 50 \ C_{211}$$

$$MP_{G322}: \ 500 \ B_{212}$$

5.3   Information accessibility

Communication lines are a highly unreliable part of a distributed computer system, partly because they are so numerous and partly because of their poor reliability [14]. The procedures of managing communication lines, diagnosing failures and tracking the repair process can be very cumbersome. Nevertheless, it is paramount to have a system that can function in spite of poor behavior of communication lines. One way to achieve this is to have some kind of redundancy for the communication. On the hardware level, this can be obtained by having multiple data paths with independent failure modes. Thus, the system designer must decide on a *redundant communications network topology* as described in [12]. This topology must ensure a degree of service in compliance with some specified, minimum threshold for communication between different production units, despite communication component failures. Some systems, for example, have adopted dual busses to solve this problem [10, 35]. Another example is provided by DEC which has adopted a dual star coupler for their VAXclusters [21].

A concept called *quasi-partitioning* for redundant communication networks was introduced in [24]. The idea is to have a redundant backup data-link with limited capacity that can be used to maintain data traffic at a reduced level when the normal data-link has been broken. Owing to the limited capacity of the backup data-link, only a relatively small amount of data can be transmitted across it, and the result is a system that has (quantitative) graceful degradation. However, this solution can cause problems to the system when different nodes with replicated data are merged after the repair of the normal data network.

While the idea of having a back-up data-link is interesting, it also has some disadvantages, e.g., the cost for the redundant data-link. This cost can be mitigated in an automated manufacturing system that has two transportation media from the start, namely, the data network and the material network. The limited extra cost that is related to this approach is then the conditioning of the material network so that it may function as a redundant data network. Our approach is thus to obtain the advantages that were pointed out in Lilien's article, without suffering some of its negative effects, by using the material network as a redundant data network.

The problem of system breakdowns caused by partitioned data networks can be difficult to handle in distributed systems. Negative effects of a network partitioning must be avoided, or at least reduced to an acceptable level. There are different ways to deal with the problem of maintaining production during data network partitionings. One approach is to use data buffers that are filled with requisite data which can be used when data network partitioning occurs. A second approach is to use AI technology to detect erroneous situations and present possible solutions to an operator, e.g., in [2], or to take advantage of previous production experience in order to find the most useful method of production. A third approach is to use redundant information links. Whichever of the three approaches is used, or whether combinations of them are used, they should be considered when designing a manufacturing system. In other words, to achieve a fault tolerant manufacturing system, it is necessary to include any fault tolerance strategy at the design stage of a system. The choice of which of the three approaches should be used depends on the application and the degree of fault tolerance required. For instance, if we use the material network to transport data, the information will be available only at certain discrete moments in time. But if it is sufficient for the information to arrive together with the material, the material network will be equally as acceptable as the data network. Another example can be illustrated by a person walking to a production unit to enter some data. The cost of that person must be considered at the design stage of the fault tolerance strategy, as must the fact that the risk of introducing errors caused by the person increases. In this paper, however, we will not deal with aspects of this kind.

The primary task of a material network is to transport material. However, a material network can also be used for fault detection or to transport information [5], which is illustrated in figure 22. It is possible to register the information via the material network in several ways, for example: observation of physical material arrival, optical recognition, barcoding and EPROMs. Some of these aspects have been described in [36].

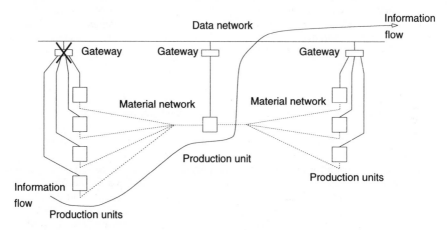

Figure 22: Information transportation via the material network

Including the material network as a medium for information transportation gives three possibilities by which the information can be transported among production units:

1. the production information is passed manually,

2. the production information is passed via the data network, or

3. the production information is passed via the material network.

The first possibility is still common and is relevant in most semi-automated manufacturing systems. The second and third possibilities can be found in many currently computerized installations. Between the three possibilities exists a range of hybrids in which some information is passed in one way while other information is passed in another.

In order to produce without any interruptions, *all* the necessary production information must be present at, or retrievable by, the corresponding production unit.

When writing *all the necessary information*, denoted $I_{n,all}$, it means:

the information that a specific production unit (e.g., a device or a manufacturing cell) needs in order to fulfill a task n *(that specifies what the production unit shall produce). Furthermore, the information has to be up-to-date.*

If a manufacturing system consists of a data network, a material network, manual transportation and data caches, the following situation exists; the information $I_{n,all}$ can be retrieved via one of the four media only, or redundantly via two or more of the four media.

The latter case is called a *fault tolerance strategy*. This means that the system will be able to deal with some types of failures, such as data network partitions and delays in the material network [5]. Furthermore, it is also necessary that a system actually retrieves $I_{n,all}$ for an adequate task n. This is necessary in order for the manufacturing system to be able to produce according to its specifications. The result, whether the manufacturing system can produce or not, is called *production liveness*.

It must be pointed out that every fault tolerance strategy is unique for every production of a specific product at a specific production unit. Also, every production liveness is unique for every production of a specific product at a specific production unit at a specific time. The fault tolerance strategy specifies *what piece of information should be sent in a redundant form, or is already present in a cache*. The production liveness describes *a minimum piece of information that is needed at a specific moment in order to maintain production*. The fault tolerance strategy is defined for a specific node in the tree structure that constitutes a hierarchical manufacturing system. Thus, in the most extreme case the fault tolerance strategy can be different at each node in the tree. A complete fault tolerance strategy for an entire manufacturing system can be obtained by joining the strategies from all nodes. The complete definition of fault tolerance strategy and production liveness can be found in [6].

5.3.1 Information retrieval

In most modern manufacturing systems information is transmitted in heterogeneous distributed systems [11]. This is one of the reasons for introducing MAP/TOP in manufacturing systems [25]. The heterogeneity aspect is also valid in the production system in the case study.

The main source of information in the case study system is a relational database using shadowing techniques, located on a VAXcluster, i.e., P_{-1}, that is placed outside the production area. There is also a local database on a PC, i.e., P_0, in the production area. In addition to this, most of the low level equipment have secondary memories but these are, however, not used as caches today. Equipment settings for the production equipment are saved on lists when values are transmitted from the VAXcluster to the PC.

Each entity in the Logical Hardware Configuration has a requirement regarding the accessibility of information. The possible physical placement of information is given by the Physical Hardware Configuration. How this placement could be done is ruled by the different physical storage capacities in the production units and the routes of information between different production units.

The different production units that have storage capacities were illustrated in figure 8. In the current production system information about one week's production is present at P_{-1}. Twice a day the information about a half day's work is loaded into P_0. This information is called a manufacturing set. One manufacturing set consists of all the necessary operations to produce one, or several, complete kits that must be produced in a sequence because of group technology reasons. For each operation the information about the equipment settings is loaded into the system from P_0 except the information to P_{21} and P_{22} which is distributed as paper listings from P_{-1}.

5.3.2 Reliability improvement

Two ways to improve the accessibility of information in a manufacturing system is to use redundant information carriers [5] and to use data caches [3]. This improves information reliability and will therefore improve the reliability of the entire system.

From an abstract point of view it is of no interest how information is transported in a manufacturing system, i.e., whether it is done via a data network, material network or through manual data transportation. The abstract network is known as an *abstract information network* [5]. How the physical transportation is done is defined in the physical information network. The information network is a useful abstraction when discussing transportation of information in intelligent manufacturing systems.

The physical information network uses a *physical data network*, a *physical material network* and *manual data transportation* to transport data between different production units. For example, the physical information network in the case study looks like:

1. the physical data network consists of Ethernets and current loops (used today),

2. the physical material network consists of roller conveyors and special handling devices at the automatic $P_{k,i}$ level (EPROMs can be introduced),

3. the manual data transportation consists of lists (used today) and data at the manual P_k level (EPROMs can be introduced).

The caches can be placed on production units which have storage capacities as described in table 8. The currently existing production system does not use the storage capacities to cache information. The reason for this is that it is considered more expensive to produce erroneous kits than to stop production entirely. However, by relaxing this demand it is possible to improve system behavior when data network partitions occur.

Table 8:   The different production units and their corresponding storage capacity

P_{111} : volatile memory	P_{112} : volatile memory	P_{113} : volatile memory
P_{114} : volatile memory	P_{121} : volatile memory	P_{122} : volatile memory
P_{123} : volatile memory	P_{124} : volatile memory	P_{131} : volatile memory
P_{132} : volatile memory	P_{133} : volatile memory	P_{134} : volatile memory
P_{141} : volatile memory	P_{142} : volatile memory	P_{143} : volatile memory
P_{144} : volatile memory	P_{145} : volatile memory	P_{146} : volatile memory
P_{11} : computer with secondary memory		
P_{12} : computer with secondary memory		
P_{13} : computer with secondary memory		
P_{14} : computer with secondary memory		
P_{21} : paper list	P_{22} : paper list	
P_{311} : small sec. memory	P_{312} : small sec. memory	P_{313} : small sec. memory
P_{321} : small sec. memory	P_{322} : small sec. memory	P_{323} : small sec. memory
P_0 : PC with database		
P_{-1} : computer center with large database		

Two examples are given to illustrate how the reliability of the production system can be improved. In the first example caching is introduced in the production system and in the second example, EPROMs are used in the material network.

Example 1 (using caches)

One possibility to avoid a production stop when the data network partitions between the PC, P_0, and the counter groups, $P_{1,i}$, is to use the storage capacities in the counter groups to store extra equipment settings and other necessary production information to be used during the partition. Under certain circumstances, it could even be economic to increase the storage capacities in the counter groups if the possibility of having this type of data network partition is high. The cached information is thus stored in P_{11}, P_{12}, P_{13} and P_{14}.

This gives the following fault tolerance strategy:

$$(I_{n,data} \cap I_{n,cache} \subseteq I_{n,all}) \Rightarrow \text{Total (double redundant) FT}$$

This means that fault tolerance, FT, exists because all the necessary information, $I_{n,all}$, is retrievable in a redundant form via the data network, $I_{n,data}$, and a data cache, $I_{n,cache}$.
The production liveness will be:

$$\exists\, n\, \{I_{n,all} \subseteq I_{p,data}(t) \cup I_{p,cache}(t)\} \Rightarrow P_p(t)$$

This means that if for at least one mission, n, all the necessary information, $I_{n,all}$, is a subset of the information that is retrievable via the data network, $I_{p,data}(t)$, and a data cache, $I_{p,cache}(t)$, at time t within time-limit Δt, then production, $P_p(t)$, is possible before $t + \Delta t$ expires.

Example 2 (using EPROMs)

If the data network partitions between the VAXcluster, P_{-1}, and the PC, P_0, it is possible to maintain production if EPROMs are used in the material network. The EPROMs would be attached to a pile of trays and the information would be valid until the data network functions correctly again or until a new EPROM arrives. The information would contain, for example, the machine settings for the bagging machines, the check weighers and the printing machines, i.e., P_{3i1}, P_{3i2}, P_{3i3} and P_{3i4}.
This gives the following fault tolerance strategy:

$$(I_{n,data} \cap I_{n,mat} \subseteq I_{n,all}) \Rightarrow \text{Total (double redundant) FT}$$

This means that fault tolerance, FT, exists because all the necessary information, $I_{n,all}$, is retrievable in a redundant form via the data network, $I_{n,data}$, and the material network, $I_{n,mat}$.
The production liveness will be:

$$\exists\, n\, \{I_{n,all} \subseteq I_{p,data}(t) \cup I_{p,mat}(t)\} \Rightarrow P_p(t)$$

This means that if for at least one mission, n, all the necessary information, $I_{n,all}$, is a subset of the information that is retrievable via the data network, $I_{p,data}(t)$, and material network, $I_{p,mat}(t)$, at time t within time-limit Δt, then production, $P_p(t)$, is possible before $t + \Delta t$ expires.

6   Acknowledgements

This chapter has been supported by the Swedish National Board for Industrial and Technical Development, no. 93-04177.

7   References

[1]   Abbott RJ, Resourceful systems for fault tolerance, reliability and safety, ACM Computing Surveys, vol. 22, no. 1, March 1990, pp 35-68

[2]   Abu-Hamdan MG, El-Gizawy AS, An error diagnosis expert system for flexible assembly systems, the 7th IFAC/IFIP/IFORS/IMACS/ISPE Symposium on Information Control Problems in Manufacturing Technology, INCOM'92, Toronto, Canada, May 1992, pp 451-456

[3]   Adlemo A, Andréasson SA, Models for fault tolerance in manufacturing systems, Journal of Intelligent Manufacturing, vol. 3, no. 1, February 1992, pp 1-10

[4]   Adlemo A, Andréasson SA, Johansson MI, Fault tolerance strategies in an existing FMS installation, Control Engineering Practice, vol. 1, no. 1, February 1993, pp 127-134

[5]   Adlemo A, Andréasson SA, Fault tolerance in partitioned manufacturing networks, Journal of Systems Integration, vol.3, no. 1, March 1993, pp 63-84

[6]   Adlemo A, Andréasson SA, Johansson MI, Information accessibility and reliability improvement in an automated kitting system, the 12th World Congress of the International Federation of Automatic Control, IFAC'93, Sydney, Australia, July 1993, vol. 2, pp 99-106

[7]   Andréasson SA, Andréasson T, Carlsson C, An abstract data type for fault tolerant control algorithms in manufacturing systems, Information Control Problems in Manufacturing Technology, E. A. Puente and L. Nemes (Eds.), IFAC Proceedings Series, Pergamon Press, Oxford, U.K., no. 13, 1990, pp 51-56

[8]   Avizienis A, Fault-tolerance, the survival attribute of digital systems, Proceedings of the IEEE, vol. 66, no. 10, October 1978, pp 1109-1125

[9]   Avizienis A, Software fault tolerance, the 9th World Computer Congress, IFIP Congress'89, San Francisco, U.S.A., August - September 1989, pp 491-497

[10]   Bartlett J, Gray J, Horst B, Fault tolerance in Tandem computer systems, Symposium on the Evolution of Fault Tolerant Computing, Baden, Austria, June 1986, pp 55-76.

[11]   Chintamaneni PR, et al., On fault tolerance in manufacturing systems, IEEE Network, vol. 2, no. 3, May 1988, pp 32-39

[12]   Cristian F, Issues in the design of highly available computing systems, the Annual Symposium of the Canadian Information Processing Society, Edmonton, Canada, July 1987, pp 9-16

[13]   Fukuda T, Tsukiyama M, Mori K, Scheduling editor for production management with human-computer cooperative systems, Information Control Problems in Manufacturing Technology, E. A. Puente and L. Nemes (Eds.), IFAC Proceedings Series, Pergamon Press, Oxford, U.K., no. 13, 1990, pp 179-184

[14]   Gray J, Why do computers stop and what can be done about it?, the 5th Symposium on Reliability in Distributed Software and Database Systems, Los Angeles, U.S.A., January 1986, pp 3-12

[15]   Gray J, Siewiorek DP, High-availability computer systems, IEEE Computer, vol. 24, no. 9, September 1991, pp 39-48

[16]   Hatzikonstantis L, et al., Interactive scheduling for a human-operated flexible machining cell, Information Control Problems in Manufacturing Technology, E. A. Puente and L. Nemes (Eds.), IFAC Proceedings Series, Pergamon Press, Oxford, U.K., no. 13, 1990, pp 445-450

[17]   Hennoch B, A strategic model for reliability and availability in automatic manufacturing, International Journal of Advanced Manufacturing Technology, vol. 3, no. 5, November 1988, pp 99-121

[18]   Hopkins AL Jr., Smith TB, Lala JH, FTMP: a highly reliable fault tolerant multiprocessor for aircraft, Proceedings of the IEEE, vol. 66, no. 10, October 1978, pp 1221-1239

[19]   Johansson MI, Johansson B, High automated kitting system for small parts - a case study from the Volvo Uddevalla plant, the 23rd International Symposium on Automotive Technology and Automation, Vienna, Austria, vol.1, December 1990, pp 75-82

[20]   Kopetz H, et al., Distributed fault tolerant real-time systems: the Mars approach, IEEE Micro, vol. 9, no. 1, February 1989, pp 25-40

[21]   Kronenberg N, Levy H, Strecker W, VAXclusters: a closely-coupled distributed system, ACM Transactions on Computer Systems, vol. 4, no. 2, May 1986, pp 130-146

[22]   Laprie JC, Dependability: a unifying concept for reliable computing and fault tolerance, Dependability of Resilient Computing Systems, T. Anderson (Ed.), BSP Professional Books, Blackwell Scientific Publications, Oxford, U.K., 1989, pp 1-28

[23]   Laprie JC, et al., Definition and analysis of hardware and software fault tolerant architectures, IEEE Computer, vol. 23, no. 7, July 1990, pp 39-51

[24]   Lilien L, Quasi-partitioning: a new paradigm for transaction execution in partitioned distributed database systems, the 5th International Conference on Data Engineering, Los Angeles, U.S.A., February 1989, pp 546-553

[25]   McGuffin LJ, et al., MAP/TOP in CIM distributed computing, IEEE Network, vol. 2, no. 3, May 1988, pp 23-31

[26]   Melliar-Smith PM, Schwartz RL, Formal specification and mechanical verification of SIFT: a fault tolerant flight control system, IEEE Transactions on Computers, vol. C-31, no. 7, July 1982, pp 616-630

[27]   Nelson VP, Fault tolerant computing: fundamental concepts, IEEE Computer, vol 23, no. 7, July 1990, pp 19-25

[28]   von Neumann J, Probabilistic logics and the synthesis of reliable organisms from unreliable components, Automata Studies, C. E. Shannon and J. McCarthy (Eds.), Princeton University Press, Princeton, U.S.A., 1956, pp 43-98

[29]   Ossfeldt BE, Fault tolerance in the AXE switching system central control: experience and development, the 13th International Symposium on Fault Tolerant Computing, Milan, Italy, June 1983, pp 384-387

[30]   Randell B, System structuring for software fault tolerance, Current Trends in Programming Methodology, R. T. Yeh (Ed.), Prentice Hall, Englewood Cliffs, U.S.A., 1977, pp 195-219

[31]   Rao TRN, Fujiwara E, Error-Control Coding for Computer Systems, Prentice Hall, Englewood Cliffs, U.S.A., 1989

[32]   Siewiorek DP, Architecture of fault tolerant computers, IEEE Computer, vol. 17, no. 8, August 1984, pp 9-18

[33]   Siewiorek DP, Fault tolerance in commercial computers, IEEE Computer, vol. 23, no. 7, July 1990, pp 26-37

[34]   Siewiorek DP, Swarz RS, Reliable Computer Systems, Digital Press, Burlington, U.S.A.1992

[35]   Taylor D, Wilson G, The Stratus system architecture, Dependability of Resilient Computers, T. Anderson (Ed.), BSP Professional Books, Blackwell Scientific Publications, Oxford, U.K., 1989, pp 222-256.

[36]   Trevino J, Kilgroe C, Computer-aided selection of automatic identification systems, International Journal of Computer Integrated Manufacturing, vol.5, no. 3, May - June 1992, pp 164-170

[37]   Wiendahl HP, Winkelhake U, Strategy for availability improvement, International Journal of Advanced Manufacturing Technology, vol. 1, no. 4, August 1986, pp 69-78

11   Integration of CAD, CAM and Process Planning

Heiko Nordloh

1. Introduction

A large number of commercial software systems are available for design, production and process planning specialized on various applications. Although much effort has been spent on the integration of individual systems under the umbrella of computer-integrated manufacturing (CIM), there are still gaps in the different steps from product idea to NC program. From the author's point of view many implementations can still be described by the Y-Model of Scheer [1]. The Y-Model distinguishes two main streams in a factory organization. On one hand order-oriented tasks have to be performed and on the other hand product-oriented tasks. Order-oriented tasks are production planning and control (PPC) functions, for example order control or capacity planning. Product-oriented tasks include product design, process planning and NC programming. The concepts presented consider the integration of product-oriented tasks as well as order-oriented tasks.

Although standard interfaces like IGES or EDIFACT transfer common data between applications, the different data models used in the individual systems still require repetitive input of the same information which can lead to inconsistancy in the models. Therefore, the implementation of common technical data models plays an important role in achieving integration on the technical instrumental level. Considerable progress has been achieved in the development and industrial application of computer-aided design (CAD) and computer-aided manufacturing (CAM) systems. Compared with CAD and CAM systems, computer-aided process planning (CAPP) did not get the same attention. Process planning tasks are the link between CAD and CAM and often consume more time and effort than the actual product design. For mass-produced products the resulting high costs of manual planning might be acceptable, because the costs can be distributed on many copies of the product. For small batch sizes and increasing product variety manual process planning is no longer economical. Therefore, one of the goals of CAPP is to reduce the costs of process planning for small batches while still achieving adequate plan quality in terms of utilization of various manufacturing resources. Two approaches of CAPP to achieve this goal will be described in the following subchapters.

Besides the utilization of common data models both approaches make use of a feature-based product description. There are various definitions of the term "feature" in literature. In the following text the definition of Shah [2] will be used. "Features are generic shapes with which engineers associate certain properties or attributes and knowledge useful in reasoning about the product." Corresponding to this definition a feature links geometry and topology information with application-oriented semantics. Different classes of features have to be used in different engineering domains, which will be expressed with terms like form, manufacturing or assembly features.

The first approach strives for fully automatic planning that can be entirely performed by a computer program. Chapter 2.4.2 outlines the implementation of an almost automatic planning system based on the results of the BRITE/EURAM project No. P2406, "Integration of CAD/CAM and Production Control for Sheet Metal Components Manufacturing"[1]. Starting from the feature-based description of two-dimensional sheet metal parts the system automatically determines manufacturing processes, selects tools, calculates the cutting sequence and generates the NC program. The chapter will demonstrate how process planning data can be derived from a feature-based workpiece description as well as process planning tasks specific for sheet metal manufacturing.

Another approach is the application of group technology (GT) principles. Planning tasks are fully described for one part and can be adapted to parts with similar properties (part families). Chapter 2.4.3 describes a system based on modified GT. It is the result of the BRITE/EURAM project BE-3528 "Manufacturing Cell Operator's Expert System (MCOES).[2] The systems application domain is the manufacturing of nonprismatic machined parts in a workshop-oriented factory, the characteristics of which will be summarized. The chapter focuses on the setup and fixture planning task and defines fixturing features to extend the part model.

1.1. Integration of CAD/CAM and Production Control for Sheet Metal Components Manufacturing

In recent years many computer programs aimed at the optimization and automation of special subtasks in design, production planning and manufacturing. The objective of the presented research was the development of a software system for design, production and process planning, NC programming as well as shopfloor monitoring and control. The system addresses requirements of small batch and one-of-a-kind production for sheet metal components. Companies specialized in the manufacturing of sheet metal parts often have different types of cutting machines at their disposal. The machines vary, for instance, in the cutting process, tools available, work area dimension or machining costs. Hence, the process planning system must be able to consider different machine capabilities and cutting processes. Furthermore, unforeseeable disturbances on the shopfloor require flexible generation of alternative process plans and NC programs just-in-time. Improvement of material utilization in small batch production demands the collection and nesting of several orders on one sheet metal blank. The system developed

---

[1]The project team consisted of sheet metal manufacturers from Belgium (Actif Industries, ETAP), Denmark (Dronningborg Maskinfabrik) and Germany (BICC-Vero Electronics, Schichau Unterweser AG), a software and consulting firm (Peter Matthiesen), and research institutes from Belgium (WTCM) and Germany (BIBA).

[2]The following nine partners formed the MCOES team: Disenño y Metodologia S.A. (Spain), Oxford Computer Services Ltd. (UK), Instituto Superior Tecnico (Portugal), Technical Research Centre of Finland, Helsinki University of Technology (Finland), Valmet-Tampella Ltd. (Finland), National Technological University of Athens (Greece) Ervin Halder KG (Germany), and Bremen Institute of Industrial Technology and Applied Work Science at the University of Bremen (Germany).

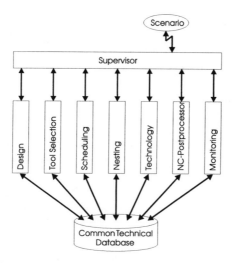

Figure 1: System Overview

will lead to increased flexibility in planning tasks as well as a reduction of delivery time and production costs. Shopfloor monitoring and control functions will allow the consideration of the actual state of the shopfloor in every planning step.

During specification and implementation of the system the following system characteristics were worked out:

- Communication of the system modules via a common database,
- Feature-based part description,
- Extendible rule base for manufacturing methods selection.

A set of individual processing modules has been developed, which can be tailored to the needs of the user. Figure 1 describes the system's structure.

The supervisor module controls the calling sequence of the system's submodules. It can be adapted to the requirements of the production site and its production philosophy like JIT and KANBAN. Disturbances on the shopfloor are taken into account by replanning from a certain level in the hierarchical calling sequence of the modules which can establish the different system goals. A more detailed description of the supervisor is given in [3].

In one-of-a-kind and small batch production the shopfloor often consists of autonomous manufacturing units. These units perform certain manufacturing operations like metal cutting, painting or assembly. The sequence of planning steps depends on the organizational structure of a company. Figure 2 gives an example for the steps to be performed for the planning of sheet metal cutting processes. The arrows on the left side of the boxes describe order dependent data generated by another module and/or have been entered by the user and is stored in the database.

330

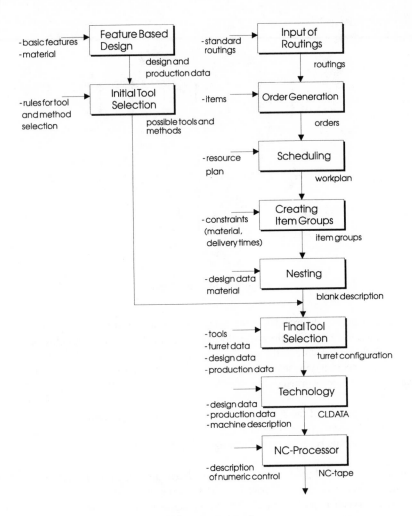

- basic features
- material

Feature Based Design

design and
production data

- standard
routings

Input of Routings

routings

- rules for tool
and method
selection

Initial Tool Selection

- items

Order Generation

possible tools and
methods

orders

- resource
plan

Scheduling

workplan

- constraints
(material,
delivery times)

Creating Item Groups

item groups

- design data
material

Nesting

blank description

- tools
- turret data
- design data
- production data

Final Tool Selection

turret configuration

- design data
- production data
- machine description

Technology

CLDATA

- description
of numeric control

NC-Processor

NC-tape

Figure 2: Sequence of planning steps

After a customer has placed an order, the parts will be described with a feature-based design system and in the initial tool selection step tools and relating manufacturing methods will be automatically calculated for the parts, taking into account the machine resources available. Parallel, routings can be selected from a list of standard routings and/or new routings for the parts can be defined. After the manufacturing orders have been created, a medium and short-term scheduling can be performed for the orders. Certain criteria like material, thickness and delivery date parts belonging to different customer orders are combined into item groups. The item groups are nested on available blanks and the turret of a selected machine can be determined within the final tool selection. The technology module will add some technology data before calculating the cutting sequence to manufacture the blank. Finally a generalized post-processor generates the NC program.

In this scenario the machines are selected during the scheduling step directly after the bill of material has been generated. In another company it might be useful to select the machine after nesting. In this case medium-term scheduling will be performed for a whole group of sheet metal cutting machines and short-term scheduling will be done for a single machine after nesting.

Existing work preparation systems often force the user to adapt the company organization to the needs of the software. When producing very small batches or even batch size one, it seems much more necessary to adapt the software to the organizational structure of the company. In the system presented here this aspect is considered by the supervisor module and the flexibility of the integrated modules.

1.2. The Database Structure

A common technical database is the fundamental mechanism for system integration. Hence, it must contain various data categories. Each of this categories consists of a fixed and of a variable data part. The categories are not isolated databases but are linked with each other very closely. In each fixed data part basic elements are defined. The fixed data part cannot be changed by the user. Every variable part consists of combinations of these basic elements. It can be extended by the user. The user cannot define new basic elements but can create new combinations of basic elements.

The geometrical part description, for example, contains a combination of form features. A form feature itself consists of a combination of basic elements and/or other form features. Basic elements can be POINTS, LINES or ARCS. Form features like rectangle or keyslot are described in the variable data part. The user can define new

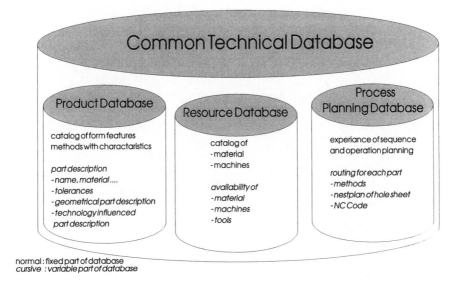

Figure 3: Database structure

form features assembled from existing elements, but it is not allowed to define new basic elements as, for example, SPLINES.

In the BRITE/EURAM project three essential categories of data have been identified (Figure 3):

- Product data describes geometrical and technical attributes of a product. The data can be divided into either form, method or item definitions.
- Resource data describes characteristics and availability of resources like tools, machines, materials, etc.
- Process planning data describes routings, orders, planning constraints, process plans, the result of the scheduling, etc.

1.3.   Description of Main Modules

1.3.1. Design

The design of a workpiece is done with a feature-based design system. Selecting a feature automatically determines possible manufacturing methods for that feature. Three types of features can be defined by the user:

- perimeter shapes, describing the outer contour of a workpiece,
- internal depressions (holes),
- external depressions (notches).

Figure 4 shows an example design of a workpiece with a complex outer contour and

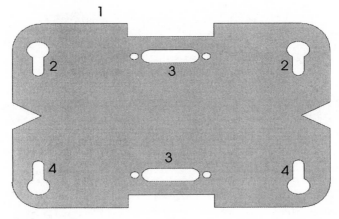

Segments

1 Contour
2 Keyslot (pattern, 1-dimensional)
3 Connector (pattern, 1-dimensional)
4 Keyslot (pattern, 1-dimensional)

Figure 4: A feature-based design

1. Complex feature

Contour (perimeter)

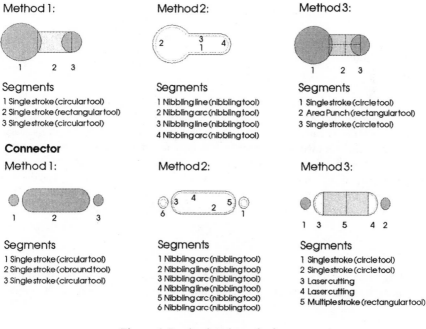

Segments

1 Arc	9 Arc
2 Line	10 Line
3 Notch (rectangle)	11 Notch (rectangle)
4 Line	12 Line
5 Arc	13 Arc
6 Line	14 Line
7 Notch (v-shaped)	15 Notch (v-shaped)
8 Line	16 Line

2. Simple features (internal)

Keyslot   Connector

3. Design data

Keyslot

E1 Length
E2 Large diameter
E3 Small diameter

Connector

E1 Size

Segments

1 Line
2 Arc
3 Line
4 Arc

Segments

1 Circle
2 Obround
3 Circle

Figure 5: Design data of a feature

Keyslot

Method 1:

Segments

1 Single stroke (circular tool)
2 Single stroke (rectangular tool)
3 Single stroke (circular tool)

Method 2:

Segments

1 Nibbling line (nibbling tool)
2 Nibbling arc (nibbling tool)
3 Nibbling line (nibbling tool)
4 Nibbling arc (nibbling tool)

Method 3:

Segments

1 Single stroke (circle tool)
2 Area Punch (rectangular tool)
3 Single stroke (circle tool)

Connector

Method 1:

Segments

1 Single stroke (circular tool)
2 Single stroke (obround tool)
3 Single stroke (circular tool)

Method 2:

Segments

1 Nibbling arc (nibbling tool)
2 Nibbling line (nibbling tool)
3 Nibbling arc (nibbling tool)
4 Nibbling line (nibbling tool)
5 Nibbling arc (nibbling tool)
6 Nibbling arc (nibbling tool)

Method 3:

Segments

1 Single stroke (circle tool)
2 Single stroke (circle tool)
3 Laser cutting
4 Laser cutting
5 Multiple stroke (rectangular tool)

Figure 6: Production data of a feature

some internal depressions. Figure 5 and 6 give an impression of a feature representation in the database.

Figure 5 describes the design data of the outer contour of the workpiece and the internal features used in Figure 4. The outer contour, called the perimeter, is a combination of the basic features, LINE and ARC, as well as the simple features, V-SHAPED-NOTCHES and RECTANGLE-SHAPED-NOTCH. Simple features are combinations of basic features. The dimensions of simple features are entered parametricly. The designer has to describe the internal feature KEYSLOT, for example, with the parameter's length, large diameter and small diameter. To define new features the user has to enter formulas to derive the basic features from the parameters of the simple feature. Furthermore, the user should specify different production methods for this feature. A combination of the following basic methods can be chosen:

- single-stroke punching,
- multiple-stroke punching,
- contour punching,
- contour nibbling,
- area punching,
- area nibbling,
- laser cutting.

Figure 6 shows manufacturing methods for the two examples: keyslot and connector.

1.3.2. Scheduling

The scheduling task was divided into three subtasks:

- order generation,
- order selection,
- scheduling.

During order generation the user defines for each item, a work order containing routings for manufacturing a workpiece, e.g. punching -> bending -> painting -> quality control -> packing. The user is free to use standard routings, already stored in the database, or can define his own routings for each item. The order selection allows the combination of different orders due to certain constraints. Possible constraints are material type, thickness of the material, or delivery date. The order selection enables a preselection of orders which can be manufactured on the same blank. If different orders are nested on one blank, this has to be taken into account by the scheduling module[3].

1.3.3. Nesting

Nesting combines workpieces of the same material and time span on blanks with the aim at minimizing material waste. This is done in two steps. In the first step irregularly

---

[3]The scheduling system PM-SIM of Peter Matthiesen is able to consider that different orders with different routings have to be on the same machine at the same time for a certain manufacturing step.

and regularly shaped pieces are interactively clustered in rectangular enclosures. In the second step the rectangular enclosures are automatically nested on blanks. The automatic nesting module selects blanks from the stock, keeps track of their availability, calculates the number of identical blanks and reserves committed blanks in stock. A more detailed description of the nesting algorithm can be found in [4].

1.3.4. Tool Selection

Tool selection automatically generates a setup of the tools in a machine's turret. The selection of tools includes decisions on the manufacturing processes for the production of the blank. Properties of the machine, for example the size of the turret and of the ability to rotate tools, have to be considered as well. Tool selection also has to consider the availability of tools in the shopfloor and allows the rejection of tools from the proposed turret, for example broken tools.

To combine the feature-based design with a rule-based tool selection, a two-stage system for the tool selection has been developed containing both a machine-independent part which is called initial tool selection, and a machine-dependent part, called final tool selection.

Initial tool selection generates a list of all tools which can be used for the manufacturing process available in order to produce a workpiece. The initial tool selection is based on parametric descriptions of the features. In a workpiece description, production information is linked to these features as well as a set of rules for the method selection.

With the rules of the initial tool selection the system can decide whether a tool of a given shape and size can be used to produce a feature or a section of it. Initial tool selection results in lists of tools linked to the manufacturing methods available for each feature of the workpiece. These lists are stored in the database.

Final tool selection generates a proposal for the turret setup of a specified machine and selects the manufacturing processes. This final step selects from the list of possible tools. The final selection step is rule-based like the initial one. The actual state of the shop floor is taken into account by this module. The rules of the final step can be divided into two categories:

- Rules for tool selection
    1. Use tools from the standard turret
    2. Use tools from the machines actual turret
    3. Minimize turret size
    4. Where possible select large tools
- Rules for the selection of manufacturing processes
    1. Select the method with the highest average priority
    2. Where possible select methods with a small number of different manufacturing processes
    3. Where possible select methods with a small number of different tools

The rules for the tool selection are static because of their implication with the algorithm. The process selection rules can be configured with the help of a table stored in the database. The following table gives an example of the process priorities.

Material	Thickness	Single stroke punching	Multiple stroke punching	Contour punching	Nibbling	Laser cutting
Steel	3.5-6.0 mm	1	4	3	0	2
Steel	0.5-3.0 mm	1	2	2	0	2
Coated Steel	0.5-3.0 mm	1	2	3	0	4
Stainl. Steel	0.5-1.5 mm	1	3	3	0	2
Stainl. Steel	2.0-4.0 mm	1	4	3	0	2
Stainl. Steel	4.5-6.0 mm	1	0	0	0	2
Aluminium	0.5-2.0 mm	1	2	3	4	5
Aluminium	2.5-6.0 mm	1	3	2	4	0

The table lists the priority numbers for the manufacturing processes for each material type. A small number indicates a high priority for the selection, a high number indicates a low priority, a zero priority locks a manufacturing process.

1.3.5. Technology Preparation

Starting with a nested blank, technology preparation has to supply as far as possible the following technical information to the automatic NC-code generation:

- the calculation of the corrected tool path for depressions and outer contours,
- which workpieces have been placed in the depressions of other workpieces,
- the detection of scrap cut out during the manufacturing process dependent on the selected manufacturing methods and tools,
- the definition of additional production segments — like bridges — to guarantee stability during manufacturing as well as read-in-lines and loops to guarantee quality of the finished parts,
- the optimization of tool changes,
- the determination of the cutting sequence according to different optimization criteria.

The generation of the NC-code based on the layout of a nested blank is performed in five steps:

1. toolpath correction
2. simulation of the manufacturing process
3. addition of technology parameters
4. tool sequence calculation
5. optimization of the toolpath for each tool.

Toolpath correction calculates the tool offset depending on the tool size to guarantee correct part dimensions. The nesting module places the workpieces on the blank with the goal of maximum material utilization. The workpieces are placed at a predefined distance depending on the tool size.

If a workpiece has a large internal depression, the material within this depression can be used by the nesting module for placing other workpieces. Workpieces which are positioned within the depressions of other workpieces have to be manufactured first, because if the depression containing the material used by the inner workpiece were to be manufactured first, the material in this depression would become free and would have to be removed. One goal of the manufacturing process simulation is to calculate the hierarchy of workpieces placed in each other.

The positioning of the workpieces done by the nesting module can result in material that is not part of a workpiece becoming free during manufacturing. This material is called scrap or garbage. To avoid damage to workpieces or even to the machine tool itself the scrap has to be removed during manufacturing. The identification of scrap is a second goal of the manufacturing process simulation.

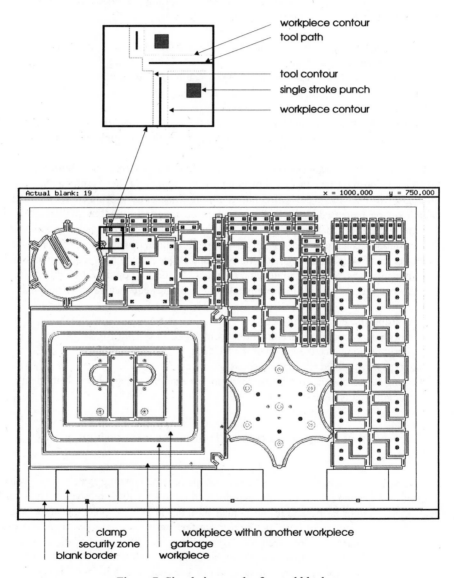

Figure 7: Simulation result of nested blank

The visualization of the manufacturing process for the user is the third goal of this module in order that the manufacturing processes and tools selected by earlier planning steps can be controlled and replanned if necessary.

Figure 7 shows an example of the simulator output. In the lower area the clamps and their security zones are drawn. At the left-hand side three levels of workpieces are nested. Different areas of free scrap can be seen.

Technology parameters are added semi-automatically. The user can position the clamps to fix the blank on the machine, define loops, read-in-lines, and insert bridges.

After manufacturing process simulation and after the addition of process-specific information like loops and bridges to the sheet, the cutting sequence can be calculated. This calculation has to consider some general constraints:

- When a workpiece is cut free, all depressions within this workpiece must already have been processed.
- Stability of the blank has to be guaranteed. During the entire manufacturing process there must be sufficient material in the area of the clamps to move the blank without vibration.
- In order that material can be removed from the machine, the last material cut has to be at a particular position. This position depends on machine dimensions and type.

Goals of toolpath sequence optimization could be, for example, the minimization of the manufacturing time and minimization of the length of the resulting NC tape. Strategies to meet these goals depend on the machine capabilities as well as on the complexity of the nested blank. Optimization of the manufacturing time can be achieved by minimization of machine movements and minimization of tool changes. Application of subroutine technique and usage of complex built-in functions of the machine controller can reduce the length of the NC tapes. Due to contrary requirements and constraints it is very difficult to develop a strategy to fulfil all optimization criteria. For this reason a set of strategies for optimizing of the cutting sequence has been adapted:

- The stripewise cutting sequence,
- Algorithms to minimize machine movement,
- Algorithms to reduce the length of the NC tape,
- Interactively defined cutting sequences,
- Cutting sequences for special sheet layouts, an example being a blank with identical rectangular parts to be laser cut.

2. Integration of Design and Process Planning for a Workshop-Oriented Factory

In this chapter the basic concepts of a prototype process planning system specifically targeted at the requirements of workshop-oriented factories will be presented. The main goal of the system is to capture the human design and process planning knowledge, store it into databases, and reuse the knowledge for nearly automatic operative process planning. The chapter is organized as follows: First, the characteristics of workshop-oriented factories are summarized. Next, the software architecture of the developed system is outlined, followed by a description of the central data model. Finally, some details on the subsystems model data preparation and process planning are given with the description of an example planning session.

2.1. Workshop-Oriented Factories

The trend towards increasing product variety and decreasing lot sizes requires not only the introduction of new information technology but also organizational changes in the production process. A particular manufacturing environment and organization is needed for the manufacturing of small batch sizes or even one-of-a-kind products.

"In contrast to traditionally highly automated factories intended for the mass production of one product or just a few limited product variants, a workshop-oriented factory is expected to be capable of producing a range of products within the scope of the recognized product families."[5] The application of the principles of Group Technology is an important aspect of workshop-oriented factories. Therefore, recognition and definition of part families is a major step in the design of a workshop-oriented factory. The part familiy members should have similarities in geometrical layout, common manufacturing processes and process sequences as well as common fixturing principles. The characteristics of workshop-oriented factories can be summarized as:

- Organization of the production flow according to the part families and therefore, grouping of machines by product families
- Usage of flexible manufacturing systems (FMS)
- Decentralized control strategies
- High planning responsibility of the machine operators
- High predictability of throughput time
- Moderate repetition of production and capacity utilization.

A comparison of the essential characteristics of functional, workshop-oriented, and traditional mass-production factories can be found in [5]. As a result, it becomes clear that a workshop-oriented factory represents an intermediate alternative combining the flexibility of a functional factory with the efficiency and controllability of an automated factory.

The structure of a process planning system also depends on the characteristics of the factory organization. For mass production a detailed process is done only once and only minor changes have to be made during product manufacturing. Therefore, the role of process planning is smaller than in small batch or one-of-a-kind production. In a functional organized factory generative process planning is preferred. Based on process knowledge process plans are generated from scratch. In a workshop-oriented factory variant process planning is preferred. Process plans are generated based on predefined process plan schemata corresponding with the product families forming the basis of the workshop design.

2.2. MCOES System Architecture

In contrast to fully automated mass production lines where decisions are made in a centralized planning department, in workshop-oriented factories many decisions are made locally. The various workshops operate autonomously and the decision-makers have to be provided with suitable tools. For the MCOES system the following end-users were identified by an analysis of the actual operations in the workshop:

- Product designer responsible for definition of the initial product family descriptions
- Method planner responsible for definition of the manufacturing process available in the workshop
- Process planner generating process plans for the product families
- Workshop technology manager providing information about the actual state of the manufacturing resources in the workshop
- Operative designer performing the operative design, creating the process plan and controlling the NC program simulation.

Figure 8 gives an architectural overview of the MCOES system implemented. The final MCOES system consists of five main subsystems:

- Feature-based design interface
- Process plan preparation system for performing strategic process planning
- Process plan generation system for performing operative process planning
- Method editing interface
- Factory modeller.

Similar to the system for sheet metal component manufacturing presented in the previous subchapter the main integration mechanism is the common data model of the involved subsystems. The data model contains part family models which are feature-based part family descriptions extended with process plan specifications. It also contains a factory model describing static factory entities and models available. Furthermore, the data model can be completed with individual manufacturing knowledge and stored in

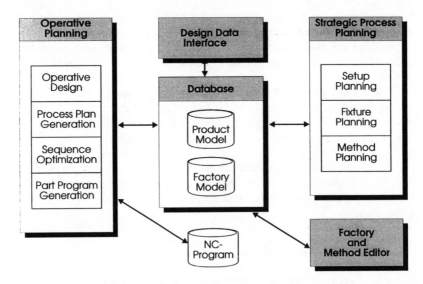

Figure 8: MCOES system architecture

the underlying database. The MCOES system allows the import of external data from tool management systems, production management systems and shopfloor control systems.

The generation of process plans and part programs with the MCOES system takes place in three stages:

- The system setup stage, whereby a factory model representing factory facilities is generated and a collection of methods representing tested, proven manufacturing processes is created. At this stage, the planning focuses on a single process.
- The strategic process planning stage, whereby a process plan specification for a new product family is created by choosing appropriate methods and setups. At this stage, the planning focuses on a single product family, comprising several processes.
- The operative process planning stage, whereby a detailed process plan is created for a given manufacturing order. At this stage, the focus is a single order, possibly comprising several instances of several product families.

"In reality, the three stages are all on-going parallel processes. That is, the system setup and the part families are continuously changed according to the changes in the factory and the products." [5]

2.3.   Central Data Model

Both the central data model of the MCOES system and the data model of the sheet metal system consists of three parts:

- the part model
- the process model
- the resource model.

This chapter outlines differences in the data model of the sheet metal system.

Part Model: The part model of MCOES has been designed with respect to the requirements for manufacturing the part families. It utilizes a feature-based representation of part families and part instances. The feature model is based on an object-oriented modelling technique where feature types are represented in a hierarchical taxonomy of classes that can inherit attributes from their upper classes. Figure 9 gives an example of a partial feature taxonomy.

Complete parts are represented by a data structure containing all the features describing the part. It is also a tree structure with the relation "has a". The root of the tree is the billet node representing the unmachined part, possibly a raw material block or a casting. The billet has a number of surfaces, on which machinable features have been defined. Feature relationship is important information needed by the process planning system to determine sequences of machining operations. Therefore, in a part family description the position, orientation and dimensions of manufacturing features should depend on the parameters of a part family.

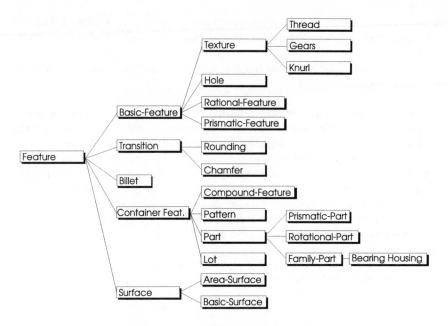

Figure 9: A partial feature taxonomy [5]

An important aspect of the feature-based modelling technique is the possibility of linking geometrical part information with technological part information. In most applications of feature-based modelling technology, form features describing the shape of a part and manufacturing features describing the machining processes and their parameters for manufacturing the part can be distinguished. The MCOES project extended this model with the definition of fixturing features. [6]

Fixturing features represent fixturing functions and a point/area of a workpiece where the fixture element and the workpiece are in contact. Fixturing features enlarge the product model with fixturing knowledge. That is useful for method planning, tool selection and collision detection. Fixturing features provide an easy way to select fixture elements. MCOES uses a parametric fixture description. Fixturing features embody a parametric description because they are positioned relative to a surface or a feature. Whenever, for example, the workpiece dimensions are changed, then the fixturing feature position and the fixture element positions respectively are changed automatically as well. Figure 10 shows the fixture taxonomy of the MCOES system.

Process Model: The feature definition contains a geometrical description of the feature as well as the definition of methods available to manufacture the feature. Alternative methods can be defined for the features. A method comprises a number of work elements. A work element is the smallest unit in a process plan and represents a single machining cycle where a tool is applied to the feature. Hence, the work element description contains references to tools available and NC macros to generate the part program.

The process plans are defined in ALPS (A Language for Process Specification) developed at NIST [7]. ALPS provides various control structures to determine, if tasks

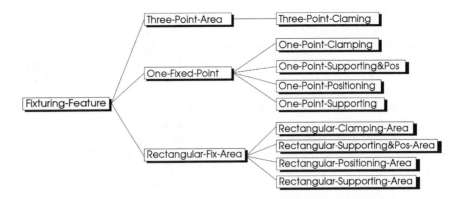

Figure 10: The fixture taxonomy

must be performed parallel, serially, or arbitrarily. A more detailed description of the ALPS taxonomy and capabilities can be found in [8]. The ALPS syntax is used for the plan specification within method and part family descriptions. For method specification ALPS terms are used to express, for example, that a hole can be manufactured using alternative sequences of work elements. Parts belonging to a part family are always manufactured using a predefined manufacturing plan. In a part family definition ALPS is used to describe, for example, the number and sequence of part setups and features machined in these setups.

Resource Model: The resource model of the MCOES system is very similar to the resource model of the sheet metal system. It is an object-oriented model of the factory resources. The model includes data such as machine capabilities, tools, fixtures, material and human resources. At the top of the hierarchy is the "factory item", which can be divided into different resources like machines or tools. Further refinement leads to the description of a resource instance. Data kept for a tool instance, for example, can include items such as dimensions, costs, order number or life-span.

2.4. A Planning Session with the MCOES system

A planning session is performed in the steps: data preparation, strategic planning and operative planning. These are not sequential steps, but in practise, these are on-going parallel processes. The part families are continuously modified according to shopfloor and the product changes.

2.4.1. Model Data Preparation

Computer-supported process planning requires extensive manufacturing knowledge which must be available in knowledge bases. The design data interface, the method editor and the factory editor of MCOES provide facilities to update and extend the process planning knowledge base.

The concepts of MCOES require feature-based design technology for the part family description. MCOES developed a native system for the creation of feature models and part families. The design data interface is implemented within a more general feature-based design system called EXTDesign, developed by the Helsinki University of Technology. EXTDesign is based on GWB, the Geometric WorkBench [9]. The GWB modeller uses boundary representation and implicitly defined polynomial surfaces. In the EXTDesign system feature-based and solid modelling operations can be utilized parallel during the design process of a part. A part design can be performed either by designing with feature types or by regular solid modelling operations. A design with solid modelling operations requires feature recognition to generate the feature model. A more detailed description of the feature recognition concepts can be found in [10]. With this approach the designer is free to choose the most convenient means for the design task. First the designer instantiates a feature class that will become the root of the instance hierarchy of the part. With modelling commands of the feature frame editor feature instances can be added to or deleted from the model. When a new feature is created, the feature type must first be selected from the list of available feature types. Thereafter, the frame of the new feature instance is displayed in the frame editor window, and the user can modify the parameters of the instance. EXTDesign allows transmutation from a part instance description to a part family description. Figure 11 shows the part family design in process. Using the structure editor for the part definition

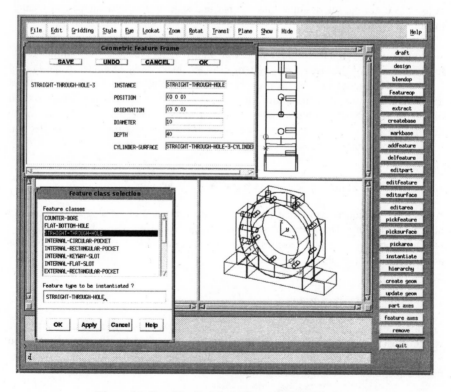

Figure 11: Part Family Modelling with EXTDesign.

the user defines the parameters for a new family. The editor provides facilities for entering formulae to compute the size and position of the part features based on the part family parameters.

Features created with the design data interface are linked to manufacturing methods. A method is a plan or description how the manufacturing features can be realized with the processes and resources available. Development and maintenance of manufacturing processes in many companies are done by an experienced planner from the shopfloor. The method editor software supports this task with the following functions:

- Definition of methods including process selection, parameter specification and sequencing.
- Definition of work elements including tasks such as tool and machining parameter selection.
- NC macro definitions.

As these tasks are very complex, they require the management of a large amount of data. Therefore, the tasks are supported by both specialized method and work element managers to maintain the method and work element knowledge bases.

Another tool for the model data preparation is the factory editor, with which the user can create the resource model described earlier in the text. It enables the user to [8]:

- import and export data to external systems
- browse the factory hierarchy
- maintain the instances of each subclass item.

2.4.2. Strategic Process Planning

Strategic process planning generates a global plan scheme for a given new product family. Because many decisions depend on the result of other decisions, there is no clear sequence of planning operations. Two examples might make this clear. Methods cannot be chosen without knowing the fixturing in order to avoid obstructing. Fixturing cannot be chosen without the knowledge of the processes chosen in order to consider the machining forces. Therefore, three assistant type systems have been developed for the strategic planning task. The modules can be executed relatively independent of each other and support quick user interaction with an evolving process plan. The steps of strategic process planning are:

- setup planning
- method planning
- fixture planning.

Setup Planning: It is the tasks of setup planning to ascertain the part orientation relative to the machine spindle and the number and sequence of setups. Setup planning depends on the geometrical attributes of the part in terms of manufacturing features as well as the relation between manufacturing features. Furthermore, setup planning has to take into account that features can be manufactured with alternative methods. This leads to a large variety of solutions. Consideration of machine tool capabilities, part tolerances and part fixturing, makes setup planning a complex task requiring long-term practical experience. The main objectives of setup planning are to:

- select features and related methods which can be machined when the part is fixed in the work area with a single part orientation
- ensure the machining of closely related features in the same setup
- minimize the number of setups.

Setup planning focuses on a single part setup for a new part family. The setup plans are specified in LISP frames which are the input for operative process planning. In operative process planning frames will be instantiated. In case of lot manufacturing operative process planning will merge several part setups in a lot setup.

Input data for the setup module is the geometrical part description in terms of manufacturing features and their selected manufacturing methods. The input data is provided by the EXTDesign module and the method planner. The output of setup planning is a part setup description including the part orientation, the features which can be manufactured, as well as possible manufacturing methods for this setup. The output data will be processed by the method planning, the fixturing planning and the NC-code generation module.

Figure 12 displays the setup planning step of a bearing house in progress. The setup planning module is a highly interactive tool which supports the user in finding optimal part orientations for a machine setup. It uses the facilities of EXTDesign for visualization.

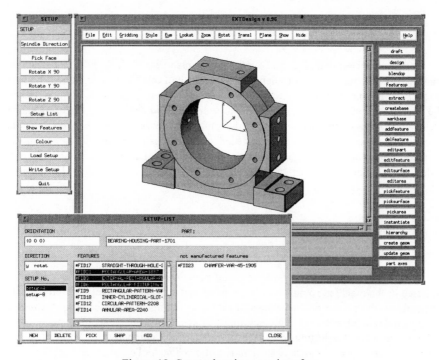

Figure 12: Setup planning user interface

For part-setup planning the following steps have to be performed.

- First the user selects the spindle direction of the machine tool. The spindle direction can be either horizontal (parallel to the y-axis) or vertical (parallel to the z-axis). The selected spindle orientation will be considered in the following steps.
- The part orientation relative to spindle direction can be modified if necessary with the buttons Rotate_x, Rotate_y, Rotate_z or Pick Face.
- A new setup will be created for each part orientation. For a new setup the system provides a list of features that can be manufactured in this setup. The system selects the appropriate features based on the orientation of the method. The features of this setup can be visualized with the button Show Features.
- Next the user decides whether all suggested features should be manufactured in that setup, or for some reason, it would be better to manufacture a certain feature in another setup. In this case the user can move the feature to the list of not manufactured features. In order to take into account that a machine tool has a rotatory table, setups with different orientations can be merged together to a single

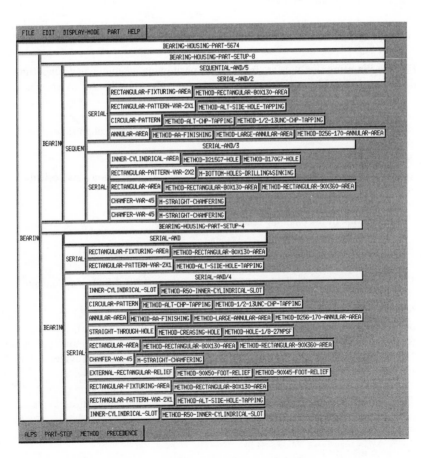

Figure 13: Method planning user interface

setup.
- Finally, the setup planning result in terms of one or more setups will be written in a UNIX file or a database and will be made available for the other MCOES modules.

Method Planning: A part description and the setup directions with a preselection of possible methods from the setup planner is the input of the method planning task. In the method planning step a complete process plan is generated for the part family. In reality it may be necessary, as already mentioned, to revisit the setup planning or fixture planning before method planning can be finished. The following subtasks have to be performed for the method planning task:

- enumerate all methods suitable for a part family,
- recognize precedences among the methods as a whole and among the work elements
- check for interference between fixturing and manufacturing features
- determine the best method within a set of alternative methods
- consider possibility of using the same tool in different methods
- generate ALPS format plan specification of the result.

The method planning user interface displaying an ALPS format plan specification is shown in Figure 13.

Fixture Planning: Fixtures are required for positioning, clamping and supporting of workpieces on a machine tool during manufacturing. The objectives of fixture planning are:

- determination of clamping, positioning and, if necessary, supporting fixturing features
- selection of fixturing elements during part instantiation
- calculation of the number of fixturing elements, their positions and orientations relative to the workpiece
- visualization of the fixturing layout
- documentation of the planning result for the shopfloor.

In a fixture planning session the following steps have to be performed.

- First, the planner selects the type of fixturing system. Special fixturing elements or a modular fixturing tool kit can be chosen for part fixturing.
- Next, for every part-setup the fixturing features are interactively defined. The user determines fixturing functions (positioning, supporting, clamping), fixturing area (rectangular, one-point, two-point) and the position for the fixturing feature. Depending on the fixturing function and the area, the system provides a list of fixturing elements available from which the user selects a suitable element.
- Then, the number of required elements as well as the accurate arrangement of the fixture relative to the workpiece and the base element are automatically calculated by the system.
- Finally, the calculated fixture will be displayed using EXTDesign.

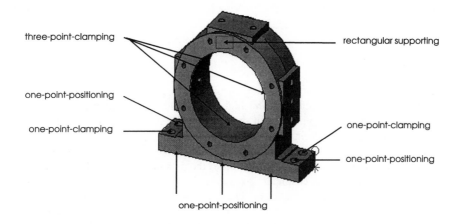

Figure 14: Definition of fixturing features

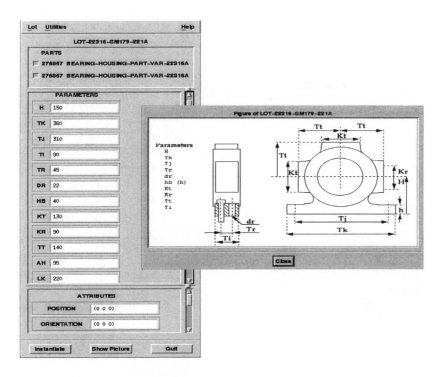

Figure 15: Interface of operative design

Figure 14 gives an example of the fixturing features defined for a bearing house part family.

2.4.3. Operative Process Planning

The overall goal of operative process planning is to generate correct process plans and NC programs in a few minutes. This high performance is a result of the preparation work done by the other components of the system. The first step of operative process planning is the "operative design".

Using the part family description of the design data interface as a basis, the operator instantiates a predefined order type and modifies the default parameter values according to the requirements of the customer. As shown in Figure 15 pixmaps illustrate the meaning of the part family parameters. The result is an instantiated family lot model consisting of part instances, each itself consisting of some feature instances. The process plan is then generated automatically by evaluating the plan specification of the lot and can be inspected with an object browser. If necessary, modifications can be made at this point. Next, a final sequencing algorithm is invoked. The algorithm tries to minimize the number of tool changes and pallet rotations considering all precedence relations defined by the process plan structure. This sequenced process plan is the input information for the automatic part-program generation. The part-program generation algorithm scans the

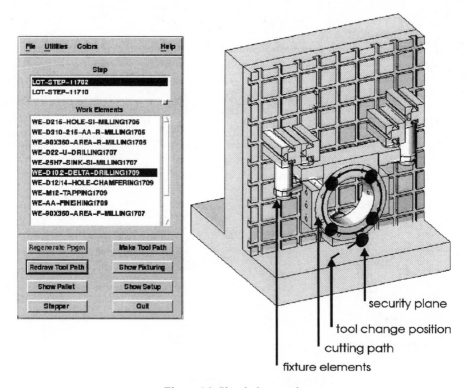

Figure 16: Simulation result

sequenced work elements of a process plan and generates a piece of part program for each. The code generation is based on parametric pieces of code stored in the work element classes. Finally, the part programs generated can be visualized using the geometric models of the parts, pallets, fixturing elements and tools. A simulation result is shown in Figure 16.

3. Summary

In the previous chapters two integrated systems covering the planning functions from product design to the preparation of the NC programs have been presented. Both systems are prototype developments resulting from research projects funded by the European Commission and have not been commercialized until now. In both systems the underlying concept for the integration of CAD/CAM and process planning are the application of feature-based design methodology and the definition and access to common data models. The system implementations presented are examples of different planning approaches.

The implementation dealing with the 2D cutting problems of sheet metal manufacturing attempts to achieve almost automatic process planning and NC program generation. A part description is composed of common geometrical features associated with various manufacturing methods. Features and manufacturing methods are described in a parametric form. The part itself is not described parametricly and therefore the positions and the parameter values of the features are absolute values. This approach means many decisions have to be made during process planning and NC program generation and requires very complex algorithms, making the application of the concept to 3D sheet metal parts with bending and forming operations difficult.

The other approach used in the MCOES system utilizes principles of group technology. Part families are described parametricly and all feature parameters describing the part can be derived from the global part parameters. Similarly, the process plan specifications including setup and fixture planning are created for the part families with a preselection of manufacturing methods available.

The evaluation of both systems in an industrial environment pointed out two general results:

- The systems presented increase the flexibility of the design and manufacturing process, specially for small batch production. With the improved data flow the throughput time can be reduced.
- These systems reduce design and planning effort and increase the quality of process plans. Product quality can be improved with the utilization of reliable manufacturing methods and parameters already tested.

The development of software tools supporting new concepts in production engineering is much faster than the application to these technologies in the industrial environment. On the one hand, the personal has to be trained to use these systems, they have to see the advantage of new methods and acquire practice in using the tools. This task can be accomplished in a short term. On the other hand, a much more difficult and time-consuming task is to change organizational structures in a manufacturing process. It was very clearly observed that when a functionally organized production was reorganized to a workshop-oriented production, such as was done during the MCOES project, organizational problems arose. In the MCOES system, concept design responsibility moves from the design department to the shopfloor. The operative design is done on the shopfloor, thus requiring good communication channels between shopfloor and design department.

The concepts presented provide tools to store design and manufacturing knowledge of individual companies electronically. Today, many companies have forward information flow from product design towards process planning to manufacturing. Information feedback from the shopfloor to the design is mostly limited to order-oriented data like "part is ready" or "delay in production process". The feedback of technology-oriented information from shopfloor to design is very rare. Feature-based design in combination with knowledge bases describing a companie's manufacturing capabilities improves technological feedback which can be used in the early design phase to test for manufacturability or perform cost analysis.

In recent years the concept of feature-based design has become increasingly important. Various leading commercial CAD systems from America and Europe like, for example, Pro/Engineer, I-DEAS, KONSYS or SIGRAPH support the design by features. Therefore, from the authors' point of view, some of the ideas presented in this chapter will be found in future industrial manufacturing systems. In this chapter integration on the technical instrumental level focusing on the automation of data exchange between computer systems in production has been described. Future research should aim at integration on a higher organizational level providing facilities for collaborative process planning and will require the application of computer-supported cooperative work principles in the process planning domain.

4. References

[1]   Scheer, A.-W., 1987, CIM - Der computergesteuerte Industriebetrieb, Berlin, Springer

[2]   Shah, J.J., 1991, Conceptual Development of Form Features and Feature Modelers, Research in Engineering Design, Vol 2 pp 93-108, Berlin, Springer

[3]   Knackfuß, P.C., 1990, A JIT Application in Sheet Metal Production. BIBA, Bremen, Germany. Proceedings of International Conference on Advances in Production Management Systems.

[4]   Schalla, A.J., Knackfuß, P.C., Hirsch, B.E., 1991, Integration of CAD/CAM and Production Control in Sheet Metal Manufacturing - An Application Area of Operations Research. Production, Planning & Control, Vol. 2, No. 2, pp 96-101

[5]   Opas, J, Kanerva, J, Mäntylä, M, 1992, Automatic Process Plan Generation in an Operative Process Planning System, Helsinky University of Technology.

[6]   Hämmerle, E., 1993, Werkstattorientierte Systeme zur Arbeitsplanung und kurzfristigen Fertigungssteuerung, Ph.D Thesis at the University of Bremen

[7]   Ray, S.R, Catron, B.R., 1991, ALPS: A Language for Process Specification, Int. J. Computer Integrated Manufacturing Vol 4 no 2 pp 105-113

[8]   Mäntylä, M., 1993, BRITE/EURAM Project 3528 Manufacturing Cell Operator's Expert System (MCOES) Synthesis Report/Overview, Helsinki University of Technology

[9]   Mäntylä, M., 1988, An Introduction to Solid Modelling, Computer Science Press, College Park, Maryland

[10]   Laakoo, T., Mäntylä, M., 1991, A New Form Feature Recognition Algorithm, Computer Applications in Production and Engineering, CAPE '91 Bordeaux, France, 10-12 September 1991, North-Holland Publ. Co., Amsterdam

[11]   Eloranta, E., Mäntylä, M., Opas, J., Ranta, M., 1989, HutCapp - A Process Planning System Based on the Integration of Knowledge Engineering. Feature Modeling and Geometric Modeling. Laboratory of Information Processing Science, Helsinky University of Technology.

[12]   Ephraim, P., Gaensmantel, G., Knackfuß, P.C., 1986, A Concept of Fault Tolerant Operation of FMS. Proceedings of the CIM EUROPE Working Conference on Production Systems, Design, Engineering, Management and Control, Bremen.

[13]   Erve, A.H. van't, 1988, Generative Computer Aided Process Planning. University of Twente.

[14]   Hammer, D.K., 1992, Lean Management: The Integrating Power of Information, IFIP Transactions, Integration in Production Management Systems, North Holland

[15]   Pels, H.J., Wortmann, J.C., 1992, Integration in Production Management Systems: An Integrating Perspective, IFIP Transactions, Integration in Production Management Systems, North Holland

[16]   Integration of CAD/CAM and Production Control for Sheet Metal Components Manufacturing. BRITE P-2406, 6-Month-Report, 12-Month-Report, 18-Month-Report, 24-Month-Report, 30-Month-Report, BIBA, Bremen, Germany.

12 Features as Modeling Entities for Intelligent Design Systems in Mechanical Engineering

Michael Schulte, Rainer Stark and Christian Weber

1 Introduction

The future success of CAD will depend on the ability of CAD systems to support the design engineer in more aspects than merely to describe and reproduce the geometric characteristics of design objects (parts, sub-assemblies, technical products) in terms of points, lines, surfaces, and/or volume primitives. One way to enhance the representation of design objects within CAD systems is the definition of features and their usage in the computer aided design process.

Today a considerable number of research projects deals with computer aided design processes based on features. In most cases, however, these features are primarily related to manufacturing processes and – in the last years – also to assembly processes. In other words: they refer to aspects which come after the design process and subsequently are aimed at supporting activities after design (e.g. operations planning for manufacturing).

These circumstances raise some interesting questions which will be discussed in this issue:

- Is there some sort of features that can be utilized to support the design process itself?
- What should these features look like?
- How can they be used?

2 Activities in engineering design

Before trying to answer these questions a brief look at the term "design" is taken. In the following some fundamental notes on engineering design will be made. They are based on the guidelines VDI 2221 [17] and VDI 2222 [18] which give a general view of the fundamental principles and methods of engineering design and are widely accepted as a useful framework in this field. (Similar conclusions can be found in [6, 7, 8, 10, 12, 13].)

The guideline VDI 2221 describes the term "engineering design" as the sum of all activities, which help to work out the information necessary for the manufacturing and the use of a technical product or system starting from the given requirements or functions the product or system is supposed to fulfill[1]. Furthermore, if the layout of the technical product is completely unknown at the start of the design process (novel design) it can be subdivided into four general phases according to the guideline VDI 2222: product planning, conceptual design, embodiment design, and detail design. In addition, the guideline VDI 2221 subdivides the four a.m. design phases into seven general working stages (cf. Fig. 1):

1. *Clarify and define the task:* This stage is necessary to clarify and define the requirements the final design has to fulfill. The result is a detailed specification (requirements' list) concerning the technical product to be designed.

2. *Determine functions and their structures:* The second working stage decomposes the overall function of the technical product to be designed into several sub-functions. Here a function represents the relationship between input(s), output(s), and state variable(s) of a system resp. sub-system, at this time still independent of the particular solution. The arrangement of individual functions resp. the relationship between the overall function and the sub-functions is expressed with the help of the function structure. Work so far accomplished is exclusively aimed at logic considerations.

3. *Search for solution principles and their combinations:* At this stage a search is made for solution principles to realize the sub-functions. In the field of mechanical engineering design physical effects have to be selected for this purpose. A physical effect shows how to transform the input(s) of the underlying sub-function into the output(s) in terms of a physical process. The physical effect can either be known empirically or it is given in accordance with a law of nature.

   Each sub-function must be satisfied by the application of a physical effect, which is mainly realized by a specific arrangement of surfaces, and a specific choice of motions and materials [12, 17]. Those parts of the surfaces of modules necessary to realize the physical effect are called "active surfaces". Examples for active surfaces are application-surfaces of forces or reference-surfaces of velocities. Further below, the term "primary functional faces" for the active surfaces is introduced, which in the here given context gives a better characteristic of what is meant.

   To satisfy the overall function the solution principles of the various sub-functions have to be combined corresponding to the function structure elaborated at the previous working stage. Its result is called the "principle solution". (As it is described in [19], it is quite easily possible to come from rather abstract "methodical terms" like sub-function, function structure, physical effect, and principle solution to the more concrete level of mathematical models for technical products.)

4. *Divide into realizable modules:* In stage four the principle solution is divided into realizable modules. This results in a module structure. It provides – in contrast to the function structure or the principle solution – a preliminary

---

[1] In the following only the term "technical product" will be used, because there is no clear cut line between "product" and "system" anyway.

356

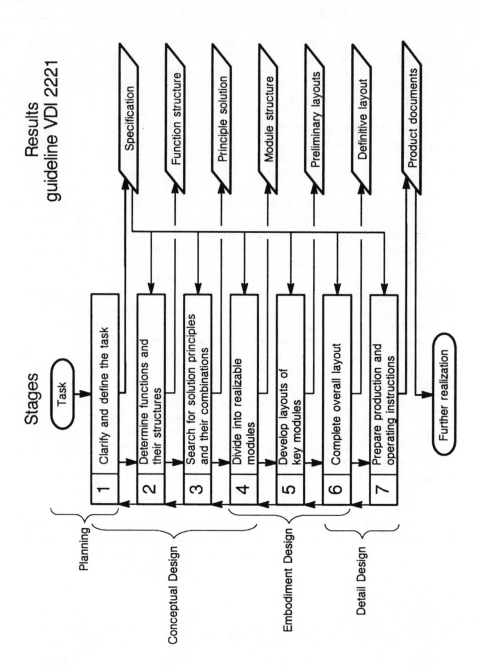

Fig. 1. General approach to design

indication of realizable groups and elements, which together with their links (interfaces) are essential for its implementation.

5. *Develop layouts of key modules:* The modules of the fourth stage will be put into a more concrete layout at this stage. The level of refinement of geometry, materials, and other details should only be pursued as far as to allow the optimum design alternative to be selected. The results of this stage are called "preliminary layouts", which can be represented as scale drawings. A complete dimensioning will not be carried out in this working stage.

Before the fifth working stage the modules have been represented geometrically mainly by the primary functional faces derived out of the underlying physical effects. To obtain the required preliminary layouts of these modules further surfaces have to be added. Within the context of the feature technology discussed here, these will be called "filling faces". Since filling faces do not have a functional meaning they may usually be modified without changing the product's behaviour.

6. *Complete overall layout:* In the sixth working stage the preliminary layouts of the modules are completed and combined to the overall solution. During this process all the modules have to be broken up into arrangements of subassemblies and individual parts, which are called "assembly units" in this issue. At this stage the aspects of manufacturing, assembly, and standardization have to be taken into consideration. Those parts of surfaces, which are the result of these separation processes of modules into assembly units will be called "secondary functional faces". It should be noted that secondary functional faces always appear in pairs[2].

Finally, the assembly units have to be dimensioned as far as their system behaviour and the conditions of strength are concerned. The result of this stage is the definitive layout containing all the essential information about the configuration necessary for the realization of the product. The main forms of representation are scale layout drawings and parts lists.

7. *Prepare production and operating instructions:* At this stage all the final production and operating instructions the design department is responsible for are prepared.

The a.m. working stages do not have to be followed rigidly during the design process. They are often carried out iteratively, returning to preceding ones, thus achieving a step-by-step optimization.

The survey of the general approach to engineering design clearly shows that the most important aspects of the design process (at least in connection with novel design) are the definition and realization of specific functions the technical product being designed is supposed to fulfill. Only after the strictly function oriented preliminary layout is found, additional aspects such as manufacturing, assembly, and standardization problems can be considered.

---

[2] The terms "primary functional faces", "filling faces", and "secondary functional faces" introduced here have a close relationship to the terms "functional surfaces", "free surfaces", and "connecting surfaces" as they are used in [1].

3 Features in design

3.1 The use of features in general

The impetus for feature research originally came from the desire to find more sophisticated ways for the definition of geometry with regard to manufacturing operations planning and NC programming. Therefore "features" were originally thought of as geometry associated with specific machining operations [5].

Today the term "feature" is used in a much broader sense. Any set of information that can be formulated by generic parameters and properties and that can be referred to as a specific entity within the reasoning process of an application can be called a feature [16]. In this issue, however, the authors support the conception that features are mainly based on geometric information.

Based on this supposition, in the last years the conception becomes more and more accepted that a feature must consist of three information components (cf. Fig. 2) [2, 4]:

1. A feature is mappable to a generic shape. In other words – as linguists say – it has a syntax. To represent the generic shape one component of a feature is a specific shape element (e.g. geometric element resp. form feature).
2. A feature has a specific meaning within the engineering context. Therefore it represents a semantic. Consequently the second component of a feature are specific semantic elements to express the feature's engineering meaning.
3. The third component of a feature are the relations between the two said components.

In many cases work so far accomplished along this line neglects the fact that very often the syntax of a feature cannot be described entirely by geometric elements resp. form features alone. Other sorts of information that may be required are, for instance, material properties, surface specifications, dimensional tolerances, and/or shape tolerances. These additional syntax elements of a feature, however, are in most cases closely related to its geometric elements, which in nearly all cases play the key role in the syntax description.

It appears plausible, that feature definitions (syntax and/or semantic meaning) differ in various engineering disciplines such as engineering design and manufacturing operations planning [2, 14]. As mentioned before, this issue does not discuss the quite common features that are related to manufacturing processes, but deals with the definition and usage of features supporting the design process itself.

Based on the conception that features for the design process should relate shape elements (syntax) to functions (semantic), the next step is to investigate these relations in detail. The results of this investigation can then be transferred to enhance CAD tools, which besides the geometric characteristics of design objects can describe and handle their functional backgrounds.

To conduct the fundamental investigation presented here, there are two possible approaches:

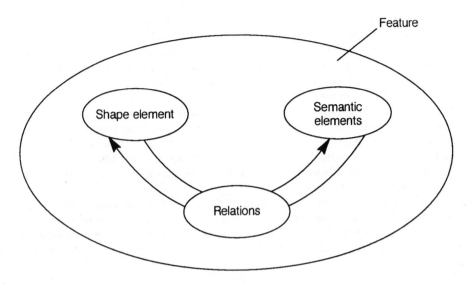

Fig. 2. Features and their components

1. "Start with the functions, consider the related physical effects and conclude the investigation by looking at the resulting shape elements." This approach might be called the synthetical one, since it follows the phases or stages of the design process as shown above in Sect. 2, that is in a top-down direction.
2. "Start with known shape features, ask for the physical effect they realize and conclude the investigation by looking at the underlying functions." This approach might be called the analytical one, since it goes backwards through the a.m. phases or stages of the design process, that is in a bottom-up direction.

In this issue, at first the "top-down" approach is presented. Features examined in this approach are called "functional features". Later on there will be a look at "design features", which stem from the second ("bottom-up") approach.

3.2 Functional features

A physical effect is layed down with regard to a specific (sub-) function and at the same time is usually characterized by plotting a certain arrangement and interrelationship of primary functional faces. Hence according to the guideline VDI 2221 (see above) it is the third stage of the design process (search for physical effects and their combinations), which during design synthesis offers the first direct relation between the functional and the geometric characteristics of a technical product.

The authors believe that it is an advantage to model the primary functional faces of a design object in larger and more meaningful units than just look at them separately. Therefore functional features are introduced as sets of primary functional faces, which embody the active surfaces of a physical effect to meet the requirements of a certain design (sub-) function [20].

Functional features are well defined if the combination of several primary functional faces leads to a generic description of the related design (sub-) function and physical effect. The primary functional faces thus represent the key elements of functional features. The syntax description of functional features mainly contains geometric information. Additionally it could be useful to combine geometric with attributive information concerning material properties, surface specifications, or tolerances.

Since design functions in the design of shafts and their adjacent components are easily comprehensible, they are taken as an example to demonstrate the modeling process with functional features in the following. Some basic functions that often have to be realized in the design of shafts in combination with adjacent components are:

- input or output of torque/speed;
- conduction of torque/speed;
- increase/decrease of torque/speed;
- realization of different kinematic conditions of adjacent modules/support of forces;
- storage of momentum energy.

Figure 3 shows an example of a simple functional feature called "frictional connection" related to the basic design function "input or output of torque/speed". As the underlying physical effect "dry friction" can only be realized by at least one pair of functional faces, the functional feature consists of two (a, c) resp. four (b) functional faces (more pairs are possible, of course). The necessary separation of functional faces, indicated in the figure by material vectors, will automatically result in separate modules and separate assembly units in later design stages. The three different types of the functional feature "frictional connection" shown in the figure could lead to some kind of a flat belt drive (a), or – after variation of the functional faces in orientation and/or number – of a vee-belt drive (b), or of a disk type clutch (c).

Another functional feature is described in Fig. 4. The underlying (sub-) function is "support of a shaft". The semantic meaning related to this function is explained in more detail on the right side of the figure.

On the left side of Fig. 4 there are shown the functional faces of one physical effect to fulfill the a.m. (sub-) function, which constitute the shape elements (geometric elements) of the functional feature examined here. The separation of functional faces, again indicated in the figure by material vectors, is necessary to allow the required rotational movement between the shaft and its housing. To transmit the radial forces the physical effect "form closure" is applied, which requires two pairs of functional faces to be horizontally oriented (perpendicular

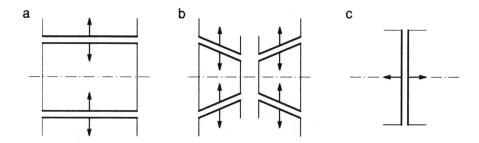

Fig. 3. Functional feature "frictional connection" and its variation

to the direction of the forces transmitted). Finally, to transmit the axial forces the physical effect "form closure" is applied also, so that another two pairs of functional faces have to be vertically oriented (again perpendicular to the direction of the forces transmitted).

The examples taken from the field of the design of shafts show two quite interesting aspects of functional features:

1. Unlike the commonly known manufacturing features, which always refer to the shape elements found in one individual part, functional features usually contain information about shape elements of several modules (parts, subassemblies) and their relations.
2. The example "support of a shaft" shows another interesting aspect: as it is indicated by the terms "loose bearing" and "locating bearing" within the form elements' part of the feature, functional features show hierarchical arrangements of shape elements[3].

3.3 Design with functional features

Generally, feature-based design can be seen as a two-step process. The first step is the modeling of the features themselves. This leads to a catalogue of predefined feature types, which can be transferred to corresponding CAD modules. The second step is the actual feature-based modeling, that is the usage of the features in CAD practice. The usage of features in the practical CAD process is a repetition of the following three actions:

1. Select a feature type from the predefined catalogue.
2. Define values for the feature parameters to scale the feature to the desired shape and size.
3. Define position and orientation of the feature, or define to which other feature(s) the new one applies.

---

[3] This can be seen in more detail when discussing results of the more analytical ("bottom-up") investigation of features later on.

362

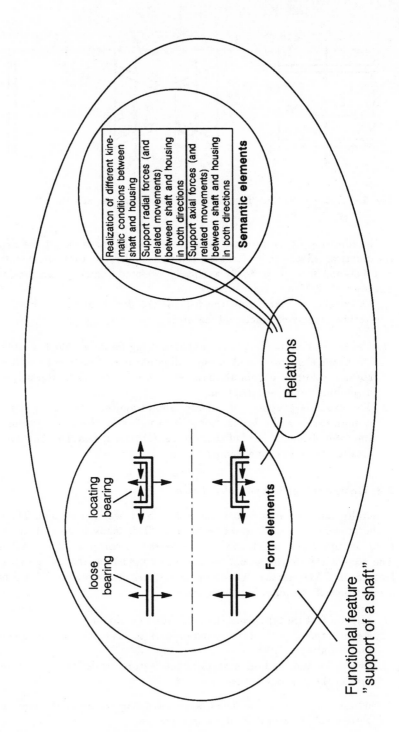

Fig. 4. Functional feature "support of a shaft"

The complete definition of a feature requires the specification of all relevant dimension and location parameters. However, not all these parameters are available, or even important, until the final stages in design. This is especially true in the case of the functional features discussed here. Therefore it is necessary to introduce the concept of "abstract features" [15].

The usage of functional features will now be described by illustrating the most significant steps of the feature-based design of an example, which in this case is a technical product for the storage of momentum energy (cf. Fig. 5 and 6).

After clarifying and defining the task of this technical product at stage 1 (according to guideline VDI 2221) the following (sub-) functions have been determined at stage 2:

- Input of torque.
- Storage of momentum energy.
- Realization of different kinematic conditions between the carrier of the momentum energy and its environment, at the same time support of forces between the two.

In the next step of the design process (stage 3 according to guideline VDI 2221) physical effects have to be chosen to fulfill the required (sub-) functions. Their representation by functional features is of interest here:

- Functional feature "frictional connection".
- Functional feature "mass momentum of inertia".
- Functional feature "support of a shaft", the most common solution which can be split into the sub-features "loose bearing" and "locating bearing" (see above).

At this stage of design the dimensions of the primary functional faces and the dimensions of the distance between the functional features are not considered, whereas the arrangements of the functional features already have to be determined (illustration 1.a of Fig. 5)[4]. It will be readily granted that this first working step on the geometric level is directly related to functional reasoning. Therefore it can be seen as direct continuation of the designer's logic considerations.

In the case of already known and predetermined solutions it is evident that more detailed functional features can be used at this stage of design. Illustration 1.b of Fig. 5 shows a possible specification of the functional features "loose bearing" and "locating bearing" by implementing rolling movements for the realization of the bearings in order to minimize friction.

In the subsequent step (stage 4 according to guideline VDI 2221) the principle solution is to be divided into realizable modules. This usually means splitting up the pairs of primary functional faces: originally belonging together in functional

---

[4] Dotted lines indicate that one "half" of the primary functional faces of the functional feature "frictional connection" are omitted in these considerations because of the system boundaries given here.

364

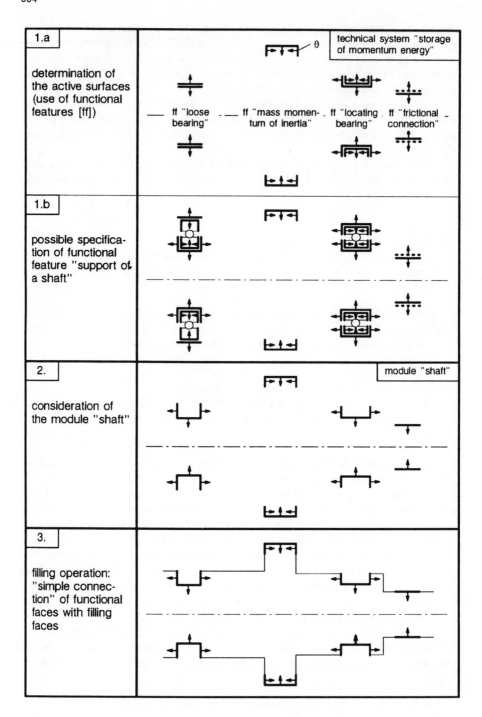

Fig. 5. Example of working with functional features in the design process; part 1

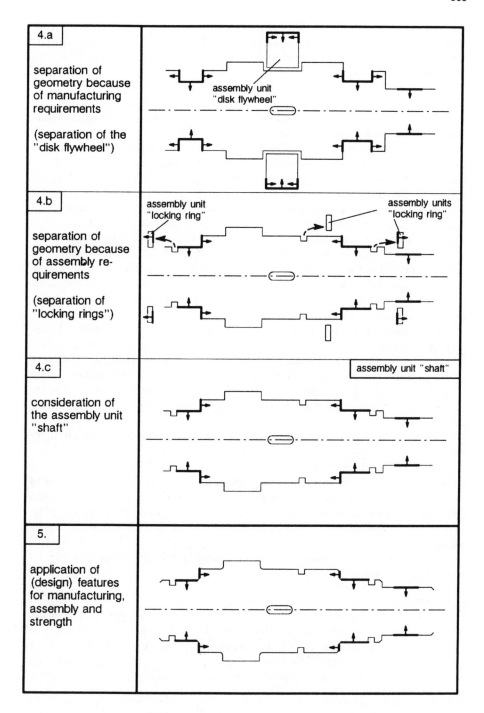

4.a	
separation of geometry because of manufacturing requirements (separation of the "disk flywheel")	
4.b	
separation of geometry because of assembly requirements (separation of "locking rings")	
4.c	
consideration of the assembly unit "shaft"	
5.	
application of (design) features for manufacturing, assembly and strength	

Fig. 6. Example of working with functional features in the design process; part 2

features to realize specific physical requirements such as different kinematic conditions (indicated in the illustrations by material vectors), they are now considered as elements of individual modules, for instance the module "shaft" shown in illustration 2 of Fig. 5.

Subsequently, the primary functional faces belonging to one module are connected by filling faces[5] (stage 5 of the design process according to guideline VDI 2221). This step, shown in illustration 3 of Fig. 5, is necessary to get a first shape design, a preliminary layout of the respective module.

Based on this first shape design now the definitive layout of the modules has to be elaborated and the modules have to be combined to the overall solution (stage 6 according to guideline VDI 2221). Because of this the designer makes further decisions concerning the realizable modules in terms of assembly units (parts, sub-assemblies) and their links. In this stage the aspects of manufacturing, assembly, and standardization are taken into account.

In the presented example it might be an advantage to separate the disk flywheel from the rest of the module in order to provide a sufficient amount of momentum of inertia and at the same time to minimize the effort of manufacturing. Another reason for the separation might be material considerations. Illustration 4.a of Fig. 6 shows the separation of the flywheel by inserting pairs of secondary functional faces.

Furthermore, the geometry of the first shape design has to be checked concerning assembly requirements. It is often necessary to carry out further separation operations in order to make assembly possible. Consequently several more assembly units are determined. In the given example the assembly units "locking ring" are introduced (c.f. illustration 4.b of Fig. 6). The separation itself is accomplished in the same manner as before, that is by inserting pairs of secondary functional faces.

The result of all the separation steps is exemplary shown for the assembly unit "shaft" in illustration 4.c of Fig. 6.

The last step in the design process is focussed on the further optimization of the assembly units with manufacturing, assembly, and standardization as well as strength considerations in mind. As it is shown in illustration 5 of Fig. 6 within this step the application of already known features for manufacturing, assembly, and strength ("modifiers") such as undercuts, chamfers, and roundings can be applied.

The way of working with functional features in the design process can be summarized as follows:

1. Starting from the functional structure of a technical product to be designed, physical effects have to be found to realize all the (sub-) functions (guideline VDI 2221 stage 3). Predefined functional features can be used to describe the primary functional faces (active surfaces) of each physical effect.
2. The division of the principle solution into realizable modules (guideline VDI 2221 stage 4) can be accomplished by combining related functional faces.

---

[5] Filling faces are indicated in the figures by the narrow black lines as opposed to the wide black lines, which characterize primary functional faces.

3. Filling faces have to be inserted to obtain a preliminary layout of each module (guideline VDI 2221 stage 5).

4. In order to meet manufacturing, assembly, and standardization requirements geometric elements (primary functional faces and filling faces) have to be separated by inserting (pairs of) secondary functional faces. Consequently individual assembly units and their relations to each other are determined (guideline VDI 2221 stage 6, part 1).

5. To meet detail requirements of manufacturing, assembly, and strength adequate modifying features (such as undercuts, chamfers, roundings) can be added to the individual assembly units. This leads to the definitive (qualitative) layout of the design solution (guideline VDI 2221 stage 6, part 2).

It appears from the preceding expositions that the concept of functional features is able to serve several purposes during the design phases of conceptual, embodiment, and detail design:

- With the help of functional features, which describe sets of primary functional faces, the designer is able to continue functional reasoning not only on the logic level of the design process, but also on the geometric level.
- The consideration of physical effects and of their primary functional faces as basis for the definition of functional features makes it possible to document the fundamental ideas of the designer in terms of geometry.
- The distinction between functional and filling faces enables the selective variation of primary functional faces to change the system behaviour. On the other hand it is possible to identify those parts of the geometry, which do not have an influence on the behaviour of the technical product.
- Functional features contain sets of geometric entities, which can easily be implemented in – current, mainly geometric oriented – CAD systems. At the same time they show the relations between these geometric entities and the underlying functions. On this basis it seems possible to come from (current) CAD systems supporting embodiment and detail design only to (enhanced) CAD systems that can support the designer also in the field of conceptual design.
- The introduction of functional features leads to a new way of designing: It is possible to separate aspects of function from aspects of manufacturing, assembly, and strength.

3.4 Design Features

In this section the more analytical ("bottom-up") approach in the investigation of features for the design process is described. As mentioned before, the features examined here are named "design features" to differentiate them from the functional features that stem from the synthetical ("top-down") approach of the investigation.

Figure 7 demonstrates the basic idea of design features, using the same example as above taken from the field of the design of shafts and their adjacent

368

components. In the figure the technical drawing of a shaft is shown on the upper left. On the upper right the installation situation of the shaft is described, because – as mentioned above – all sorts of features for the design process should be related to functions and consequently cannot be restricted to the examination of single parts. In the lower half of Fig. 7 the design features are listed, to which the designer can break down the examplary shaft in his mind. Each design feature combines shape elements (partially including surface and/or tolerance attributes) and design functions.

The short statements in Fig. 7 characterizing the design functions show that the design features describe the functional information on a much more concrete level than the functional features discussed in the first part of this issue. The design features shown in Fig. 7 as well as the above described functional features combine shape elements with functions. Nevertheless it is obvious that the functions in the context of design features are of another type than those functions that represent the engineering meanings within functional features.

Looking at the literature, the functions described in the context of the synthectical approach – from now on called "methodical functions" – are related to systems [10, 17]. (Systems are characterised by the fact that they have boundaries that cut across their links with their environments. A methodical function describes the relationships between input(s), output(s) and state variable(s) of a systems – as a.m.). Furthermore, a methodical function corresponds to a physical effect. A function in the context of the button-up approach – it will be designated as a "technical fuctions" from now on – can also relate to a system and to a physical effect but it does not have to.

Another difference between design features and functional features is that in the geometric resp. syntactical part of the description of the design features it is not unequivocally possible to distinguish between functional and filling faces as it is done with the functional features.

As mentioned above, features relevant for the design process usually are characterized by hierarchical arrangements of geometric elements. This is shown in Fig. 8 where a feature-based description of the examplary shaft discussed here is given. Two basic terms are used: "single features" and "compound features". Single features are the lowest order canonical forms supported by a feature-based system. It should be noted that there are two classes of single features: "generic features" and "modifiers" [11]. Generic features describe the rough outlines of shapes, while modifiers determine local changes in the topologies and geometries of the generic features. Examples for modifiers are roundings, undercuts, and all sorts of grooves, slots, and pockets. The dividing line between generic features and modifiers, however, is not always clear-cut.

In the upper half of Fig. 8 the examplary shaft discussed here is shown as an "elementary feature-based model": here only the generic single features are represented explicitly, while the modifying single features are added as attributes (albeit in parametric form). In the lower half of Fig. 8 generic features are characterized by square frames, whereas circular frames point to modifiers.

A compound feature is a combination of more elementary features, which may be single features or compound features, within a specific context. (In the

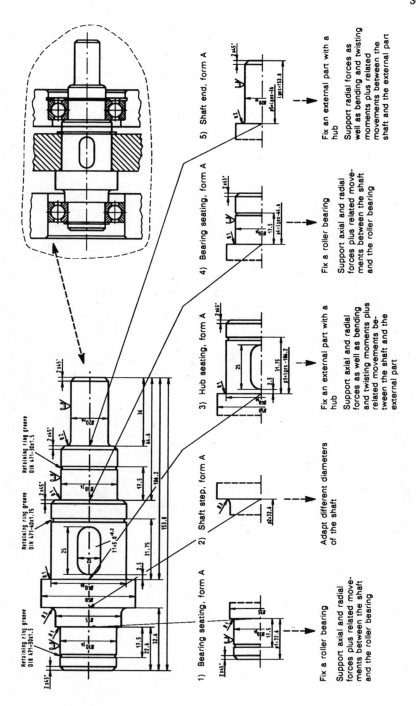

Fig. 7. Compound features of a shaft

370

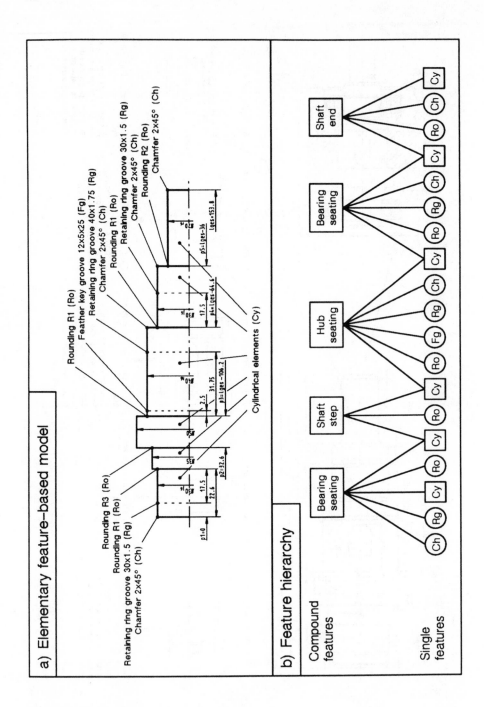

Fig. 8. Feature-based description of a single part

example presented in Fig. 8 the compound features are built up on single features only.) One asset of the concept of compound and elementary features is the possibility of generating and manipulating features at multiple levels: a related group of single features making up a compound feature can be manipulated as an unit rather than working on each single feature individually; at the same time the definition of a compound feature determines certain relationships of its constituents. Here the authors are more interested in an additional point: compound features represent more complex technical functions than single features.

Figure 9 shows the single features of a compound feature "bearing seating" and their related technical functions. Technical functions of single features are seen from four points of view: (i) (basic) design function, (ii) strength, (iii) assembly, and last but not least (iv) manufacturing. These low-level features are combined to the compound feature "bearing seating" because a more complex technical function can be associated with this design feature. The engineering meanings of compound features are just seen from the viewpoint of design function.

Without going into detail there are two reasons why features like "bearing seating" are splitted into single features. Firstly, from the syntactical point of view, if all features stored in a feature modeler are built up on a small number of elementary features, the feature-based system just has to be able to express the single features explicitly; all other, more complex, features, can be presented implicitly. Secondly, from the semantical viewpoint, if designers have to describe the features of a single part, first of all they think in terms of compound features with a complex engineering meaning. Then, if they have to determine more elementary design functions that have to be warrented to realize the complex design function of a compound feature, they split it into its elementary features. So, the technical functions listed in Fig. 9 as the semantics of the single features are necessary to realize the design function of a compound feature "bearing seating". Looking at these necessary resp. elementary design functions it should be noted that they are important at different phases of the product's life cycle:

- The technical functions listed in the columns "assembly functions" and "manufacturing functions" in Fig. 9 apply to the production phase.
- Contrary to the a.m. technical functions the elementary functions in the two columns "(basic) design functions" and "strength functions" are related to the operation phase.

To come to feature-based models of technical products feature-based description of single parts – an example is presented in Fig. 8 – have to be combined. For this purpose "connection features" are used. In Fig. 10 a connection feature "support of a shaft" is illustrated. In this example two bearing seatings of a shaft, two bearing seatings of a housing, two roller bearings, and three retaining rings are combined to one feature. Therefore connection features couple feature hierarchies of several single parts – in the given example the feature hierarchies of the shaft and of the housing – and they include additional parts as a whole – in the given example the roller bearings and the retaining rings. The parts included completely might be represented with features as well.

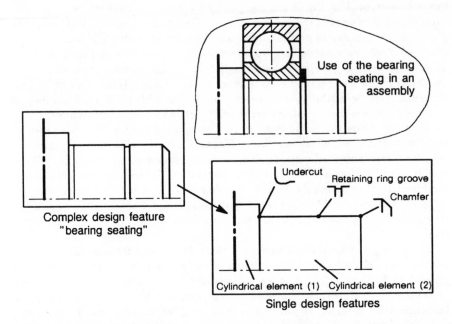

Use of the bearing seating in an assembly

Complex design feature "bearing seating"

Undercut   Retaining ring groove   Chamfer

Cylindrical element (1)   Cylindrical element (2)

Single design features

Single features	(Basic) design functions	Strength functions	Assembly functions	Manufacturing functions
Cylindrical element (1)	Support of axial forces and related movements		Stop of motion during the assembly operation of the roller bearing	
Cylindrical element (2)	Constructive length Support of radial forces and related movements			
Chamfer			Centering the roller bearing during the assembly process	Realisation of a definite form of the edge
Retaining ring groove	Seating a retaining ring to support axial forces and related movements		Realisation of a detachable connection with the roller bearing (with the help of a retaining ring)	
Undercut		Decreasement of the fatigue strength reduction factor with regard to sharp-edged steps		Run out for tools necessary for grinding

Fig. 9. Single features of a "bearing seating"

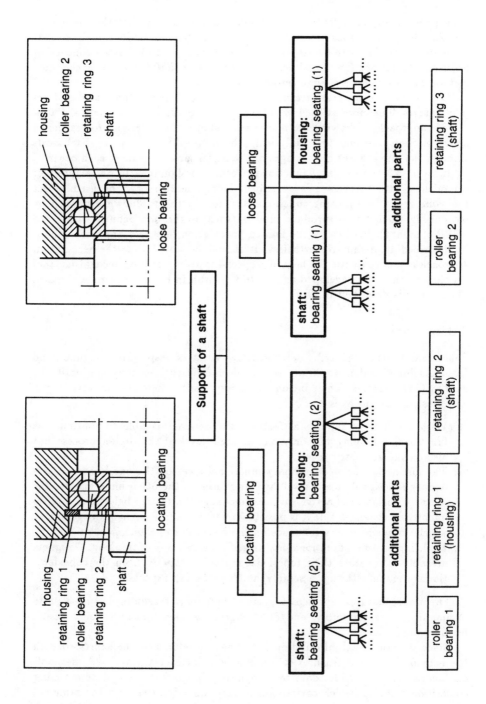

Fig. 10. Connection feature "support of a shaft"

Whereas compound features are defined in such a way that they represent complex design functions connection features are build in such a manner that methodical functions can be associated with them. Figure 11 shows a connection feature and a related functional feature "support of a shaft", both have the same methodical function resp. engineering meaning.

Consequently connection features are the bridge between functional and design features. Therefore at the level of connection features the synthetical ("top-down") approach – following the phases resp. stages of the design process shown above in Sect. 2 – and the analytical ("bottom-up") one – going backwards through the a.m. phases resp. stages of the design process – meet each other.

In Sect. 3.3 the usage of functional features designing a technical product – modeling with features – is described. Design features can be used to get the same result as by using functional features in the top-down approach by recognising them out of product models (such as the one defined within the STEP standard). Because of the limited space given here this approach cannot be described in detail. Nevertheless, it should be noted that the recognition approach is realized with the help of a special type of a feature recognising tool – a parser – that was developed at the DFKI GmbH in Kaiserslautern, Germany. (The parser is described in [9].)

4 Conclusion

Functional Features describe sets of active surfaces resp. primary functional faces. As functional features are sets of geometric entities they can easily be used in CAD systems. Their introduction resp. usage offers several advantages. The most important ones are:

1. With the help of functional features the designer is able to continue his functional reasoning not only on the logic level of the design process but also on the geometric level.
2. The distinction between primary functional faces on the one hand and secondary functional faces as well as filling faces on the other hand enables the selective variation of active surfaces to change the system behaviour. Consequently on the other side it is possible to identify those parts of the geometry which do not have any influence on the behaviour of a technical product.
3. Modeling based on functional features offers a new way of thinking. Designers are able to separate their functional reasoning strictly from the considerations of manufacturing, assembly, and/or strength aspects.

Functional features represent aspects of functional reasoning but no aspects of manufacturing, assembly, and/or strength. For these purposes only design features can be used.

Since the functional information of technical products can be expressed with the help of functional features it is possible to extract the realized methodical functions from product data by recognising design features and determining connection features (which correspond to functional features with the same engineering meaning).

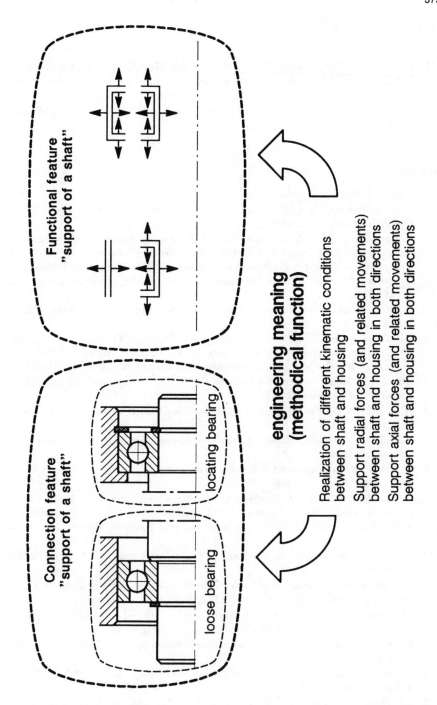

Fig. 11. Connection feature and corresponding functional feature

5 Outlook

At the Lehrstuhl für Konstruktionstechnik/CAD (Institute of Engineering Design/CAD) of the university in Saarbrücken features are used in the context of two research projects: (i) automatic classification of design objects and (ii) feature-based tolerance representation and analysis. Results will be published next year.

References

1. M.M. Andreasen, S. Kähler, T. Lund, *Design for Assembly*, Springer Verlag, Berlin, Heidelberg, ..., 1983.

2. A. Bernardi, C. Klauck, R. Legleitner, M. Schulte, R. Stark, "Feature-Based Integration of CAD and CAPP", *CAD 92 – Neue Konzepte zur Realisierung anwendungsorientierter CAD-Systeme*, Edited by F.-L. Krause, D. Ruland, and H. Jansen, Springer Verlag, Informatik Aktuell, Berlin, Heidelberg, ..., 1992, pp. 295-311.

3. N. Cross, *Engineering Design Methods*, Chichester, New York, ...: John Wiley & Sons Ltd, 1989.

4. J.J. Cunningham, J.R. Dixon, "Designing with Features: The Origin of Features", MDA Technical Report 3-88, *Proceedings of the ASME International Computers in Engineering Conference and Exhibition*, San Francisco, CA, 31.07.-03.08.88.

5. A.R. Grayer, *A Computer Link Between Design and Manufacturing*, Ph.D. Thesis, University of Cambridge, 1976.

6. W.E. Eder, V. Hubka, *Principles of Engineering Design*, Springer Verlag, Berlin, Heidelberg, ..., 1987.

7. W.E. Eder, V. Hubka, *Theory of Technical Systems*, Springer Verlag, Berlin, Heidelberg, ..., 1988.

8. R. Koller, *Konstruktionslehre für den Maschinenbau*, 2nd ed., Springer Verlag, Berlin, Heidelberg, ..., 1985.

9. C. Klauck, J. Mauss, "A Heuristic Driven Parser for Attributed Node Labeled Graph Grammars and its Application to Feature Recognition in CIM", *Proceedings of the International Workshop on Structural and Syntactic Pattern Recognition (SSPR)*, 1992.

10. G. Pahl, W. Beitz, *Engineering Design, a systematic approach*, Springer Verlag, Berlin, Heidelberg, ..., 1988.

11. M.J. Pratt, P.R. Wilson, *Requirements for Support of Form Features in a Solid Modelling System*, R-85-ASPP-01, CAM-I, Arlington, Texas, 1985.

12. W.G. Rodenacker, *Methodisches Konstruieren*, 4th ed., Konstruktionsbücher Vol. 27, Springer Verlag, Berlin, Heidelberg, ..., 1991.

13. K. Roth, *Konstruieren mit Konstruktionskatalogen*, Springer Verlag, Berlin, Heidelberg, ..., 1982.

14. J.J. Shah et al.: *Current Status of Feature Technology*, R-88-GM-04.1, CAM-I, Arlington, Texas, 1988.

15. J.J. Shah, "Conceptual Development of Form Features and Feature Modelers", *Research in Engineering Design*, Vol. Feb. 1991, pp. 93-108.

16. J. Tikerpuu, D. Ullman, "General Feature-Based Frame Representation for Describing Mechanical Engineering Design Developed from Empirical Data", *Proceedings of the ASME Computers in Engineering Conference*, San Francisco, CA, 1988.

17. Guideline VDI 2221: *Systematic Approach to the Design of Technical Systems and Products*, VDI-Verlag, Düsseldorf, 1987.

18. Guideline VDI 2222, Sheet 1: *Konstruktionsmethodik, Konzipieren technischer Produkte*, VDI-Verlag, Düsseldorf, 1973.

19. C. Weber, "Ableitung von Rechenmodellen für mechanische Systeme aus Funktionsstrukturen und Gleichungen physikalischer Effekte", *Proceedings of the International Conference on Engineering Design (ICED 91)*, Zürich 27.-29.08.1991, Switzerland, WDK 20, Heurista Verlag, Zürich, 1991, Vol. 2, pp. 873-884.

20. C. Weber, M. Schulte, R. Stark, "Functional Features for Design in Mechanical Engineering", *Proceedings of the 8th International Conference on CAD/CAM, Robotics and Factories of the Future*, Metz, France, Vol. 1, pp. 179-192, 1992.

This article was processed using the LaTeX macro package with LMAMULT style

13   Implementation of JIT Technology

A.S. Sohal and L.A.R. Al-Hakim

1. Introduction

In the late 1970s and throughout the 1980s, most US and Western Manufacturers faced intensified competition from Japanese manufacturers. Competitive analysis revealed the ability of Japanese manufacturers to produce and deliver higher quality products but at a lower cost. Just-in-Time (JIT) manufacturing was identified as the main contributor to this. Since then JIT manufacturing has received much publicity and has become one of the most important topics in the production and operations management literature. Surveys of the utilisation of production techniques in the USA [1], the United Kingdom [2] and in Australia [3] reflect the popularity of JIT manufacturing systems by comparison with the traditional MRP (Material Requirements Planning) or its version MRPII systems.

1.1 What is JIT?

A popular, and more pragmatic definition offered by Schonberger [4] likens JIT to a production system which aims to:
> "Produce and deliver finished goods just-in-time to be sold, sub-assemblies just-in-time to be assembled into finished goods, fabricated parts just-in-time to go into the sub-assemblies and purchased materials just-in-time to be transformed into fabricated parts".

Schonberger's interpretation is often the basis for other definitions, the number of which attests the all-embracing nature of JIT. Hall [5] explores the possibility that:
> "JIT is not confined to a set of techniques for improving production defined in the narrowest way as material conversion. It is a way to visualise the physical operations of the company from raw material to customer delivery".

In the APICS Dictionary [6], the JIT concept is defined as more of a philosophy than a production technique; achieving excellence in a manufacturing company based on the continuing elimination of waste. Wantuck [7] extends this definition and Schonberger's description by including another element - concern for people. He describes JIT as:
> "A production strategy for continuous improvement based on two fundamental principles, both of which we believe, but do not necessarily practice very well: the total elimination of waste, and a high respect for people".

Based on Wantuck's definition of JIT, Sohal et. al [8] summarise the multifarious benefits of JIT to include:

* Reduce lead times, space, materials handling, work-in-progress and defects.
*Improve productivity, flexibility, employee involvement and market performance.

1.2 Implementation Difficulties

The implementation of JIT manufacturing systems is primarily seen as a technical problem requiring changes, with different degree and complexity, in the entire organisation.

  Although the changes are initiated by the experts (usually external consultants), many managers who implement these systems in their organisations have often been disappointed with the results. The inability to cope significantly with the variation in market requirements and to reduce production costs drastically may affect the organisation's market share that will ultimately cause these systems to collapse completely [9].

1.3 Factors Contributing to Lack of JIT Success

A study carried out by researchers at the California State University shows that the majority of organisations do not have the means to reorganise their facilities at the cost of lost production (Plenert and Best 1986). This raises the problem of modus operandi, especially for an organisation that already has practised an MRP system [10].

  A similar research conducted in Australia [3] indicates several difficulties confronting companies in operating their JIT systems. Factors identified as having a major influence on the outcome of their JIT systems were "Lack of management commitment and leadership", "Supplier difficulties", "Insufficient resources", "Departure of the JIT instigator/enthusiast", "Lack of employee training and involvement", "Lack of perseverance and the propensity to revert to traditional practices when difficulties encountered", "Employee scepticism and resistance to change", and "Need to introduce Total Quality Management program to complement JIT".

  Lorinez [11] provides a summary of a survey carried out in the United States and concludes that a high degree of confusion and uncertainty exists regarding implementing JIT. Another study by Al-Hakim et. al [12] indicates that many Australian managers are reluctant to publicise their JIT implementation failure and some managers are claiming using JIT system while they operate a MRP system.

1.4 Elements Of JIT

Researchers and practitioners [13, 14, 15] have identified several requirements of JIT operation to include:

- small batches;
- set-up time reduction;
- cellular manufacturing;
- multiskilled workers;
- quality at the source;
- preventive maintenance;
- Kanban/pull production scheduling;
- JIT purchasing/delivery.

  Hall [16] adds another requirement, that is, uniform plant loading and describes it as the most difficult requirement to implement in a JIT system. The term "uniform plant loading" is the core of an article by Park [17]. Park mentions other requirements such as simplification and quality circle activities.

Most of the researchers in their explanation of JIT emphasise the need for high respect for people and the elimination of waste. It is a mistake to imagine that a unique prescription for action is possible, and therefore no attempt is made to provide one. Rather, the aim is to show how needs may exist that can be met by filling the gap created by the reformed objectives through systematic and continuous reduction of waste, how these needs may make themselves felt and, hence, how they may be identified and the problems put right.

2. JIT and Organisational Objectives

Organisational performance has been linked to organisational objectives. It is therefore clearly important to know what these objectives are. Improper definition of the organisational objectives will put the behaviour changes achieved by training in improper directions.

In-depth study of JIT manufacturing systems reveal that they differ from the traditional approaches to production especially with respect to the means by which they manage waste, rather than how to produce more competitive products [18]. By "waste" here we mean anything that adds no direct value to the product. This viewpoint is necessary finally to understand, redefine and reform;

    1. the organisational objectives,
    2. factors affecting achievement of these objectives.

Indeed, any organisation which does not change its objectives in line with the environment could be described as unhealthy, since factors affecting achievement of the objectives very often alter so rapidly that failure to change will cause stagnation or even bankruptcy. Here we are not facing the problem of objective achievement alone but also the precise definition of the main objective synchronised with the installed (or about to be installed) production system.

Traditional organisational objectives such as profit maximisation, increased productivity to a certain level or producing more competitive products, so long the central assumption of economic theory, are no longer considered viable in relation to newly-installed production systems. This is not the place to enter into a lengthy discussion on this particular matter, but the inherent weaknesses in the traditional theory can be seen immediately one asks the questions;

    a. what are the factors affecting achievement of profitability?
    b. how is the profitability measured?

There is ample evidence, however, that the key parameter for competitive advantage is not profits but something else. The minimising of waste or the maximising of long-term waste reduction would certainly be a more realistic organisational objective. The phrase 'long-term' here implies that the waste reduction is not a one-step process but a continuous one. Most Japanese industries and certain Western companies provide the most instructive examples, not because they are necessarily unique but because they are capitalising on waste reduction as a critical source of competitive advantage [18].

3. The Turning Point: Overproduction

Overproduction is created by producing components and products over and above those required by the market. When this happens, more raw materials and resources are consumed, more time spent, and more wages are paid for unnecessary work, thereby creating unneeded work-in-process (WIP) and inventory. This in turn requires additional space and cost to hold the inventory, a more sophisticated inventory control system and more resources to monitor the WIP and inventory. It also requires more paper work, more materials handling as well as the additional investment in components and products that are not needed. This is besides the quality problem created by excessive inventory and WIP.

Furthermore, WIP and excessive inventory lead to confusion on the shop-floor in scheduling and assigning priority for production. Since resources (people, machines and equipment) appear busy and all facilities (including warehouses) are occupied unnecessarily, additional investment in resources and space may be allocated on the mistaken assumption that it is needed. As a result, overproduction is considered one of the worst wastes and should be eliminated. This requires a small lot size which in turn necessitate a complete study of all sources of waste.

3.1 JIT and Lot Size

Under traditional manufacturing, lot size decisions are based on the minimising of Economic Manufacturing Quantity (EMQ). The EMQ formula is usually expressed as a function of the holding cost and set-up cost (see Fig. 1a). Both carrying cost and set-up cost are considered as waste in JIT manufacturing and need to be minimised. If set-up cost is reduced as presented in Fig. 1b, the EMQ lot size is reduced from Q1 to Q2. Further reduction in set-up cost will reduce further the lot size to even one for very small set-up cost. Reducing the lot size will cause a reduction in holding cost since the average inventory will decrease. The result of reduced holding cost causes an increase in lot size to Q3 as presented in Fig. 1c. However, the result is a decrease in lot size from Q1 to Q3. The JIT objective of continual reduction in waste helps the continual reduction in the lot size to a level suitable to achieve JIT.

Smaller lot sizes mean frequent production runs. This in turn causes a major scheduling problem. Here, we face two options. Either to install a sophisticated scheduling system which ultimately creates more waste, or to change the whole manufacturing philosophy. The later was behind the motivation of the manufacturing pull system instead of the classic push system.

3.2 JIT and Pull System

A pull system is a new philosophy for running the manufacturing system. The basic difference between pull and the classic push, as explained by Karmarker [19], is that a pull system initiates production as a reaction to present demand, while push initiates production in anticipation of future demand. JIT, as indicated in the literature, is a pull system.

At the JIT operational level, every workstation considers the preceding workstation as a customer. Starting at the shipping dock and working back through the whole processes, the system pulls the necessary "products" from the previous workstation only

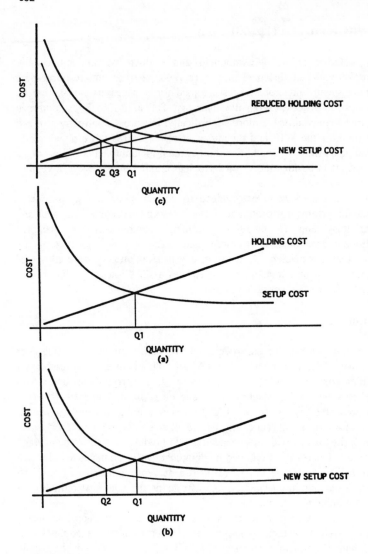

Figure 1. Effect of the setup cost on the lot size.

as needed. Accordingly, the workstation initiates production upon receiving a demand from the preceding workstation. Under condition of minimum waste, there is no significant delay.

To facilitate communication and scheduling the Kanban system was developed. Kanban is a Japanese word, one interpretation of which is "card". It is also indicates a "signal". A Kanban system is fairly simple. The cards are used to initiate transactions. The issuance of one Kanban card causes the production, vending, or conveyance of one unit (or one container) of the required product. The issuance of two Kanban cards causes production, vending, or conveyance of two units (container), and so on. Of course such a system requires smooth flow with a minimum lead time. A tight connection with the supplier and stable demand of the finished products are also necessary.

4. Zones of Waste

Unfortunately there is too much waste in our environment. While we talk about waste and explore the importance of productivity, export, and competitiveness, we tend to ignore the opportunities for improvement through waste elimination. One must re-orientate the thinking to identifying the different zones of non-value added costs of the product. This is in order to recognise the real levers for improving factory performance such as set-up time, lead time, work-in-process, materials movement, defect levels and so on. Zones of non-value added costs are (See Fig. 2)

1. Work added by unnecessary design.
2. Work added by poor quality.
3. Work added by inefficient methods of manufacturing.
4. Work added by inefficient facility planning.
5. Work added by set-up and changeover.
6. Overhead and carrying cost.

Waste zones as illustrated in Fig. 2 are consistent with the meaning of waste discussed by Suzaki [20]. That is " anything other than the minimum amount of equipment, material, parts, space, and worker's time, which are absolutely essential to add value to the product". In a very simple term " If it doesn't add value, it's waste" [21].

Indeed, the traditional approach tries to reduce these types of waste to a certain level but here we are working to drastically reduce the waste to a level of a single digit minute set-up time or zero stock level or zero defects, etc. This is quite different from the way we have traditionally managed the production system, for even quality inspection and overproduction can be considered as a waste. Obviously, JIT manufacturing have some very different concepts that we need to understand. Clearly, the more waste reduction is achieved the more productivity and quality can be raised. A certain attitude, knowledge, skills and behavioural changes are required to orientate the production direction toward successful achievement of this objective (see Fig. 3).

We may emphasise that the educational institutions base their management education on the development of analytical skills alongside the traditional management skills. This increases the knowledge on decision-making and problem solving processes which are highly needed to identify the waste zones. However, understanding and identifying the waste zones will not necessarily eliminate the waste. While the waste zones may be similar in many respects from one situation to another, unique waste elements will frequently be present. Thus, there is a gap between "learning how to do it" and learning how it is done". This gap in the technical knowledge is but one more aspect of waste.

We may fall into yet another trap if we consider the reduction or elimination of any element of waste as an organisational objective by itself. For example, many people think that work-in-process or inventory reduction is the objective of many new production systems such as JIT. Emphatically it is not. The basic requirement of JIT is to ensure that the right quantities are purchased and manufactured at the right time and in the right quantity with absolutely nothing to spares and that there is "no waste" [22]. The elimination of all unnecessary inventory is but one step towards waste reduction rather than an objective itself. Of course, elimination of inventory as an end or main objective may cover other elements of waste and may have a big effect on productivity, but it is not the whole story.

384

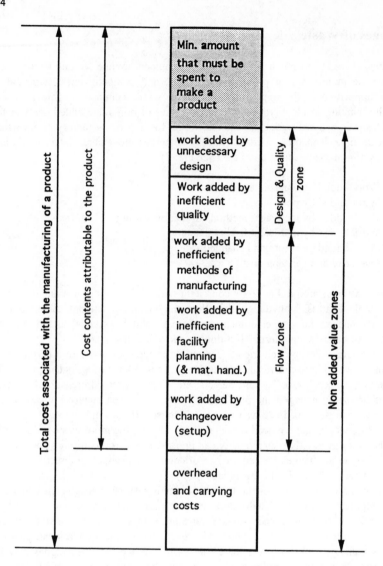

Figure 2. Zones of the total cost associated with the manufacturing of a product.

5. Environment of JIT

Elimination of waste is really the linchpin of the entire JIT phenomenon. It is not an element of JIT but the whole philosophy of JIT. It is an umbrella term under which all the elements (techniques and approaches) discussed earlier are used to accomplishing the elimination of waste within the zones of waste mentioned in Fig. 2.

By closely examining the waste zones, one can group these and the related elements into three interrelated and equally important categories. These are personnel performance, quality, and flow. However, waste elimination is not a static process, it is a continuous activity within the entire organisation. The prerequisites for continuous improvement are:

1.   Simplification, and
2.   Personnel Performance.

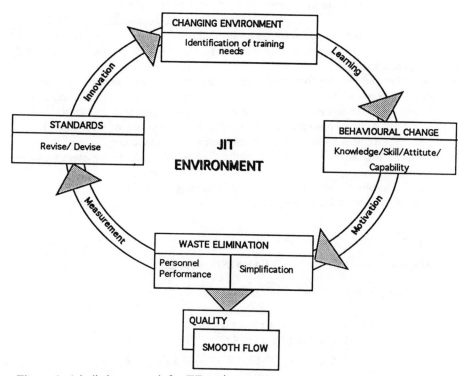

Figure 3. A holistic approach for JIT environment.

To put this into action, we argue that continuous improvement is essential. This includes a holistic approach (see Fig. 3) by all players exercising responsibility for revising the standards and for the wide range of training issues. It involves not merely a more interactive personnel function but the active participation of top and line management at every interface.

5.1 JIT and Personnel Performance

As the main organisational objective and the factors affecting it are recognised, it may seem reasonable to relate the personnel performance level with the level of occurrence of these factors. But that is not enough for two reasons:

(a)   WIP, set-up times, breakdowns, and defective products, etc., are interrelated and are not the only source of waste in themselves. They are also a source of confusion.

(b)   Without making a commitment to ongoing reduction in waste and ongoing learning, a firm will gain no more from any training than a one-time boost in performance.

Responding to new demands and new opportunities requires change, but much confusion comes from manufacturing units facing such pressure. Poorly qualified managers always seem to be caught by surprise, operate haphazardly and "leapfrog" from one crisis to the next. Much of the confusion in their plants is internally generated and, in such a factory, new production technology will only create more confusion and not more productivity.

On the other hand, we are convinced that a firm's learning rate - the rate at which its managers and operators learn to make it run with more waste reduction consciousness (ie., run with less manufacturing cost) - is at least of equal importance to its current level of productivity. A firm whose total productivity is lower than that of a competitor, but whose rate of learning is higher, will eventually surpass the competitor.

Thus, the two essential tasks for management of the newly installed JIT system are to:-

(a)   create clarity (reduce or prevent confusion), and
(b)   accelerate the learning rate.

MAXIMIZE   CLARITY + LEARNING RATE

SUBJECT TO :

1. MARKET CONSTRAINTS
2. CAPACITY CONSTRAINTS
3. CLARITY * LEARNING RATE > 0

Figure 4. Personnel performance problem formulation.

Clarity and the learning rate are a powerful combination to measure personnel performance for the purpose of training needs. They need to be maximised and oriented according to market and capacity constraints (see Fig. 4). Without such a combination, JIT creates more confusion not more productivity (the third constraint of the training formulation of Fig. 4 ensures that the rate of both clarity and learning need to be positive). Needless to say that every member of the organisation - from the shop floor to the senior management - has a part to play in the waste elimination.

5.2 Simplification

Experience shows that there is no "perfect method" in manufacturing. In fact there are always opportunities for improvement.   By contrast to Taylorism and Scientific Management the Japanese reception was open and widespread. The major reason for this was in the recognition by Japanese management of the benefits of simplification.

.Scientific Management in the West is limited to the study of work method and time measurement, while in Japan it is oriented towards simplification of both methods and design and in connection with quality. Simplification has endless opportunities to:

1.   Reduce the number of components in the products.
2.   Reduce the amount of assembly required.
3.   Combine operations and components
4.   Reduce the number of steps/operations in the process flow.
5.   Change the sequence of steps/operations for less handling and better utilisation of resources.
6.   Simplify the operations.
7.   Reduce the number of components in the fixtures and tooling.

With the above improvements, all aspects of the manufacturing system becomes simpler, taking up less time, costing less and much more reliable with better "easy-to-see" quality. There can be no doubt that the new, simple design of products and systems will work. The continuous process of simplification aimed towards further reduction of waste creates an environment where attitudes are changed permanently.

6. JIT and Quality

The link between JIT and quality is simple. While quality does not need JIT to be successful, JIT can only be successful in a company that is producing quality products. That is, quality makes JIT possible.

Determining a statistical sample and judging the whole lot size according to the acceptability of that sample is standard procedure in the traditional approach to quality control. Sampling tables and lot acceptance sampling inspections are no longer viewed favourably in the JIT environment. Quality in a JIT environment is not a control system and if viewed this way it will never be substantially improved. In sampling tables the acceptable level expressed is a percentage of defectives, ie., number of defects per lot of 100 units. JIT rejects any level of defects as acceptable.

Inspection is regarded as a non value added process. It is a waste. In the JIT environment an effort should be made to inspect every part. This is as part of the manufacturing process itself rather than through traditional methods of inspection. The kind of quality required for a JIT environment is quality at the source. That is *"do the right things right the first time"* with the possibility to *"detect errors before they become defaults"*. Quality control in JIT may serve as "facilitator" only. Here, JIT brings a number of unique elements and techniques to quality management. Some of the significant contributions of JIT are robust design, quality at the source and Andon system.

6.1 Robust Design

The underlying theme of continuous quality improvement is supported by the work of Taguchi [23]. Taguchi directs quality to the design phase of the product and therefore is referred to as an off-line quality control method. Taguchi introduced a philosophy in which quality is measured by the deviation of a characteristic from its nominal target value. Two types of factors causes these deviation. Controllable factors and uncontrollable (noise) factors. Since the elimination of noise factors is costly and often impractical, the Taguchi method seeks to reduce the product's sensitivity to variation by adjusting factors that can be controlled in a way that minimises the effects of factors that cannot be controlled. This results in what Taguchi calls a "robust" design.

6.2 Quality At The Source

100% quality is the ultimate target for all organisations. Inspection is a costly operation and does not add value to the product. It is waste. Unless we can develop a low-cost method to ensure 100% quality, 100% quality will not be possible. Therefore, it make sense to control quality at the source. The key concept of quality at the source is to achieve 100% quality at each process by preventing defects at the source. This should significantly reduce or even eliminate inspection totally.

Poka-yoke [24] is a Japanese technique which interpret as "foolproof mechanism". The concept is to develop an autonomous control mechanism that requires the least supervision. Whenever a defect is produced, the mechanism will stop the machine or give a signal requiring a correction to be taken. This allows an operator to concentrate on his\her work.

Besides defect prevention, further consideration should be given to process capability. Quality at the source concept pays extra attention to developing a reliable process capability of the machines. To do this all the quality parameters should be monitored over time by using appropriate method. In order to make such a system more effective, the Andon system is developed. The Andon System (or stopping the process) system consists of placing one or more coloured lights above the assembly line or machines to signal the assembly line or machine needs repairs or to signal the existence of a defect. It may also be used to signal that the machine or assembly line is out of materials or components.

7. JIT and Method Study

Method study is part of Scientific Management which was originally developed by Taylor in the late 19th century. Method study is an analytical systematic approach in which the manufacturing system is studied for the purpose of improving the system. This includes an analysis of each step in the manufacturing process or system. Charts are commonly used for such analysis. The process chart is one the oldest and most commonly used for analysing the materials flow in a manufacturing system. The basis of the process chart is in the process symbols developed during the 1920's which includes operation, transportation, inspection, storage and delay. Another version of the process chart is the flow diagram which is the graphical record of the steps in a process made on the layout of the manufacturing plant under considerations. This helps to visualise the area of excessive material handling and WIP. Table 1 illustrates the processes, movements, inspections, storage and delays that need to be eliminated or minimised in a JIT environment.

Table 1. Activities that need to be minimised or eliminated in JIT environment.

Symbol	Activity	Action: Technique
Operation	* Ordering	* Minimise: Forward planning; Master schedule; Kanban.
	* Operating	* Minimise: Method engineering; Work simplification; Work design.
	* Setup	* Minimise: Method engineering; Single-minute exchange of die (SMED); One-touch exchange of die (OTED).
	* Search	* Eliminate: Method engineering.
	* Loading	* Minimise: Work design.
	* Unloading	* Eliminate: Point-of-use delivery.
	* Unpacking/ Repacking	* Eliminate: JIT delivery.
	* Receiving	* Minimise: JIT delivery; Point-of-use delivery.
	* Rework	* Eliminate: Zero defects; Quality at source.
Transportation	* Transport loaded	*Minimise: JIT delivery; Point-of-use delivery.
	* Transport unloaded	* Eliminate: Focused factory; Cells formulation; Flow design.
	* Move WIP	* Eliminate: JIT manufacturing.
Inspection	* Incoming inspection	* Eliminate: Supplier certification.
	* Final Inspection	* Minimise: Zero defects; Quality at source; Off-line quality control. (Taguchi method).
	* Counting	* Eliminate: Small lots; Poka-yoke method.
Storage	* On-site	* Minimise: Focused factory.
	* Off-site	* Eliminate: Long-term contracts; Certified suppliers; JIT delivery.
	* WIP	* Eliminate: JIT manufacturing; Kanban.
	* Finished goods	* Minimise: JIT manufacturing.

Table 1. (continue)

Symbol		Activity	Action: Technique
Delay	Delay	* Batching parts	*Eliminate: JIT manufacturing.
		* Waiting for material	* Eliminate: Flow design.
		Batching WIP	* Eliminate: JIT manufacturing; Kanban; Flow design.

8. JIT and Materials Flow

There are general guidelines, principles, and techniques that, if followed, may lead to an effective facility layout that has smooth flow of material with minimum handling and hence a more successful JIT environment. Some of these techniques and principles are described below:

8.1 Focused Factory Organisation

Skinner [25] suggested the idea of establishing the focused factory organisation, that is `plants within a plant' or `factories within a factory', as a potential solution to increasing competitiveness. One objective of the focused organisation is to organise new factories of the smallest size practical within an existing factory. These compact entrepreneurial units are called subplants. The related major issues are the nature of the subplant and how to reorganise existing operations within it. A second objective is to organise manufacturing units along either product-family or component family lines. The later applies when the capacity of a machine or cell greatly exceeds that needed for any product family. Each subplant of the focused factory requires simple systems to handle and retrieve information and transactions. Moreover, factory support services are often provided by machine operators and assemblers. Repairs, preventive maintenance, and housekeeping are part of their responsibilities. The above environment eases the implementation of JIT.

8.2 Uniform Process Modules

Under this, processes are designed into modules such that each module has one common dimension, `N', or a multiple of `Ns' (see Fig. 5). Inputs and outputs of each module are organised to be closest to the plant aisle in a way that every aisle serves modules on both its sides. In these modules, the machines and assembly lines are organised in U-form or serpentine line shapes. Serpentine and U-form shapes are fairy easy to compress and expand to fit the `N' dimension modules. The modules (cells or workstations) should be designed;

1   So that the operator can pick up and discharge materials without walking or making long or awkward reaches.
2.   To minimise the time spent manually handling materials.

In uniform modules layout, it is possible to slide the process back and forth to fit the new layout requirement or to insert a new module, etc.

8.3 Regular Aisle System

The uniform modules layout requires a regular systematic framework of aisles. The symmetry of aisle system often indicates the quality of the layout. This can be achieved by the following (see Fig. 5):

1. Avoid irregularities.
2. The aisle should service production area of `N' depth.
3. Every aisle serves production areas (modules or stations) on both its sides.
4. Every aisle should originate at some point on the building's perimeter.
5. The entry aisles points for a rectangular building should be on its longer sides.
6. The aisle system is designed primarily for one-way traffic.

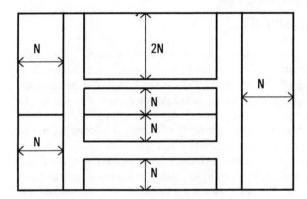

Figure 5. Uniform process modules layout.

8.4 An Example

This example illustrates the relationship of manufacturing methods and facility design in creating a successful JIT environment. Fig. 6 illustrates the flow structure process of assembling a major product that needs several parts which in turn are assembled by several sub-assembly lines and stations. The speed of each line and the machinery and the technology needed for each line is determined such that:

1.   Receiving parts from any sub-assembly signals the preceding sub-assembly to start production (as a Kanban).
2.   The speed, manpower, machines and the kind of technology needed for each line are determined such that the main assembly line receives parts from the sub-assembly lines at the time needed and in the necessary quantities.

By this example, we can recognise the important of cooperative recovery in producing components and products. It also illustrates the simplicity and effectiveness of the pull scheduling system in the JIT environment. With this type of flow structure, there is no necessity to produce extra work (WIP) by any sub-assembly line.

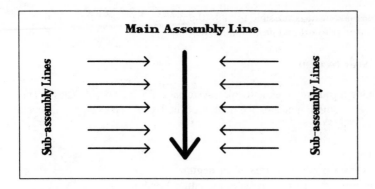

Figure 6. Flow structure for an assembly process.

9. Comparisons: MRP with JIT

At the planning level, MRP may be considered as a pull system, since all the required computations are based on a master production schedule which in turn is derived from forecasts or orders [26]. However, this is not the case at the operational level, where MRP is a push system. Push implies that the schedule introduces the job to the production process to meet the due date.

At the first sight, MRP logic matches almost perfectly with the JIT concept of having the required quantity at the right time and the right place, but MRP deals with fixed market environment with predictable demand and takes into consideration fixed lead times and existing set-up times. Parts are processed in batches and batch sizes are usually too large due to high set-up costs. Obviously, increasing the batch size and lead time of manufacturing increases WIP as well as finished goods. JIT places particularly heavy emphasis on the reduction of set-up times and offers attractive incentives for lead time management. This facilitates reduction of batch sizes and WIP to a minimum, preferably one and zero, respectively.

Another major difference is that MRP requires accurate data for its inputs (master production schedule, bill of materials and inventory status files). This also requires sophisticated computer software, and therein, Belt [27] argues, lies its greatest danger. Time and again, companies become obsessed with the data processing hardware and software and forget that MRP software does absolutely nothing but add, subtract and print - the rest is up to people. Data accuracy is not important in JIT production and there is no need to use computer software to schedule production.

Since lead times are predictable, MRP works well, but so does JIT. The latter tends to be cheaper because JIT does not require computerised production data processing systems. However, there seems to be a paradox here. MRP inherently aims to be a JIT system while pull systems do not really recognise the future events that are supposed to be just-in-time [19]. Furthermore, Karmarker [19] explains, as we move to more dynamic and variable contexts, MRP become invaluable for planning and release of orders. Pull techniques can not cope with increasing demand and lead-time variability. Shop floor control requires higher levels of thinking and scheduling sophistication. Materials flow is too complex for strict JIT. Monden [28] concludes that JIT is difficult, or impossible to use when there are:

(1) job orders with short production runs, or
(2) significant set-up, or
(3) scrap loss, or
(4) large, unpredictable fluctuations in demand.

Krajewski et. al [29] have used a large scale simulation model to show that JIT performs poorly in an environment which has low yield rates, worker inflexibility and lack of product standardisation. MRP, on the other hand, can be used in almost any discrete part production environment. JIT seems to produce superior results when it can be applied [30].

10. Case Studies

To complete this chapter, we present three case studies on the implementations of JIT. The first case, taken from Al-Hakim, Okyar and Sohal [12] study illustrates the effect of operational environment, eg. improved material flow and reduced set-up times, on the implementation of production systems. Case Study 2, taken from Samson, Sohal and Ramsay [31], illustrates the integration of a Kanban system with a Flexible Manufacturing System. This case demonstrate that successful implementation is very much dependent on how the work-force is managed. The third case study, also taken from Samson, Sohal and Ramsay [31], again illustrate the importance of the management of people.

10.1 Case Study 1

As part of studying the effect of operational parameters on the performance of MRP and JIT systems, we consider a manufacturing unit in a batch production company. This unit is one of several manufacturing facilities producing a range of diverse domestic appliances. The unit under consideration contains five hydraulic presses, two plastics moulding machines, two guillotines, and one assembly line. The layout is a traditional one with extensive flow of material. The unit produces 10 different sizes of a domestic product, ie., brands. In general, manufacturing of these brands involves similar operations on same machines and assembly on the same assembly line, though it may require different dies and machine set-ups.

The company has been using an MRP system for planning and controlling the production. Due to high set-up times required by some machines, especially the plastic moulding machines, the batch sizes are large. This brings about the problems associated with high WIP, eg. space and defects. The management requires a JIT system without changing the present layout or set-up technology.

To persuade the management on the consequences of introducing JIT without environmental changes, we developed a simulation model for the relevant unit using SIMAN simulation language (Pegden 1984). We approximated the processing times to fit a range of practical values of the normal distribution with adequate mean for each workstation (machines and assembly line). The set-up times are approximated to fit to uniform distribution. In our simulation program we disregard the effect of WIP on

394

quality and the machine breakdowns. There are two situations experimented with:

Situation 1:   Existing layout, ie., present material flow and set-ups.
Situation 2:   Cell formulation, ie.; improved material flow, with no set-up considerations. Suitable room for WIP is assumed when MRP is simulated.

The parameter used for the evaluation of each situation is the service level. Service level is the fraction of jobs whose flow time, ie., the time between job release and its completion (cycle time), is not greater than their lead time. It is our opinion that service level (in percentages) is an effective and more understandable parameter to the company's management. Our analysis shows the following:

(1)   Introducing a JIT Kanban system with the existing layout, material flow and set-up times affects the service level considerably (see Figure 7). MRP gives a better service level with larger brands to be manufactured.
(2)   Both MRP and JIT yield the same service level with cell formulation and no set-up consideration (see Figure 8).
(3)   Considering the second situation with 10 brands manufactured, the performance of JIT is highly sensitive to the changes in the variance of the processing time (see Figure 8)
(4)   While WIP is bounded in JIT, it grows to a dangerously high level (see Figure 9) with MRP system.

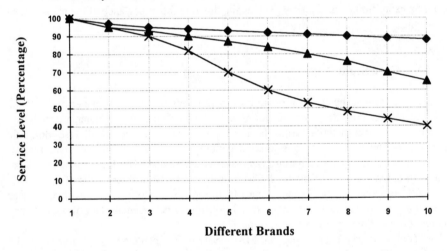

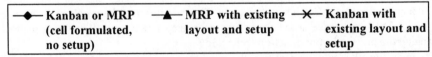

Figure 7. The effect of product's variability on service level.

Regardless of the effect of WIP, MRP seems to be superior to JIT when it is desirable to provide a higher service level. On the other hand, we know that cutting WIP in JIT leads to lower rejection rates and a rapid feedback when a process starts to malfunction.

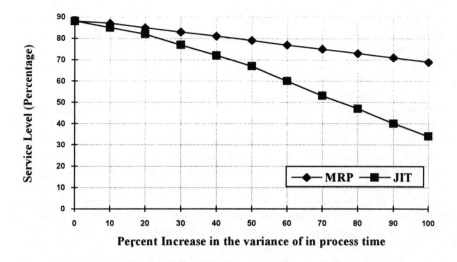

Figure 8. The effect of in process time on service level.

Excess WIP is not the only problem. It is also a source of confusion [18]. However, the experiment suggests that improvement of the operational environment, eg., improved material flow (cell formulation) and reduced set-up time is more important than the selection of the technique itself. It is the prerequisite for efficient utilisation of any system.

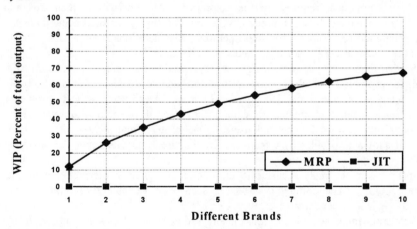

Figure 9. The effect of product's variability on WIP level.

10.2 Case Study 2

Dowell Remcraft, employing around 200 people, is a manufacturer of wooden window and door frames. The company has invested nearly $1.4 million in a new Flexible Manufacturing Cell (FMC) to replace existing technology which was 15 to 20 years old. The FMC, comprising of 27 cutting spindles, has enabled the company to reduce it's stock volume by half as well as provide increased flexibility. This has improved the company's ability to respond to the changing needs of its existing local customers and has enabled the company to market its products in overseas markets. Set-up times have been reduced to 90 seconds compared to five to six hours with the old technology. The new technology has resulted in better surface finish and better accuracy on length and profile of components.

In addition to the FMC, the company has also implemented a computerised Kanban system to provide better production and inventory control. This system includes a portable bar-code reader which enables fast and accurate identification of component parts. The company is currently integrating the FMC with the computerised kanban system and other computerised managerial control systems.

Before the new system was delivered to the site the manufacturing manager set about proposing an implementation program. During 1987 and 1988 considerable training was undertaken with all second-level supervisory staff attending a short course to recognise how they could contribute in their own areas of production. A formal meeting time was established for all supervisors to discuss current problems and future strategies. Representative trade unions were consulted and reassured that there would be no retrenchments of existing staff. The unions were told that a program of retraining would be embarked upon and the company would allow natural attrition to reduce staff members.

One month was set aside for installation of the new equipment. This was complicated by the fact that the old manual system had to run concurrently to satisfy existing orders. A significant factor in the success of the implementation has been attention to the human element of this project. The kanban cards are produced on site, on a local area network management information system. Plant operators use bar codes printed on the kanban to identify the component required to the computer. An otherwise complicated and exacting task to accurately program the CNC computer, that is repeated many times per shift, is simplified to a rapid error free operation, that additionally collects significant data on the machines performance for feedback to the operators.

Profiles that require manual tool changes are prompted to the operator, even from shift to shift, for example to verify that tooling has not been removed overnight. Software has been carefully designed and debugged well before the end user discovers problems. Changes have been implemented to extend the ease of operator use, many from suggestions put forward by the end users. One change implemented extended the number of allowable profiles by a factor of one hundred and fifty times the computer capacity.

Information is made freely available to the operators and their interpretation of data collected sought. The computer system is seen as a valuable tool rather than a constantly vigilant management spy.

The successful implementation of the manufacturing system has relied heavily in the staff accepting change as a positive and constructive process. This has been achieved by involvement in the decision making process, and having access to the critical information base from which the decisions must be made. Trust and openness are critical to constructive involvement. Management is seen as a guiding authority rather than issuing decrees and ultimatums in isolation. Consensus on priorities, methods and desirable outcomes and their effect on staff and the business unit are reached before implementation.

Careful selection of achievable objectives and getting several morale boosting successes on the board in the first few attempted tasks was extremely important. This helped to alleviate the frustration of failure and prepare staff to accept that some decisions would not be successful in practice whilst looking good in theory. Team building was also enhanced by common successes and the resultant post mortems on the failures. Fear of failure is discouraged and an atmosphere where even seemingly ridiculous ideas can be put forward and tested by "the team". Goal setting and priorities are discussed and every team member encouraged to participate and lead individual projects through to solution. Traditional artificial barriers put up by craft or union demarcation practices are minimised with genuine consultative negotiation on skill building and personal development. Training and multi-skilling is accepted as beneficial to the employee just as much as the employer.

A careful balance between the technology aspects, commercial and human aspects is necessary. One very significant problem at Dowell Remcraft has been to sustain motivation of key operational personnel. As the mechanical and electrical problems associated with the machine commissioning were overcome, there were long periods of repetitive loading components into the machine with much of the intellectual challenge diminishing. Rotation of job tasks within the team and the creation of a second team also provides a welcome break in the routine. The additional team also provides a degree of friendly rivalry and competition in addition to safeguarding against the vulnerability of losing key personnel and increasing the points of view in the best method of operating the line. Training and the regular reviews are achieved without disruption to work schedules. A third team is well advanced and all key operators, technicians, supervisors and the manufacturing manager have received detailed training in programming the computerised Line.

This marriage of people, the Kanban System and computerised machines has provided a valuable advantage in reducing inventory and operational costs, component obsolescence and enhancing manufacturing flexibility and responsiveness. Many of the key elements in promoting efficiency and quality and most importantly.. business viability.

The system provides numerous opportunities for extending the new technology into other areas of the company. A project is now underway to extend the computer system into areas such as time keeping and attendance, complete work-in-progress tracking and factory reporting. This will increase control over operations and further reduce the labour content slightly. The company is also considering marketing some of the software and technology developed in-house to other overseas manufacturers.

10.3 Case Study 3

Trico (Australia), currently employing around 260 people, manufactures windscreen wiper assemblies for the automotive industry. It was established in 1954 and operated very successfully with little competition from overseas until the early 1980's when it found itself in direct competition with the Japanese. For the first time in its history, the company made an operating loss. Factors contributing to this were identified as inefficient production, a high level of inventory, an inflexible manufacturing system and poor quality.

To improve their competitiveness the company adopted the JIT manufacturing strategy in mid-1984. Improvements through JIT manufacturing at Trico have been impressive. Inventory turns improved from 2.5 in 1983 to 7.5 in early 1989, and business had increased 30-40 percent. Whereas the company was threatened with closure by its US parent company in 1983, in 1988 Trico Australia was exporting 50,000 wiper assemblies per month to the USA.

The Trico success resulted from positive changes in attitude by all personnel. Management had to learn to accept involvement from shop-floor employees on day-to-day problems while the employees had to accept long-term goals, such as job security and job involvement rather than the more obvious short-term goals of wage increases.

With only four levels of management from the Managing Director to the shop-floor and with employees working in groups (manufacturing cells), information networks and communication flows were shortened allowing greater focus on the production processes. Groups of workers were formed into improvement teams which held regular meetings. During 1985, there were 41 such meeting. Because of the consultative nature of problem-solving, changes which were implemented were generally accepted by all concerned and implemented fully.

The result of employee involvement and improved communication between management and employees has resulted in higher productivity and competitiveness. Much of the initial employee resistance to change was overcome with education and training in the JIT process and related benefits. Regular briefing were held on changes that were to be implemented under the JIT scheme and these briefings have continued to date.

New products to be introduced to the range in the future are the subject of considerable discussion with personnel from all areas, including the shop-floor employees, in order to achieve optimum results.

The implementation of JIT at TRICO (Australia) has resulted from the following people-related success factors:

- realisation by everyone in the organisation that change is necessary.
- the total commitment of the top management team and shop-floor employees. The middle and lower management team (manufacturing manager and supervisors) acted as champions who lead the JIT push.
- everyone in the organisation having an open mind and sought assistance when needed. The ability of the people to question each change was also an important factor in accepting the change.

References

1. White RE. An empirical assessment of JIT in U.S. manufacturers. Prod & Invent Manage J 1993; 2nd Quarter: 38-42.
2. Oakland J, Sohal A. Production management techniques in UK manufacturing industry: usage and barriers to acceptance. J Oper Prod Manage 1987; 7 (1): 8-37.
3. Ramsey L, Sohal A, Samson D. Just-in-time Manufacturing in Victoria. University of Melbourne, Melbourne, 1990.
4. Schonberger R. Japanese manufacturing Techniques: nine hidden lesson in simplicity. The Free Press, NY, 1982.
5. Hall RW. Zero inventory. Dow Jones-Irwin, Homewood, IL, 1983.
6. American Production and Inventory Control Society. APICS Dictionary, 6th ed. Falls Church, VA, 1987.
7. Wantuck K. Ken Wantuck's concept of just-in-time manufacturing techniques. Tecnology Transfere Council, Melbourne, 1988.

8. Sohal L, Ramsay L, Samson D. JIT manufacturing: industry analysis and a methodology for implementation. Int J Oper Prod Manage 1993; 13 (7): 22-56.

9. Al-Hakim L, Jenney B. Training for new production technology. Proceeding of 1990 industry training conference, Melbourne, 1990.

10. Al-Hakim L, Jenney B. MRP: an adaptive approach. Int J Prod Econ 1991; 25: 65-72.

11. Lorinez JA. Suppliers question approaches to JIT. Purchasing Word J 1985; 29 (3): 42-63.

12. Al-Hakim L, Okyar H, Sohal A. A comparative study of MRP and JIT production management systems. In: Sumanth D, Edosomwan J, Poupart R, Sink D(eds) Productivity & quality management, IIE Press, NY, 1993. pp 547-556.

13. Baldwin RE. Adaptive effects on purchaser-vendor relationships resulting from Japanese management techniques. Procedding of the national purchasing and materials management research symposium, 1989, pp 42-54.

14. Celley A, Clegg W, Smith A, Vonderembse M. Implementation of JIT in the United States. J Purchasing & Material Manage 1986; 22 (4): 9-15.

15. Pegler HC, Kochhar AK. Rule-based approach to just-in-time manufacturing. Comp Integ Manuf Sys J 1990; 3 (1): 11-18.

16. Hall RW. Leveling the schedule. Proceeding of zero inventory philosophy & practices seminar, Falls Charh, VA: APICS, 1984.

17. Park PS. Uniform plant loading through level production. Prod & Invent Manage J 1993; 2nd Quarter: 12-17.

18. Hayes RH, Clark KB. Why some factories are more productive than others. Harvard Busness Review J 1986; Sept-Oct: 66-73.

19. Karmarker U. Getting control of just-in-time. Harvard Business Review 1989; Sept-Oct: 66-73.

20. Suzaki K. The new manufacturing challenge, The Free Press, NY, 1987.

21. Ford H. Today and tomorrow, Garden City Publishing, NY, 1926.

22. Aggrawal SC. MRP, JIT, OPT, FMS, Harvard Business Review, 1985: 8-16.

23. Taguchi G. Introduction to quality engineering. Asian Productivity Organisation, Tokyo, 1986.

24. Schonberger R. Zero quality control; source inspection and the poka-yoke system. Productivity, Inc., Stanford, 1986.

25. Skinner W., The focused factory. Harvard Business Review 1974; May-June: 113-121.

26. Lambrecht M, Decalume L. JIT and constraint theory: the issue of bottleneck management. Prod & Invent Manage 1988; 3rd Quarter: 61-65.

27. Belt B. MRP and kanban- a possible synergy?. Prod & Invent Manage 1987; 1st Quarter: 71-80.

28. Moden Y. Toyota production system: practical approach to management. Industrial Engineering Management Press, Norcross, GA, 1983.

29. Krajewski L, King B, Ritzman L, Wong D. Kanban, MRP, and shaping the manufacturing environment. Manage Sci J 1987; 33 (1): 39-57.

30. Spearman M, Woodruff D, Hopp W. CONWIP: a pull alternative to kanban. Int J Prod Res 1990; 28 (5): 879-894.

31. Samson D, Sohal A, Ramsay E. Human resource issue in manufacturing improvement initiatives: case study experiences in Australia. Int J Human Factor in Manuf J 1993; 3 (2): 135-152.

Index